Crossing Boundaries, Building Bridges

Studies in the History of Science, Technology and Medicine
Edited by John Krige, CRHST, Paris, France

Studies in the History of Science, Technology and Medicine aims to stimulate research in the field, concentrating on the twentieth century. It seeks to contribute to our understanding of science, technology and medicine as they are embedded in society, exploring the links between the subjects on the one hand and the cultural, economic, political and institutional contexts of their genesis and development on the other. Within this framework, and while not favouring any particular methodological approach, the series welcomes studies which examine relations between science, technology, medicine and society in new ways e.g. the social construction of technologies, large technical systems.

Other Titles in the Series

This book is part of a series. The publisher will accept continuation orders which may be cancelled at any time and which provide for automatic billing and shipping of each title in the series upon publication. Please write for details.

Crossing Boundaries, Building Bridges

COMPARING THE HISTORY OF WOMEN ENGINEERS 1870s–1990s

Edited by

Annie Canel
*École Nationale des Ponts et Chaussées,
Paris, France*

Ruth Oldenziel
University of Amsterdam, The Netherlands

and

Karin Zachmann
Technische Universität, Dresden, Germany

hoap

harwood academic publishers
Australia • Canada • France • Germany • India • Japan • Luxembourg • Malaysia
The Netherlands • Russia • Singapore • Switzerland

Amsteldijk 166
1st Floor
1079 LH Amsterdam
The Netherlands

British Library Cataloguing in Publication Data
A catalogue record of this book is available from the British Library.

Crossing boundaries, building bridges: comparing the history of women engineers, 1870s–1990s. – (Studies in the history of science, technology and medicine; no. 12)
   1. Women in engineering – History – 19th century 2. Women in engineering – History – 20th century 3. Women engineers – Social conditions
   I. Canel, Annie II. Oldenziel, Ruth III. Zachmann, Karin 620′.0082′09

ISBN: 90-5823-069-4 (soft cover)
ISSN: 1024-8048

Cover. Three women draftsmen trained as engineering aides posing in classic Rosie-the Riveter government propoganda style in 1943. Permission of Schlesinger Library, Radcliffe College, Harvard University, Cambridge, MA, USA.

Contents

List of Illustrations

List of Tables

Ruth Schwartz Cowan

Foreword: Musings About The Woman Engineer As Muse

A few years ago my husband and I spoke to a literary agent about a book we were planning to write, based on interviews with women engineers. This particular agent specialized in feminist books, so we thought she might be interested in our project. She wasn't. 'I can't imagine a more boring subject,' she exclaimed, 'than women engineers.' We were horrified – both by the speed of her unthinking response and by her blanket (and profoundly un-feminist) dismissal of a whole group of women.

Women engineers are not boring at all; in fact, they are fascinating – as a group and also as individuals. One reason is that they are rare. As many of the chapters in this book demonstrate, there is still a lower proportion of women in engineering than in any other profession. Why should this be so? No one knows for sure, although several scholars have proposed theories (about which more below). I raise the question here only to point out that in most fields of inquiry that which is rare is inherently worthy of study. As it is with prehistoric clay pots, so it should also be with women engineers: the fact of their scarcity is itself both interesting and instructive.

Women engineers are also fascinating because of the work they do. Engineers are the people who invent, develop, maintain, and supervise the technological systems without which modern societies could not exist. Without chemical engineers we could not process large quantities of food-stuffs or produce medications on a mass scale. Without civil engineers we would not have roads, bridges, and the multi-story buildings in which most of us live. Life and work would crumble. Without aerospace and naval engi-neers we would not have national systems of defense or peace-keeping forces. Without electrical engineers we could not power the machines, large and small, with which we do our daily work – whatever it may be – and without electronic engineers we could not communicate over a distance. Engineers, to put the matter another way, make it possible for us to feed, clothe, and house ourselves, which means that engineers, engineering and the engineer-ing sciences are central to our lives and crucial features of our economies – whether those economies are industrial, post-industrial or in the process of development. Of course, since I am an historian of technology, it is hardly surprising that I find engineers (and their work) intriguing. Other scholars (and policy makers also) might be equally fascinated if they could get beyond the stereotypes that say all engineers are boring people and engineering is work suitable only for technical drones. For reasons which will, I hope, become clear below, the study of women engineers can help us put many of these stereotypes to rest.

Women engineers are also brave women, which is a third reason to find them interesting. They are mold-breakers, people who have had the courage – both in youth and in middle age – to violate very strong social norms about gender roles. When they were in early adolescence, a time when most young people favor gender conformity, women engineers were willing to display mathematical skills that many modern cultures have defined as 'unfeminine.' Later on in their youth, many women engineers courageously defied institutions and demanded that barriers to their education be lowered. When necessary, many had the ingenuity to circumvent institutional barriers in order to obtain the specialized training they needed and deserved. Many women engineers, in the present as well as in the past, have also possessed the strength of character to persevere in the face of the slings and arrows of harrassment. Finally, as they moved into middle age, those women who sustained careers in engineering did so in the full knowledge that many of their peers regarded the work that they did – always technical, sometimes dirty, occasionally heavy, often supervisory – as stereotypically male work and, therefore, for women, inherently unfeminine.

Perhaps one reason that there are so few women engineers is that there are very few people who have the courage, especially in adolescence, to challenge pervasive gender stereotypes. Several articles in this collection suggest that similar gender norms have existed in most industrialized countries during most of the twentieth century, thereby helping to explain why the scarcity of women engineers appears to be a cross-cultural phenomenon. Those of us who would like the world to be populated with more courageous and creative people than it currently holds might be well advised to conduct more systematic studies of family environments and other conditions that produce women engineers. I wish I had had the presence of mind to think of this aspect of their lives when conversing with that literary agent; imagine her surprise if I had suggested to her that women engineers might have been among the first gender-benders – or, alternatively, that women engineers can teach us that there is more than one way to bend and twist a gender!

Interestingly enough, although women engineers have challenged gender norms just by the fact of their existence, many of them are not feminists, at least of the standard political variety. Political feminism has meant advocacy of many different things in different nations in the last two centuries – voting rights, divorce reform, protective labor legislation, educational opportunities, equal pay laws, and legalized abortion among them. As we learn more about the history of women engineers we learn that whatever has been defined as feminism in any particular time or place has been unlikely to attract them. Most women engineers tend not to engage in organized political activity and not to identify themselves as feminists. Perhaps a person can only rock one boat at a time. Having challenged so many gender stereotypes, women engineers may find that they need to remain apolitical (like so many of their male colleagues) in order to achieve professional acceptance. In any event, the fact that so many women engineers do not identify themselves as feminists ought to suggest to those of us who are feminists, and students of feminism, that we need to reorganize our thinking. Perhaps one can live a feminist life without necessarily being politically active. Perhaps, as some of the papers in this volume suggest, there is more than one form of feminist politics.

The papers collected in this volume – the first effort to look at women engineers systematically in an historical context and across cultures – raise a host of questions great and small, scholarly and practical. Some of these questions are directly related to women's participation and equality in the labor force. Others are indirectly – but very concretely – related to issues of economic development and growth. From these papers we can begin to assess whether it is appropriate, or wise, or perhaps misguided to treat women in general, and women engineers in particular, as a reserve labor force, dispensable except in military or economic crises. Thinking through some of the events narrated in this volume can help us to determine whether policy initiatives that worked in one nation or time period (for example, the establishment of technical colleges exclusively for women) should be tried in other contexts.

Some of these papers provide reason to believe that women engineers have different class and regional backgrounds than their male counterparts. In the United States, at least, most studies suggest that male engineers come from a lower socio-economic stratum than other professionals (such as scientists, for example) and that they are more likely to have been reared in rural, rather than urban or suburban, families. This does not appear, however, to hold true for women engineers; their families of origin tend to be better educated, more prosperous, more urban – and, possibly, less politically conservative. None of the studies collected in this volume contains definitive sociological or historical research on this issue, but they do provide leads to the archival materials that would make such an inquiry feasible.

Similarly, some of these studies suggest that engineering sub-disciplines have their own patterns of gender segregation, but that this does not mean one framework explains all cultures. For example, in one country women may tend to congregate in electrical engineering, while in another they are more likely to be found in chemical engineering. Or women may be entirely excluded from mining engineering in one national tradition, while in another they may find extensive opportunities in metallurgy. Clearly, in many countries it should be possible to follow up with careful research the suggestive evidence presented here in these preliminary studies.

Gender segregation in the engineering workforce is a topic worth pursuing because a larger theoretical issue is at stake: the issue of biological determinism.

If gender segregation in the engineering workforce turns out to be both local and contingent, then those who would argue that biology is destiny have a weaker foundation for their argument. My reading of the papers in this volume suggests that both biological and economic determinists will find little comfort between its covers. No author has been able to demonstrate conclusively that women engineers think or design differently than their male colleagues. Several authors have, however, provided us with grounds for believing that, in many places and times, men have gone to considerable lengths to prevent women from becoming engineers: something they would surely not have needed to do if women inherently lacked either the mathematical abilities or the personal traits requisite for success in the profession. Indeed women have had to overcome immense hurdles to obtain training and – once trained – have found difficulties in obtaining employment. This appears to have been true in good economic times as well as bad, in state-controlled socialist economies as well as in Free market economies, in countries in which most engineers work for the government as well as in countries in which most engineers work for private enterprises. The comparative evidence in this volume, suggests that the factors which explain the relative paucity of women in engineering are neither biologic nor economic.

Technology has, quite obviously, changed profoundly since engineering began emerging as a profession a century and a half ago. This has meant that the training engineers require and the work that they end up doing has also shifted radically. Several scholars have noted, in addition, that certain cultural stereotypes about engineers have undergone considerable change over this period: people who were once regarded as Romantic, Promethean heroes in the industrial era came to be seen as drones, functionaries, and robots in post-industrial decades.

Yet one aspect of these cultural stereotypes has remained remarkably consistent. From its beginnings, engineering has been gendered *male,* so that a women engineer has been, almost by definition, a non-conformist, an iconoclast, *the other*. In this sense, women engineers are the alter-egos to economically crucial actors, which means that they can teach us about the societies into which we were born and in which we currently live. In addition, almost by necessity, the effort to study people who are non-conformists, moldbreakers, and gender-benders, requires that we stretch our imaginations and reconstruct the categories into which we usually pigeon-hole our knowledge. To put it another way, it requires scholars to imitate their subjects and become mold-breakers themselves.

How foolish of that literary agent to conceive of such an enterprise as boring!

Ruth Schwartz Cowan
Professor of History and Chair of the Honors College
State University of New York at Stony Brook, USA.

Notes on Contributors

Boel Berner is professor at the Department of Technology and Social Change, Linköping University, Sweden. Her publications in Swedish, *The State of Things* (1996), *Gendered Practices* (1997), and *Perpetuum Mobile* (1999), all deal with the social and gendered construction of technical expertise.

Annie Canel received her Ph.D. degree in industrial organization and sociology. She has been working first as senior researcher in social sciences at the *École Nationale des Ponts et Chaussées*, Paris, France, and currently at the Ministry of Transportation. She is on the executive board of the Society of ENPC Graduates. Since her Fullbright scholarship in 1992, she has been working on the history of engineering and gender cross-cultural perspective.

Konstantinos Chatzis has been a researcher in the social sciences at the *École Nationale des Ponts et Chaussées*, Paris, France since 1993. His specific interest is the long-term development of large socio-technical systems of the nineteenth and twentieth centuries, the history of engineering education, and nineteenth-century history of mechanics. He has published numerous articles, and a book on the history of storm water drain systems entitled, *La pluie, le métro et l'ingénieur* (Paris: Harmattan, 2000).

Irina Gouzévitch is a researcher and ingénieur d'études at the Alexandre Koyré, Center, at the *École des Hautes Etudes en Sciences Sociales*, Paris, France. Her research is on the social and comparative history of engineering in the eighteenth and nineteenth centuries.

Dmitri Gouzévitch is a senior researcher at the Institute for the History of Science and Technology at the Russian Academy of Science, St Petersburg, Russia and a researcher affiliated with the *Centre d'Histoire des Techniques du Conservatoire National des Arts et Métiers*, Paris, France. His research is on the history of the engineering practice in Russia and Europe from the seventeenth to the nineteenth centuries.

Moniko Greif received her Ph.D. degree in mechanical engineering from the Technical University of Darmstadt, Germany, where she wrote her doctoral thesis in production technology and machine tools. After working for 13 years in applied research and automotive industry she is now professor of production technology, quality management and computer aided techniques in production and quality assurance at the Department of Mechanical Engineering of the Fachhochschule Wiesbaden (University of Applied Sciences) and also became its vice-president. She is co-founder of *Women in Natural Science and Technology* (NUT), former chair of *Women in Engineering* section of the Association of German Engineers and has published widely on high speed milling, tools, fatigue of welded structures, women in engineering, and feminist technology.

Juliane Mikoletzky is university archivist at the Vienna University of Technology and lecturer at the Department of Social and Economic History at the Vienna University in Austria. Her research interests include the history of engineering profession and education, of higher education, and of the labor market and technological change. She recently edited a book on the history of women engineers in Austria.

Efthymios Nicolaïdis is senior researcher in history and philosophy of science at the National Hellenic Research Foundation in Athens, Greece. He has published books and articles on the history of science of the Greek world during the Byzantine and Modern periods.

Ruth Oldenziel received her Ph.D. from Yale University in American history (1992) and is associate professor at the University of Amsterdam, the Netherlands. She has been working on engineering cars and household technology. She coedited *Technology and Culture's* special issue on gender and is the author of *Making Technology Masculine: Men, Women, and Machines in America, 1880–1945* (Michigan: Michigan University Press, 1999).

Carroll Pursell is Adeline Barry Davee Professor of History and Chair of the Department of History at Case Western Reserve University, Cleveland, Ohio in the United States. Over the years he has written several articles on gender and technology. Most recently, he is the author of *White Heat: Technology and People* (Berkeley: University of California Press, 1994) and *The Machine in America: A Social History of Technology* (Baltimore: The Johns Hopkins University Press, 1995).

Ruth Schwartz Cowan is Professor of History and Chair of the Honors College at the State University of New York at Stony Brook, New York, USA. A former president of the Society for the History of Technology, she is the author of *More Work for Mothers* (New York: Basic Books, 1983) and of *A Social History of American Technology* (Oxford: Oxford University Press, 1997).

Annette Vogt is a research scholar at the Max Planck Institute for the History of Science in Berlin, Germany. She has published articles on the history of mathematics, ties between Russian/Soviet-Union and German science communities, and Jewish and women scientists in Germany during the nineteenth and twentieth centuries.

Karin Zachmann is a researcher in the Institute for the History of Technology and Technical Sciences at the *Technische Universität*, Dresden, Germany. She is completing a study of women engineers in the state socialist society and teaching the history of technology. She has published widely on German industrial history in the nineteenth and twentieth centuries and on the gender division of labor.

Ruth Oldenziel, Annie Canel, and Karin Zachmann

Introduction
Crossing Boundaries, Building Bridges: Comparing The History of Women Engineers, 1870s–1990s

Whenever women have entered the engineering profession they have crossed cultural and social boundaries. This book is about when and why they crossed boundaries, what passports they needed to do so, and what strategies women employed to demand their passage into the male world. *Crossing Boundaries, Building Bridges* presents studies from the US, East Germany, Czarist Russia, and several European countries. The articles offer a panoramic view of the experiences of women engineers from nine different political and economic situations. Each nation underwent industrialization at a different moment, but all needed engineers to staff and supervise the process. Cross-cultural comparison shows that women engineers were not a rare species, but always part of history: they were important economic and political actors within the process of industrialization. Women engineers have occupied their places in the world of engineering long before the 1960s and as early as the nineteenth century. Bringing these diverse histories together in one volume allows us to see how varied were the passports they needed and the means by which women sought to cross the borders. Women faced male engineering institutions associated with the military throughout the nineteenth century, yet sometimes women were allowed to enter the male bastion with special visas when the engineering elite saw the need for women's labor. Allowing women to study or work in engineering was often part of expanding the engineering agenda to meet new industrial needs. Competition among different parts of the male elite also created small windows of opportunity for women. Singly and collectively they worked with reform-minded men to enter the profession in any way they could. War economies, hot and cold, offered other opportunities, but women were often shut out after peace was restored. In many countries, the presence of a women's movement and the mobilization of a feminist intellectual legacy helped women enter engineering, but success depended partly on how women engineers and feminist groups were able to build bridges between them or other reform movements. All of these things – the industries and circumstances of war, competition and social ties among male engineers, and women's collective organization – helped to shape the barriers women encountered and the means by which they forced their way through.

Recognizing these common factors is useful, but it is not the whole story. Their convergence happened differently in each place, which helps to explain why the timetables by which women entered engineering deviated so dramatically. Why did American and Russian women graduate from engineering programs before the first World War, though Austrian women only managed this a few years later after the collapse of the Habsburg Monarchy? Why are women in Greece now at the forefront of engineering education, even though, on the face of it, Greece used to be a traditional, patriarchal culture until recently? Why did women find employment in the war economies of Nazi Germany, East Germany, and the United States from the 1930s through the 1950s, but not in France? All of these differences, some of them quite surprising, are explained if we understand how industrialization, women's emancipation movements, and the development of male professional culture happened in each country. This helps to clarify why early women engineers in Britain specialized in electrical engineering, while in Greece it was architecture and civil engineering, and in France aeronautical engineering. And it helps us comprehend how women came to organize their own engineering society in Great Britain as early as 1919 but did not do so in the United States (ostensibly having a similar professional and educational structure) until 1949.

The history of engineering is incomplete without consideration of women and gender. The creation of a professional engineering identity depended on the erection of boundaries and their maintenance. Both education and work experience were integral to this process. Several articles in this volume demonstrate conclusively that such a process is never just about class status, but always has a component of gender: engineering was not only a male profession, but it was also coded as masculine. This had profound implications for the ways in which engineering education was structured to include and exclude various social groups. Historians of technology have discussed at great length the cultural differences between engineering training in the workplace and that which takes place in the school room. Looking at women's experiences alters the debate. Women were often able to use education as a temporary means of entry into the profession. Yet when the engineering elite insisted on the importance of workplace experience – a form of training often tied to management and coded male more than the one based on formal knowledge of science and math – this became a mechanism of exclusion. Histories of the profession have also suffered by taking a top-down approach, ignoring the experiences of the rank and file who staffed industrialization throughout the nineteenth and twentieth centuries. Focusing on women's experiences within the profession helps us to correct this imbalance in the historiography and better understand how the world of engineering works. Over the past decades, scholars of gender have mapped women's entry into numerous occupations, from teaching to medicine. In comparison to these, engineering has seemed entirely male-dominated and women's efforts to enter it a catalogue of lost time and wasted efforts, with only a few individual heroines. Based on this incorrect view one can see why, as Ruth Schwartz Cowan says in her foreword to this volume, engineering has seemed boring to feminists. Nevertheless, this book demonstrates that, although engineering had fewer women than other professions, its development was similarly complex and worthy of study. It is necessary to go beyond counting the number of women in a particular occupation, and to put individual stories in historical context. Doing this illuminates the importance of political movements and systems, and it also helps us see the structure of the labor market more clearly. Previous scholarship of women and the professions has been much concerned with the question of whether women enter an occupation because it is losing prestige, or whether their presence in fact causes such a decline. This volume cannot offer a definitive answer to this chicken-and-egg question, but it proves that the two developments occurred simultaneously. While it is difficult to chart a direct causal relationship between the women's entry into specialized areas of engineering and the status of those fields, case studies here suggest that the two developments often went hand in hand.

These historical insights should also be of use to those interested in policy issues. The historical record shows that there is no limit to where women will work: women have been employed from the engineering trades to theoretical ballistics, and from positions on shop floors, building sites, and laboratories to supervisory work in consulting and research. Good working conditions, too, foster the willingness of women to stay in a job. This is shown clearly in the case of the German Democratic Republic, explored in the chapter by Karin Zachmann. Compared to other countries women there had excellent prospects for career advancement for a whole series of structural reasons: increased employment of engineers, the establishment of state-organized child care facilities, the weakening of a tax system that had been based on the single family breadwinner, and the upgrading of women as engineers. In contrast, unified Germany only pays lip service to women's advancement, and women engineers have lost much ground. If a state wants to encourage women to enter engineering and stay there, it must see that its policies are sufficiently systematic and sustained to be effective.

The book thus focuses on the specific circumstances under which women entered engineering in nine different countries.

Engineering in the eighteenth and early nineteenth centuries was an elite male profession and, in many countries, closely linked to the military. Young men who trained in engineering were groomed to build the infrastructure for the new nation states; their career paths followed military lines. However, advocates of modernization, who sought to implement policies that would promote industrial capitalism, challenged the nature of the profession. In response, the engineering elite realized that engineering needed to bring new groups into the technical world in order to foster economic growth. Therefore, they reached out to the industrial and agricultural classes below them. Whether they looked to ethnic groups, colonial subjects, or women to become the new rank and file of the profession depended on the country's political landscape and demographic composition. But there was always much at stake, for the old elite wanted to preserve its position as best it could under these new circumstances. The nature of the political system and the specific way engineering work and education was organized determined how each nation dealt with the integration of new groups while pushing for industrial policies.

Centralized countries such as France and Czarist Russia invested a great deal in the military for their national identity. Their educational systems for engineering emphasized science, math, and theoretical training, and groomed a small sample of young men to be leaders. As Irena and Dimitri Gouzevitch and Annie Canel demonstrate, in both countries, the engineering elite then sought ways to maintain its high status and to preserve a strong core culture by segregating 'the other',' including women. It is significant that only Russia and France had single-sex engineering schools for women: the St. Petersburg Polytechnic Institute for Women (1905–1917) and the French Ecole Polytechnique Feminine (1925–1993). These polytechnic institutes for women stressed the hands-on practical and experimental training of the labs for applied-science and industrial research. Despite the similarity between the two, however, these women's schools were established under very different political circumstances. The St. Petersburg Polytechnic Institute For Women was born in the wake of the revolution of 1905 and facilitated by a temporary alliance between the Russian feminist movement and reform-minded male engineers. By contrast, as Canel shows, neither a revolution nor a feminist movement gave birth to the French school. One pioneering woman, Marie Louise Paris, responded to raised expectations for women's job openings during the First World War and the dwindling opportunities in peacetime. She argued for a separate institution that would meet the needs of young French women aspiring to prestigious engineering work in industry. In short, creating separate educational institutions for women was one way to let those who were not part of the elite enter the profession. Although separate, these schools proved to be important training grounds for women engineers. In contrast, the multi-ethnic and autocratic Habsburg Empire successfully kept women out of education until it faced a political collapse. Nothing could break the resistance of the Austrian administration to women's higher technical education: not the women's movement, which demanded technical education for women as early as 1908, and not the liberal engineering elite at some Bohemian technical education colleges, who pleaded for the admission of women from 1903. Instead, as Juliane Mikoletzky describes in her case study of Austria, the educational ban against women in engineering was lifted only when the political system fell apart.

By contrast, in Britain and the US engineers did not have a central, state-run agency to certify them. In both places the educational system was connected to upper-class economic and political power, not to the state. The success of these two countries in fostering industrial growth has therefore been attributed to the ability of the political and industrial elite to merge into one powerful class and to integrate 'the others' more easily into their system. This certainly explains why Ruth Oldenziel finds women graduating from an engineering school in the United States as early as 1867. It also allows Carroll Pursell to describe how British women engineers became numerous enough to form a fully-fledged association in 1919. The success of the British women was due in part to organizing themselves across class lines, uniting women in the engineering trades with the daughters who worked in family firms. British engineering culture was based on small companies, craft traditions, working-class associations, and kinship relations. Some daughters of family engineering firms also had links to the burgeoning women's movement. The association that resulted in 1919 was an inspiring example to women engineers in United States, Sweden, and Germany. American engineering, however, was tied to growing corporate and federal bureaucracies. Hence it became a mass, middle-class occupation with a hybrid form of professionalism, increasing emphasis on school training, and an almost knee-jerk aversion to traditional blue-collar unions.

Despite the relatively openness of the British and American systems, women who were not the daughters of small male-led family firms had difficulties. They tended, despite having high-level academic training, to end up in segregated labor markets within the large federal and corporate bureaucracies. Professional engineering organizations also turned out to be mechanisms by which male engineers excluded women. Whenever an organization strove to follow a classic professional model, with a high level of occupational control, women lost out to exclusive male jurisdiction and privileges. Women had a relatively easier time being accepted in those engineering societies associated with a business culture and lower professional standards such as the US Institute of Mining Engineers. Still their numbers remained small. Thus the relative openness of political and economic systems was no guarantee of women's full access to engineering: professional standards and segregated labor markets were serious obstacles.

Sweden is an interesting case study through which to study how a political system that was both open and closed integrated or excluded new groups. The Swedish elite was able to recruit from the agricultural hinterlands and were relatively successful in incorporating new social groups and thus industrializing the country. But Sweden also had one central agency certifying engineers and was thus better able to preserve a male identity which was closely linked to the nation state. The reproduction and maintenance of the male culture of engineering there was particularly remarkable given the long tradition of women's participation in the labor force. Boel Berner argues that, because Sweden did not participate in the two world wars, a shortage of labor did not challenge segregation in work as it did elsewhere. She illustrates how the Swedish Royal Institute of Technology groomed men for state service and industry and reproduced a specifically male elite. The Swedish women who entered the Royal Institute were too few to organize collectively or to make a dent in this vigorous masculine culture. In many countries, including Sweden, the different efforts to maintain the character of an elite culture while facing new challenges, engineering was eventually transformed into a mass occupation. The emergence of the twentieth-century industrial-military complex also generated the reorganization of engineering work both in industry and government. The alliance of corporate research and the state created niches for women to enter engineering work in countries that were gearing up for war, hot or cold. New industries created the need for new engineering knowledge and upset old categories within the engineering elite. Women began to enter the newer laboratory-oriented branches of engineering (chemical, electrical, and aeronautical) in part because these special areas were not yet firmly coded as masculine. They were still in the process of gaining professional status and needed new recruits more than the established fields, such as civil and mechanical engineering. This industrial-military complex took shape in all the countries under review here at different times and under different political systems. This had profound implications for women's engineering work and professional accreditation.

Wars – cold, hot and heating up – pushed the integration of 'others' into the system to a higher and more intense level. But they also intensified the development of long-term patterns in gender roles and the shape of the military-industrial complex. The First World War is an excellent example. Of course, the need for men to serve at the front meant there was a window of opportunity for women in technical fields. In Britain, for instance, where general engineering firms had employed 134,600 women as engineers during the war, as many as 800,000 took up munitions work of various kinds. They worked in the shipyards, aircraft factories, foundries, machine shops, shell-packing plants, and explosives factories. The First World War was a watershed in another sense: it marked the transformation of engineering from an elite to a mass occupation, though in each country this took a slightly different form.

The world wars both created and reproduced normative gender practices. For example, during the Second World War, it was an innovation of American federal policy to coin the term 'engineering aide' for the hundreds of thousands of women working on different engineering programs. But the label 'engineer' was employed only for men. The job title of 'engineering aide' thus worked to reinforce existing gender patterns. Titles were, of course, related to pay, so women remained badly paid and segregated within the profession. They were also temporary. Women had been encouraged by government propaganda agencies and the women's movement to think that education would lead to jobs, but in the aftermath of war they always faced very difficult times. Just after the war, in countries as diverse as Austria, Sweden, and Greece, women graduated from engineering programs, but were faced with the harsh realities of peacetime conversions and crises of the economy during the inter-war period. In most countries, with the possible exception of Nazi Germany, the 1930s were not good times for a woman to get a job in engineering. The brilliant aeronautical engineer Irmgard Flügge Lotz has been celebrated as a notable American for her work in the 1960s, but Annette Vogt shows how important Nazi army research was in her pre-emigration years. There was a cohort of such women whose history Vogt tells. In Germany, the Kaiser Wilhelm Association, which had been founded to aid the country's industrialization through prestigious research programs in 1912, offered positions to some qualified women scientists. The institutes led research in the new fields of applied science, such as chemical and aeronautical engineering. This case study shows that it was not enough that new fields be open to women. To succeed, women also needed the patronage of powerful men. In Germany, certain leading scientists were offered almost unlimited power to build programs of their own, and to hire and fire researchers at will. Some of them recognized the abilities of academically trained women and helped them to advance their careers. Essentially these institutes were patriarchal fiefdoms, but they proved a training ground for those women who were not the target of Nazi purges. Hence they were soon drafted into the Nazi war programs where they sometimes had prestigious supervisory positions. Unlike elsewhere, the late 1930s thus offered an opportunity for some German women scientists in the emerging military-industrial complex.

In countries such as the US and the Soviet Union, the 1950s saw a massive build-up of the military-industrial complex. Here, too, women were drafted in to help the progress of the Cold War. Because France had its own defense program from the late sixties, women entered in this cold-war equation much later. The battle between capitalism and communism generated a call for technical personnel to meet the needs of industry in America and the Western block. This was a panicked response to the technological euphoria of East Germany and the communist world, which sought to mobilize science and technology for the new communist system. In East Germany it was express policy to mobilize women within the restructuring of the society and economy. Such efforts resulted in a jump of women's participation in the engineering student population from four per cent to thirty per cent. Special women's classes in engineering were set up, child-care facilities instituted, and novels honored women engineers as newly important social actors. Yet even in this restructured society, scores of East German women found themselves in segregated labor markets – whose existence had been reinforced, not eradicated, by the new state socialist policies.

Mobilization for war granted some women passports to pass male gate-keepers, but the existence of an active women's movement was also crucial. The feminist movement in Russia, which crossed class lines, was instrumental in shaping radical and reform politics throughout the nineteenth century. It also helped to create a separate women's engineering institution whose graduates went to pursue long and successful careers. British women engineers started to organize collectively to counter the efforts of patriarchal unions and male engineering societies to push women out of the profession in 1919. The British women were successful not only because of the cross-class alliance, but also because some of the middle-class women involved had connections to the feminist movement. Successful collective organization depended on building bridges between women in engineering and women in the feminist movements. American women engineers failed in a similar effort in the same year because middle-class identity prevented them from organizing across class boundaries. It would take another war and another peacetime purge before they had enough members to organize themselves in 1949.

The connection between the entry into engineering and women's social and intellectual movements is not always easy to establish. Most authors here note the women took the step to enter engineering during the 1920s when the women won the vote in most countries under examination. The great exception is France, where the women's movement had been active much earlier, long before male engineers were looking for ways to expand industrial growth, but where the vote came to women as late as 1944. Therefore the French case is a story of missed connections between women in engineering and those in the women's movement. It was the First World War that set the stage for one strong-willed individual to make a difference. Her actions rather than the women's movement paved the way for many generations of women to enter engineering, even though they had to do so through a separate educational system and never made a dent in the French male engineering culture.

Finally, we consider the intriguing case of East Germany. While communist policy makers, like their capitalist counterparts, saw women as a reserve labor force, remarkably, they also celebrated special programs for women engineers as a triumph of socialist thinking. Communist ideologues argued that the advocacy of women's equal rights was unnecessary because socialism would insure women's equality. But at the same time, anything that showed women's equality was also a sign that socialism was really working. The celebration of women engineers was a co-optation of feminist ideals into communist policy, which then became an inspiration for feminists in the West.

Recent developments in Greece and other countries show that historical connections are complicated. As Kostas Chatzis and Efthymios Nicolaïdis explain, the demilitarization of Greek engineering, the political upheavals of the First World War, and the activities of the women's movement all contributed to a climate of rising expectations for women in the 1920s. Greek women, like their counterparts elsewhere, were beginning to enter engineering, but they followed rather than led a trend. However, today Greece leads: in the nineties, one third of all graduating civil and chemical engineers are women. And these figures continue to climb. This is despite the fact that there are no special government or industry sponsored programs, as there were in East Germany, to encourage women. Nor is there a vocal women's movement. However, these impressive figures come at a time when the prestige of engineering is declining. We are left with the chicken-or-the egg question: did women cause this loss of prestige, or have they merely suffered from it? Traditionally, women engineers in Greece were employed in federal bureaucracies that paid less than the jobs in the private sector, which was dominated by male engineers. Because they worked for the government, their income did not drop when the economy was poor, and they made much higher wages than women in other occupations. Recently, however, male engineers have moved into these public sector jobs because of dwindling opportunities in private construction. What does it mean? Women make up one third of graduating engineers in some fields, yet should this be considered a Pyhrric victory? Rather than phrase it as winning versus losing, though, we must use this case and others to understand the complexity of women's opportunities. The social, economic, and political structures that have shaped the organization of the engineering profession in each of these nine countries were diverse, and comparisons among them have limitations. The reasons individual women were motivated to enter engineering work were (and are) complicated too. Taken together, these nine case studies suggest we must not jump to easy conclusions. Women engineers were never only victims nor heroines in the struggle against male oppression. Rather they were active agents of history, facing real issues and dilemmas. We hope this collection of essays can illuminate the difficulties they faced and the choices they made.

Acknowledgements

Crossing Boundaries, Building Bridges is the product of a cross-cultural partnership. A major challenge that we faced in conducting this project was to bring authors together to meet and articulate their ideas. This challenge could not have been met without the kind support of many people and institutions.

We are particularly indebted to Dorothea Schmidt who supported our project from its inception; her helpful comments and suggestions helped to clarify and sharpen our thinking. We would like to thank Carroll Pursell and Ruth Schwartz Cowan for offering their expertise and for making our meetings so fruitful, Karin Hausen for providing us with helpful insights, and Agneta Fisher for asking some tough questions on doing comparative work. Our thanks, also, to David Hounshell for his encouragement, and to Ulrich Wengenroth for his kind support and wise advice. We want to offer special thanks to John Krige, our editor, who had the confidence to support and encourage us even though we approached him with a project that was in its very early stages. His trust helped impel us from ideas to a finished product.

We also wish to thank Tobias Dörfler for his assistance in gathering illustrations. So too we extend our thanks to Raymond Hofman for general assistance. We are particularly grateful to Caitlin Adams and Renee Hoogland, who took on the difficult task of producing a consistent style throughout the book. Our thanks also go to the authors for their willingness and patience to accept stylistic changes and rearrangements.

This book benefitted greatly from the financial and institutional support of the University of Amsterdam's Belle van Zuylen Institute, the University of Amsterdam Affirmative Action Fund, and the Dutch National Science Foundation (NWO), who provided the funding to get the project under way. We also wish to thank the French EPF (former *Ecole Polytechnique Feminine*) for its generous support during the writing of this book. The Adenauer Foundation of Germany offered us the great opportunity to present our results to a broad audience by funding and organizing a conference on *Frauen(t)raum Technik* in Berlin in January 1999.

We also want to acknowledge the help of industrial firms who strongly supported this project. We are particularly indebted to Cofiroule and L'Oréal for their generous support of our work. Our project also benefited a great deal from the support of Schlumberger. We are grateful for their interest in our topic, and for their financial support, which made this book possible.

Ruth Oldenziel

1. Multiple-Entry Visas: Gender and Engineering in the US, 1870–1945[1]

In 1978, almost hundred years after the first woman had graduated from Iowa State University's engineering program, Samuel Florman wondered in *Harper's* magazine why women failed to take the same existential pleasure in engineering that he advocated for men.[2] Florman, a spokesman for engineers, claimed to have found the answer when visiting the lofty halls of Smith College, a private women's college with an elite reputation. He could imagine the smart women students – all well trained in math and the sciences – 'donning white coats and conducting experiments in quiet laboratories.' But he could not see these sensible, bright young women becoming engineers. '[I]t is "beneath" them to do so,' he said. 'It is a matter of class.'[3] He believed that, given the option, Smith women with strong science and mathematical abilities would choose the sciences rather than the low-status profession of engineering. Florman then made a call to feminists to solve the problem of status in engineering by entering the profession. Unfair, to say the least! But his observation pushes us to take a closer look at the importance of class relations in women's entry into engineering.

In the US, engineering changed from an elite profession to a mass occupation, grew the fastest of all professions, and developed specialist areas at a great pace after the 1890s. American engineering was also a deeply divided and segmented profession. Engineers could be found working anywhere from board rooms to drafting departments, mechanics workshops, and chemical labs. They worked as executives, managers, designers, draftsmen, detailers, checkers, tracers, and testing technicians. American engineering did not become a closed profession associated with science as it did in France, where the state groomed a small elite for leadership positions. Nor did it resemble British engineering culture with its small firms, craft traditions, working-class associations, and kinship networks. Instead, American engineering evolved into something between the French and the British models: a mass middle-class occupation with a hybrid form of professionalism and an almost automatic distaste for blue-collar union organization.[4] Compared to the profession in France and Russia, engineering in America was a relatively open affair: doors were opened early to newcomers from the lower classes and different ethnic backgrounds to staff rapid industrialization. There was neither a central agency or professional organization to certify engineers, nor a national educational system. American engineers were trained either informally on the work floor or formally at a variety of engineering institutions.[5]

The openness of America's engineering system had its limits, however. Engineering advocates in the US were eager to maintain professional prestige, which proved fragile. Here gender and class were linked: even though in the United States the profession was divided and diverse, professional engineering organizations and educational institutions often used sex discrimination as a means by which to preserve class distinctions. At the same time, the fact that the profession lacked central organization nevertheless opened a small window of opportunity for some women. Both engineering educators and the state mobilized women trained in the sciences for their own purposes in times of war (cold and hot). As they faced changing coalitions of men, and old-fashioned hiring and promotional practices, women engineers developed many individual and collective strategies to force their way into the profession.

Bridges of Sisterhood: Shaky at Best

Few women engineers publicly rallied to the feminist movement or its causes.[6] This disinterest was mutual.[7] The American women's movement focused on the sciences rather than on engineering because the latter lacked cultural authority. Unlike their Russian, Austrian, and French sisters, American philanthropists and advocates of women's education neither paid any special attention to the engineering profession as a vehicle for women's equality, nor helped to establish separate engineering institutions.[8] Although in the United States, it was rapidly becoming a mass occupation and the few women in engineering who entered came from a higher class background than their male counterparts, engineering failed to attract the many young elite women looking for a suitable vocation. Raised on the propaganda of the First World War, when the government desperately tried to recruit more technical personnel, a generation of academically-trained women engineers grew up in an era when suffrage had been won and when professional women seemed to make headway. They adhered to a belief that professional status and advancement were based on merit, having nothing to do with gender.

Among American women engineers, only Nora Stanton Blatch (1883–1971) envisioned as mutually enriching the two sides of her life as an engineer and a feminist. (Figure 1) As a third-generation suffragist, Nora Blatch could see herself in a feminist genealogy. She descended from a line of famous activists (her mother was Harriot S. Blatch, and her grandmother was Elizabeth Cady Stanton) and campaigned for suffrage at Cornell University, where she had chosen civil engineering as her major: because, she said, it was the most male-dominated field she could find. Her generation of women engineers grew up in the nineteenth century, when the bond of solidarity among women was more firmly

Figure 1. Nora Stanton Blatch, civil engineer and descendant of two generations of women right activists on suffragist campaign on horseback in New York State in 1913, challenged the engineering establishment in 1916 on charges of discrimination. Reproduced from Civil Engineering *(1971). Courtesy of Delft University of Technology, Delft, The Netherlands.*

entrenched, but Blatch went a step further than her contemporaries. She accused the American Society of Civil Engineers (ASCE) of sex discrimination when in 1916 it tried to bar her from full membership; moreover, she campaigned for pay equity between men and women through the National Woman's Party for many years. Where Blatch found literary and organizational models for professional women in the feminist movements of her mother and grandmother, she found none among her women colleagues in engineering.[9]

Most American women engineers ignored any link between the women in the technical field and those working in the woman's movement. They honored instead an alternative model offered by Lillian Gilbreth (1878–1972). Gilbreth borrowed Rudyard Kipling's rewording of the biblical allegory of Martha ('simple service simply given') as the inspiration for women engineers. Gilbreth, holder of a PhD in psychology, had acquired her technical knowledge and her legitimacy in engineering through her husband – a 'borrowed identity,' as Margot Fuchs terms it, that she expertly managed. She was actually a widow for much of her working life, but she projected an image of herself as a married career woman.[10] Avoiding the open confrontation that Blatch chose, Gilbreth advocated a professional strategy for women engineers based on hard work, self-reliance, and stoicism.

Such a strategy amounted to what Margaret Rossiter described as 'the classic tactics of assimilation required of those seeking acceptance in a hostile and competitive atmosphere, the kind of atmosphere women heading for bastions of men's work encountered at every turn.'[11] These were: 'quiet but deliberate over-qualification, personal modesty, strong self-discipline, and infinite stoicism.' Indeed, American women engineers maintained their loyalty to male models of the profession at great personal cost. Vera Jones MacKay, a chemical engineer who had managed to find work on pilot plants for fertilizer production with the Tennessee Valley Authority, recalled painful memories when looking back on her career in 1975: 'it is hard to discuss my working days as an engineer without sounding like one of the most militant of the women's libbers.'[12] Jones's public admission of the personal costs involved in her career choice is unusual because most women engineers kept a stiff upper lip in the constant struggles before affirmative action. Most women engineers preferred Lillian M. Gilbreth as their role model to her contemporary Nora Blatch. Hence, women engineers drew their literary models and organizational forms from their engineering fathers rather than from their feminist sisters. Before the Second World War, American women engineers – unlike their British counterparts – cultivated silence as a survival strategy and ventroloquized discontent without ever directly articulating it. As rank-and-file members of the profession working for corporate and military establishments, American women engineers not only became invisible to themselves as a group, but also to history, not least because of the failure to build bridges with women of the women's movement or those of the Progressive movement who helped shape the infrastructures in cities.

Surrogate Sons and the Family Job

A great many American women found their way into engineering through what might be called patrimonial patronage and matrimonial sponsorship. Most of them never appeared in any statistics because their ties to engineering were through their fathers, brothers, and husbands. Supported by kinship ties, such familial patronage and sponsorship often offered relatively easy access but also resulted in what we have already seen in the case of Lillian Gilbreth: a borrowed identity. Most women in such circumstances did not have female role models but looked to their fathers and brothers. Their identity as engineers was therefore largely 'on loan,' even if some, like Gilbreth, managed to stretch the terms of the loan to the limits of social approval.[13]

Formal training was an important credential for continental European countries such as France and served as a wedge into other professions in the US. But it did not play such a decisive role in employment opportunities in American engineering before 1945. By the end of the forties and beginning of the fifties, only fifty-five per cent of American men – and even fewer women (twenty per cent) – who worked in engineering had completed such training. In the nineteenth century and probably well into the twentieth century, a few hundred women continued to manage their late husbands' engineering work, having received enough informal technical training to call themselves engineers.[14] Others besides Gilbreth became well-known after learning their trade through family and husbands. They acquired technical knowledge on the job or through an informal system of education within family firms without ever attending a specialized school. Emily Warren Roebling (1841–1902), who kept the family firm going when illness kept her husband housebound, acted as his proxy throughout most of the building of New York's Brooklyn Bridge in the 1870s and 1880s. Trained in mathematics, she learned to speak the language of engineers, made daily on-site inspections, dealt with contractors and materials suppliers, handled the technical correspondence, and negotiated the political frictions that inevitably arose in such a grand public project. The Brooklyn Bridge had been a Roebling project on which the family's fortunes depended, and Emily Roebling served as her husband's proxy for decades. Most wives, however, worked anonymously in family businesses. As late as 1922, a woman active in civil engineering wrote to the editor of *The Professional Engineer* saying that she greatly appreciated that the journal finally acknowledged the wives of engineers without specialized degrees: 'My training in engineering began with marriage and I have filled about every job … from rodding and driving stakes to running a level party, or setting grade and figuring yardage in the office.'[15]

Most such women would be forgotten. Lillian Gilbreth, however, became America's most celebrated woman engineer, in part because she used her borrowed identity to do the widest possible range of work while still appearing to conform to social norms. Frank Gilbreth's untimely death in 1924 must have been devastating for a mother with twelve children but, aided by a team of domestic hands, it also allowed Lillian Gilbreth to enjoy considerable freedom in her role as a widow for nearly fifty years. She expertly managed her image, fostering publicity that cast her in the role of a married career woman. This public persona provided perfect protection against possible disapproval of her career ambitions. Similarly, long after Frank Gilbreth's death, she allowed her marriage to be publicized as a partnership to be emulated: the perfect, most efficient match between business and love. Such a partnership was analogous to her husband's advocacy of performing tasks in the 'one-best-way' for maximum benefit.[16] As Gilbreth's strategy shows, family connections might guarantee work, and they could also serve to protect women engineers from public scrutiny. Newspaper reports and government propaganda played up women engineers' strong family ties to men as a way to ward off any possible threat from these female incursions into the male domain.

The effort to 'domesticate' women's talents into familiar categories prevailed during the Second World War, when war propaganda emphasized women's family ties to engineering. Most of the available biographical information on the social background of women engineers was generated as part of this campaign to attract women engineers. Historical narrative sources on women engineers therefore tend to overexpose women with family connections. Nevertheless, it is clear that formal engineering education, with or without a degree in hand, could be useful for some, such as the daughters of proprietors of small manufacturing firms. For example, Beatrice Hicks trained at the Stevens Institute of Technology, going on to become chief engineer, then vice-president, and finally owner of her father's Newark Controls after his death. After her graduation in engineering Jean Horning Marburg supervised the plant construction for her family's mining property in Alaska. Florence Kimball was another graduate who worked at her family's elevator firm, drafted plans for the remodeling and building of its real estate property, and drew several blueprints for patents – the most exacting of all draftsmanship. Small family firms like the Kimball and Horning companies not only tried to maximize production and profits, but were also in the business of building and maintaining a family legacy.[17]

Succession in patriarchally organized family firms was exclusively an affair between fathers and sons, but circumstances sometimes pushed daughters into becoming surrogate sons. The most celebrated and best-documented case is that of Kate Gleason (1865–1933), the oldest daughter of William Gleason. He had started his own toolmaking shop, the Gleason Gear Planer Company in Rochester, New York, which later became one of the largest of its kind. Encouraged by the example of early feminist Susan B. Anthony, and prompted by the early death of her half-brother, Kate Gleason began to take courses in mechanical engineering at Cornell University and the Mechanics Institute in Rochester. Her training followed the course of many sons of other family manufacturing firms, who no longer were expected to master a craft completely but had to have a working knowledge of all the various aspects of the firm. Gleason was her family firm's business manager for many years while the business grew dramatically, becoming a major player in the industry.[18] Patrimonial patronage thus encouraged daughters like Kate Gleason to seek formal education with or without completing a degree because it fit into a family business's strategy. The link between business sense and family interests could make such education and work acceptable.

For similar reasons husbands encouraged their wives to seek formal training. Such matrimonial sponsorship not only gave legitimacy to women's engineering accomplishments, but also offered them the hope of establishing a firm in partnership with their husbands. The pooling of resources of man and wife in an enterprise afforded the opportunity for a partnership of business and love. Sometimes, however, engineering marriages could turn into a liability. Many women met their partners at college or in the field of engineering, allowing them to enter into male social and study circles otherwise closed to them. But because of the inherent power inequality in such mentorship, such relationships could turn into a distinct disadvantage for the wife's career advancement later if she questioned the terms of her borrowed identity or the matrimonial sponsorship.[19]

As a young feminist activist and engineering graduate Nora Blatch and her husband, the engineer and inventor Lee De Forest, shared in the excitement of new emerging technologies such as the radio at first. But in the end they disagreed about who was to shape and direct the possibilities of these developments. On their first meeting, Blatch 'tremendously admired' the young radio inventor Lee De Forest and recalled that 'a life in the midst of invention appealed to me strongly.'[20] For his part, De Forest thought 'destiny' had brought her to his door and pursued her relentlessly. In desperate need of money for various ventures, he accepted funds from his future mother-in-law Harriot S. Blatch, while Nora's technical training, her love for music, and the connections with the New York powerful, brought enormous technical, financial and social resources to his flagging career. No doubt seeing an opportunity to fulfill her life's goal of combining career and marriage, Nora Blatch took extra courses in electricity and mathematics with Michael Pupin, a well-known New York electrical engineer, and worked in De Forest's laboratory on the development of the radio. Together Lee and Nora were able to air the first broadcasts of music and conversations in the New York area. On their honeymoon to Paris, the newlyweds seized the opportunity to promote their wireless phone by a demonstration from the Eiffel Tower, organized through Blatch's family connections.[21]

Both Blatch and De Forest shared an excitement about participating in the new technological developments with their contemporaries. To Harriot S. Blatch (and no doubt to her daughter as well) technologies such as the radio were new tools for women to use for their own ends. At one of the promotional experiments for the 'wireless phones' in New York in 1909, Harriot, Nora, and Lee were positioned at the Terminal Building, while a group of women's students from Barnard, their physics professor, and some male interlopers from Columbia stood listening to the transmission at the Metropolitan Life Building. 'I stand for the achievements of the twentieth century,' Harriot Blatch declared in the first message sent. 'I believe in its scientific developments, in its political development. I will not refuse to use the tools which progress places at my command … not forgetting that highly developed method of registering my political opinions, the ballot box.' Since the transmitter was only a one-way communication, she continued uninterrupted – although a male student from Columbia protested that 'that is a mean way to talk at a poor chap when he can't say anything.' Believing that technological modernity was inextricably and inevitably linked with politically progressive ideas, she continued: 'Travel by stagecoach is out of date. Kings are out of date: communication by canalboat is out of date; an aristocracy is out of date, none more so than a male aristocracy.'[22] The speech was used by De Forest and his business agent to sell stock of his Radio-Telephone company to suffragists and their supporters.[23]

Even if Nora Blatch and Lee De Forest shared in the excitement of the new technologies, disagreements emerged over the financial status of the firm once they had married and their child was born. De Forest opposed his wife's management opinions and her insistence on continuing to work in engineering. This caused their separation in 1911. Explaining his divorce, De Forest told reporters of a national newspaper that 'his matrimonial catastrophe was due to the fact that his wife … had persisted in following her career as a hydraulic engineer and an agitator [for women's suffrage] … after the birth of her child.'[24] He warned other men against employing their wives, conveniently omitting all mention of Blatch's technical and financial participation in his ventures. Eventually, Blatch started her own architectural firm with family capital. It allowed her to remain independent from partners like De Forest and from the corporate employers she had earlier learned to avoid. De Forest and other husbands were interested in joint ventures but not an equitable partnerships with their wives. De Forest, who insisted that he wholeheartedly supported suffrage for women, had admired Nora's intelligence, and employed her technical training. Yet his wife's greatest offense was that after marriage and motherhood she had rejected a loaned identity and continued to assert her feminist beliefs and heritage.

School Culture and the Strategy of Overqualification

Family businesses were based on a form of engineering knowledge which linked them to the patriarchal authority of the traditional workplace. Formal education, by contrast, was to be a more democratic form of knowledge accessible to all, but it was still in need of establishing and reproducing its own male model of authority. In the decade following the American Civil War, diversity and openness characterized American engineering education, but nevertheless it also used sex and race in exclusionary ways. Hailed as the landmark legislation that pushed co-education to unprecedented levels, the Land-Grant Morrill Act of 1862 helped establish several schools of engineering at land-grant state universities, colleges, polytechnic institutes, and private universities throughout the land. Its drafters had intended it for education of the children of farmers and industrial workers, but had not stipulated the character of 'agricultural and mechanic arts.' At places such as MIT, women, workers, and farmers attended courses in the early days. Industrialists had been the first to support education of workers and women, viewing them as potentially well-disciplined work force.[25]

This broad interpretation of the Act changed over the course of the century. 'The agricultural and mechanic arts' often came to mean industrial rather than agricultural education, technical rather than artisanal training, and school-based engineering rather than a British-style apprenticeship. Engineering educators began breaking with the traditions of vocational training, latching onto scientific rhetoric. The push to upgrade the field through professional ideals resulted in the masculinization of higher engineering education, sending women into separate fields of chemical lab work or home economics.

Before this closure, American women were welcomed as special tuition-paying students when engineering educators sought to increase their enrollment figures in new programs. Compared to their sisters in other countries, American women had free access to primary and secondary education and came relatively well prepared. The newer institutions in the US were more welcoming than the established ones. Thus the co-educational land-grant institutions and state universities showed a more favorable attitude towards women's higher education in engineering than privately owned and sex-segregated institutions such as denominational colleges, military academies, and high-status private schools. The state sponsored land-grant institutions (e.g. Purdue, MIT, Iowa State, Ohio State, Cornell, Berkeley, and the Universities of Washington, Illinois, Colorado, Michigan, and Kentucky) and many municipal universities (the Universities of Cincinnati, Louisville, New York, Houston, and Toledo) thus led the way in engineering education for women. Even some mining schools admitted women to their engineering departments.[26]

In the pre-professional era, engineering institutions and occupational clubs had not yet achieved prestige comparable to that in other professions. Nor had home economics yet been established as a separate field for women interested in technical fields and applied sciences. Pioneering women students therefore began to graduate in engineering from the 1870s onwards. Even so they received mixed messages. Engineering educators searching for higher enrollments might admit some women to their programs, but women faced outright discrimination at every turn, and staying the course required tremendous stamina. A complete set of data on the enrollment and graduation figures of three schools (Ohio State University, University of Alabama, and Stanford University) suggests – not surprisingly perhaps – that the dropout rate for women was 25 per cent higher than for men, 50 instead of 40 per cent of those enrolled.[27] Even women who managed to complete their course work did not always receive the official recognition they deserved. The experiences of Lena Haas at Columbia, Eva Hirdler at the University of Missouri (1911), and Mary and Sophie Hutson at Texas A&M (1903), are telling examples of women students who satisfied all their requirements without receiving the appropriate degrees. Engineering educators were trying to raise academic standards to compete with colleagues in the humanities by playing down the achievements of their women recruits who boosted enrollment figures for their newly established programs.[28]

Facing discrimination, women engineers paired stoicism with a strategy of overqualification. The experienced mechanical engineer Margaret Ingels warned in the 1930s that a woman engineer 'must in many cases work even harder than a man to build up confidence.' Two decades later another woman found the situation unchanged and concluded that 'a dedicated woman can succeed [but has to] run twice as hard as a man just to stay even.'[29] Women who were willing to fit into the tight-knit male world of engineering could force the doors slightly further ajar by concentrating on their math abilities and doubling their efforts. Many women opted for multiple degrees.

If women engineering students in the nineteenth and early twentieth centuries faced formidable difficulties, lack of preparation in mathematics does not seem to have been one of them. In high school, for one, American girls and boys received an equal amount of instruction in calculus and geometry.[30] Moreover, because women who entered engineering tended to come from higher social strata than their male counterparts, they often had a better general education. In the 1970s, sociologist Sally Hacker argued that the high standards of math in engineering education effectively served to exclude women. But at the point where professional standards were just beginning to be formed, math offered a brief window of opportunity for those women interested in a technical education. Before the Second World War an understanding of mathematics was required for practicing engineering, but in America it did not form the kind of obstacle or rite of passage that it would become later as educators sought to raise the standard of engineering. On the contrary, many women who went into engineering could claim superior ability and knowledge in mathematics. The increased importance that engineering educators placed on mathematics as a means of upgrading the profession might have been a major hurdle to many engineering students with average ability – both women and men – but it also helped brilliant women in a school culture that stressed academic skills over hands-on experience. Exceptionally competent women like Elsie Eaves (1898–1983), Alice Goff (b. 1894), Dorothy Hanchett (1896–1948), and Edith Clarke (1883–1959) used their mathematical skills and multiple degrees as a wedge into engineering work and mobilized them as a shield against outright discrimination.

Educational reformers such as Robert H. Thurston who sought to upgrade engineering training with a new emphasis on mathematics, history, and the humanities faced a dilemma. Their form of engineering knowledge was not linked to the patriarchal and class authority of the workplace, but was based on the new cultural authority of science and math. Not only were academic engineers often accused of failing to prepare their students to face the reality of the production floor, but academic ideals threatened to become associated with gentility and femininity. In balancing these elements, engineering educators became the most articulate purveyors of an academic male ethos that stressed hands-on experience and a slap-on-the-back kind of manliness. Many engineering educators tried to imitate 'the methods and manners of real shop-life' in college workshops that housed steam engines, blacksmith tools, foundries and the like. Here hands-on experience could be acquired in association with academic ideals. The problem with the college shops, however well-equipped, was that it was impossible to simulate or test a confrontation with the attitudes of independent workers and bullying foremen. The ability to 'handle men' – to be a professional manager – remained the true hallmark of the successful engineer. In the schools of engineering, this managerial ideal balanced precariously between working-class manliness and academic gentility.

In these environments, women students were encouraged to take math classes but often excluded from taking shop classes or field trips to factories that were required for graduation. When Nora Blatch's classmates prepared for a photograph showing them at work in the field, they arranged for a male friend to take Nora on a date on the day of the photo session so that she would not be in the picture. They thus deliberately excluded her from the rite of passage and erased her from the visual historical record. In 1925, MIT professors prohibited Olga Soroka from participating in a field trip required for graduation in civil engineering. Her professors organized a special internship with the New York subway system for her instead, considering it more appropriate to then-current ideals of women's public behavior. Anna Lay Turner, a chemical engineering student at Rice University in 1924, recalled that women were tolerated in academic environments, but barred from mechanical laboratories. Simply to don overalls was to challenge prevailing social codes.[31]

The workshop and building sites thus functioned as a way of screening out women. '[For] it must be clearly understood,' as one critic of their presence in engineering and other technical occupations wrote in 1908, that 'the road to the drafting board and the laboratory of the engineer lies through the workshop, and workshop practice means hard work and blistered hands, not dilettante pottering and observation.'[32] Women might be competent in drafting, calculation, research, and analysis, as employers attested in the 1920s, but sweat, dirt, and calluses made the engineer a real man. Or, in the words of one scholar, 'if science wears a white lab coat, technology wears a hard hat and has slightly dirty fingernails.'[33] Ideally, middle-class men belonged on the production floors and building sites where they managed other men, while women dealt with more technical details in respectable environments. But of course this was only an ideal. Had it been achievable, there would have been no need constantly to reassert its importance.

Foot Soldiers of Bureaucracy

Ever since the First World War, most women engineers were employed by the emerging military-industrial complex. Women also found their way into engineering through corporate and federal apprenticeship, particularly when the government and the private sector worried about the shortage of technical personnel in times of war and competition with foreign countries. Among entry-level jobs, women made the most headway in the laboratory-oriented and newer fields, which required more academic skills and where gender coding was not yet fixed: chemical analysis, electrical, and – after the Second World War – aeronautical engineering. If small businesses provided a way into engineering for women with family ties, the large emerging corporations did so for women without family capital or resources, to whom mechanical engineering – which was steeped in craft traditions – remained a closed shop. This is not to say that women could not be found working in mechanics shops: during the First and Second World Wars, corporations hired working-class women as lathe and punch press operators and as assembly workers. Mechanical engineering implied a different level of work than the engineering trades, however, one that involved supervision.[34] These encouragements of women's engineering work were all temporary, relatively low level, and deeply ambivalent.

For employees of large corporations without family connections or capital, an engineering job held the promise of promotion, even if this became more a vision than a reality over the course of the twentieth century. Formal and bureaucratic rules made gender discrimination endemic. But they also helped to secure better opportunities for women engineers than the informal rituals of firms, whose shop-floor culture encouraged male patterns of advancement. Only a few women, such as Kate Gleason, could crack the male code of the shop floor by invoking an authority that stemmed from family ownership. The growing importance of formal rules and the move toward professionalization in the twentieth century therefore proved to be a double-edged sword for women engineers who entered the profession without capital or connections. The two world wars – and the state – offered opportunities, but not full-fledged careers, while the war economy institutionalized old discriminatory practices and created new ones.

Women might have found formal education a viable means of access to entry-level engineering jobs, but those who excelled in the academic setting did not fare well in their subsequent careers. Highly trained women including Elsie Eaves, Olive Dennis, Patricia Stockum, and Mabel Macferren, all of whom had earned two or three degrees and showed the stamina to succeed, found that this initial advantage turned into a liability once they entered the workplace. In the new environment, male codes of managerial command and hands-on experience determined one's professional standing, rather than academic excellence. Many overqualified women ended up either as high school teachers in mathematics and sciences or as calculators in corporate offices and at research institutions. Dorothy Tilden Hanchett first trained in civil engineering at the University of Michigan ('17). No doubt she believed that gaining additional M.A. and Ph.D. degrees at Columbia University ('27) and Logan College ('45) would help advance her career. Instead, she ended up at Battle Creek High School as head of the math department. In the aftermath of the First World War, Hanchett and many other highly qualified women found that government propaganda had been empty rhetoric. They were forced to accept temporary teaching jobs in elementary and high schools, teaching instructorships at engineering colleges, or editing positions in professional organizations.[35] The tactic of obtaining multiple degrees did not guarantee employment.

In times of economic downturn, only government bureaucracy and highway projects could offer employment (badly paid) to academically trained women. Proportionally more women trained in civil engineering than any other engineering specialization. But in this already over-crowded labor market, they earned the lowest salaries, ending up in low-level positions and finding fewer employment opportunities than in any other branch of engineering. Still, while these jobs might have been demeaning for young men who expected management positions, for women they offered relatively high wages compared with other jobs available to them. The drafting departments of the State Highway Commissions gave temporary jobs to Esther Knudsen in Wisconsin and Elsie Eaves in Colorado during the 1920s, to Myra Cederquist in Ohio in the 1930s, and to Emma Crabtree in Nevada in the 1940s.[36] The large corporations such as Westinghouse, General Electric, and Boeing also offered women an avenue to technical training through a kind of corporate apprenticeship. At Westinghouse, Bertha Lamme found ample opportunity to use her superior mathematical knowledge and her engineering skills to design motors and generators for over ten years, until she had to relinquish her job when she married a co-worker in 1905. Finding the door to engineering slightly ajar during the war in 1917 as a young

Figure 2. Women trainees in aeronautical engineering posing for propaganda photograph for the Curtiss-Wright Corporation's Cadette Program, an aeronautical engineering crash course for women during the war in 1943. Courtesy of Archives of Women in Science and Engineering, Iowa State University, Ames, Iowa, USA.

civil engineering graduate from the University of Michigan, Hazel Irene Quick established a long career that lasted until 1950. As a fundamental plan engineer, she was in fact the only woman employed by the Michigan State Telephone Company. Nevertheless these were mostly individual women who advanced through the ranks at a much slower rate than their male colleagues.[37]

Even if the World Wars offered opportunities to women, employers also responded to the modest increase in the number of women by setting up clear boundaries against women and by creating separate social and spatial arrangements for their female employees. To deal with the small increase of women, employers instituted sex-segregated offices and drafting departments where some academically trained women could move into supervisory but temporary positions. After graduating from Iowa State University in civil engineering in 1894 and doing some graduate work at MIT, Alda Wilson (b. 1873) worked in architectural firms in Chicago and New York for over ten years before she found a managerial job – as superintendent of the women's drafting department at the Iowa Highway Commission in 1919. Unable to find an engineering position after the First World War, the overqualified and brilliant Edith Clarke spent several years training and supervising women in the calculation of mechanical stresses in turbines. This was in a separate women's department that existed within the Turbine Engineering Department at General Electric in Schenectady, N.Y. Such separate female spaces might have offered women a temporary niche but rarely a solid stepping stone for full-fledged careers as designers, executives, or managers.[38]

Before the era of affirmative action, neither the federal government nor the corporations offered true alternatives to the kind of patriarchal patronage experienced by daughters in family firms like Gleason. At the end of her career in 1947 when the government campaigned for women's return to the home, the experienced Olive Dennis still believed in meritocracy. '[W]e certainly do not want to discourage the ambitious young woman with the right qualifications for an engineering career,' she said, but she warned, 'anyone pioneering in this field must be made to see that, outside of the lowest levels of clerical and manual work, there are almost no standard [management] jobs for women.'[39] Dennis knew what she was talking about. A whole generation of qualified women engineers and scientists had weathered the storm during the Depression in order to continue their careers. As Nora Blatch observed when she worked as an engineering inspector for the Public Works Administrations in Connecticut and Rhode Island, federal sponsorships of women were limited during the New Deal. The investments of Roosevelt's public-work administration in major building programs provided engineering work for men only. The reforestation, highway, building, and reclamation projects were all closed to qualified women while they mobilized men for lower wages. (Ironically these men then had to travel far away to their work, leaving their women as heads of household.) Moreover, the National Recovery Board still specified lower minimum wages for women than men.[40]

These highly educated women of the post-war era waited for better times and looked for jobs in teaching, drafting, or editing and secretarial positions with engineering firms and professional organizations. Elsie Eaves, for example, located herself strategically as manager of the business news department of the journal, *Engineering News-Record*. A graduate of the University of Colorado ('20) and from a prominent family, she provided mentoring and career guidance to many young women engineers during the 1930s and 1940s. She counselled them on how to get through the Depression, and encouraged them to acquire stenographic and secretarial skills in the hope of 'a position with a fine engineer.' But she warned, 'I never encourage a girl to study engineering on the theory that if she wants it badly enough she will do it in spite of all discouragement.'[41] During the Second World War, when younger men went to the front and others moved up to fill their positions, women stood ready to take on the new jobs that were opening up. But these never materialized. Instead of recruiting among the experienced women already available, the federal government chose to train new young women still in college. The state thus helped to institutionalize old patterns, best-illustrated by the gender coding of the federal job of engineering aide.[42] This was used during the Second World War to define women as non-engineers.

American women, like those in Europe, were encouraged for the first time to seek training in technical work for the war effort. Under the auspices of the federal government and in cooperation with universities, large corporations urged young and bright women to apply for engineering jobs. Federal agencies, large aircraft companies, and engineering schools pushed over 300,000 women through various kind of engineering programs ranging from three-month courses to college engineering curricula condensed into two years. Georgia Tech University in the South, which like many other well-established schools had been hostile to any hint at coeducation before the war, opened its doors to women for a special training program sponsored by the US Chemical Services when shortages of technically trained personnel threatened the war industries. The aircraft corporation Curtiss-Wright sponsored a course for women engineering students at several American universities including Iowa State, University of Minnesota, and Rensselaer Polytechnic. (Figure 2) The women received an engineering certificate after completion of the course which included work in engineering methods, mechanics, drafting, and processing. (Figure 3) Many of these specially trained women, who had been the best and the brightest in their high school and college, ended up in drafting, testing, and routine lab work, however. Juliet K. Coyle trained in medical technology and biology as a young college student in 1943, when she

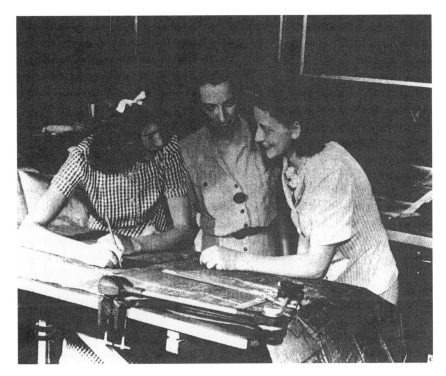

Figure 3. Three women draftsmen trained as engineering aides posing in classic Rosie-the-Riveter government propaganda style in 1943. The photo campaign promoted by the Department of Labor's Woman's Bureau advertised women workers' importance to the war industry. Permission of Schlesinger Library, Radcliffe College, Harvard University, Cambridge, MA, USA.

was recruited by an aircraft company for a short course in engineering where she learned to read blueprints, drafting, statistics, and mechanical practice.(Figure 4) When the war ended and she finished her studies, however, the company had no idea what to do with her and her equally well-trained women colleagues. The firm not only moved them around through different departments and paid them less than their male counterparts, but gave them explicit instructions to avoid giving orders to workers on the production floor – the male avenue to further promotion on the managerial ladder.[43]

Coyle's experience illustrates that none of these educational efforts, either during or after the war-time labor shortages, were meant to turn women into full-fledged engineers. All programs were clearly intended for women's temporary employment as technical assistants to the various engineering departments, despite propaganda claims to the contrary.[44] During the Second World War, vocational literature attempted to assure female recruits that such engineering work would lead to full careers, but, Margaret Barnard Pickel, an adviser to graduate students at Columbia, questioned that promise and advised women to prepare for a backlash in peacetime. 'Are the educators of women justified in encouraging their students to start on the long, arduous and expensive training for an engineering degree with the expectation of a career at the end of it?' she asked. After taking an inventory of the barriers women would face, she concluded, 'it seems hardly honest to hold out such a prospect as a pro-

G-E Campus News

GIRLS, GIRLS, GIRLS

INASMUCH as only one-third of the 12,000 engineers who will graduate in 1943 will be available for private industry, General Electric is hiring young college women to do work formerly done by male engineers.

Forty-four "test women" are on the job now, and others will report each week until the quota (150) is reached. The girls will make computations, chart graphs, and calibrate fine instruments for use in the machine-tool industry.

Miss Virginia Frey (U. of Michigan), one of the 12 women in the country who received engineering degrees this year, is the only graduate engineer in the group. However, each of the others has majored in either mathematics or physics and has received training in both.

Although no one expects these girls to become full-fledged engineers, most of them will be given the Company's famous "test" course.

HI-YO, SILVER!

BROTHER, can you spare a dime?

Manufacturers don't really need it yet, but they are using more and more silver as other metals become increasingly difficult to get. G-E engineers, for example, are using silver in the manufacture of electric apparatus in order to conserve tin, copper, and other scarce materials.

There is now at least a little of the precious metal in almost every motor, generator, transformer, and other piece of equipment built by General Electric for the war.

In many cases the use of silver adds to the cost—a consideration secondary to production at present. Here its use is probably temporary.

But in current-carrying contacts and in brazing alloys, the use of silver results in an improvement in quality sufficient to justify the greater cost. For these purposes, silver will very likely be used in even greater quantities after the war.

TEST PILOT

THE versatile electronic tube has now become somewhat of a test pilot. On test flights, it goes along and writes a complete record of the strains on certain structural parts of the plane as it dives and twists and streaks across the sky.

When a fighter plane goes into a power dive at 500 miles an hour, for example, it has to withstand terrific strains. How great a strain is a vital question to the designer, who wants to know whether he can reduce the weight of the plane to give it greater speed.

Here's how the electronic tube helps furnish the answer to that question: strain gages measure minute changes in dimensions, converting them into tiny electric impulses which electronic tubes amplify sufficiently to drive a highly sensitive oscillograph galvanometer; the galvanometer makes a permanent record of the impulses on a photographic film. *General Electric Company, Schenectady, N. Y.*

GENERAL ⊕ ELECTRIC

Figure 4. Propaganda cartoons, as published by General Electric's Campus News, 1942–1943.

fessional possibility for women.'[45] Olive Dennis, often quoted by the Baltimore and Ohio Railroad in their public-relations literature during the war, also pointed out that 'Women engineers have been ignored or else glamorized with newspaper publicity that is harmful to serious advancement in their work.'[46] The investigation by the Women's Bureau into employment opportunities in peacetime was equally realistic and cautious. In explaining their research project, the officials at the Women's Bureau stated in their correspondence that 'it is our opinion that the increase of women [in technical and scientific work] during the war has been greatly exaggerated because of the publicity presented to attract them. However, we want to find the facts through first-hand contact with professional organizations ... those who employ women in technical and scientific jobs, and with training centers.'[47]

Indeed, Margaret Pickel had been justified in sounding the alarm, as the Women's Bureau found. Strategies to contain women – most of whom were college educated – took various forms. Not only was a woman's job given the title of 'engineering aide,' but an implicit division of labor by sex relegated women to drafting departments and laboratories, while men were assigned jobs on the production floor that enabled them to advance to managerial positions. In short, the federal policy created the term 'engineering aide' to refer to women engineers, while it continued to use the title of 'engineer' to denote men. The job title of 'engineering aide' thus forcefully drew a line between technical expertise and management, and both reproduced and helped to create standard practice.[48]

In the United States, as elsewhere, understanding the formulation of a male professional identity requires attention to the relationship between management and the technical content of engineering as an occupation. Overqualified women found that the initial advantage of academic education would turn into a liability once they entered the workplace, where male codes based on managerial qualities and hands-on experience determined one's professional standing. The gender division of engineering labor – between production floors and drafting offices – hinged on the very same (class) distinctions by which male engineers sought to distinguish themselves from self-trained foremen who had risen through the ranks. The division of labor between the laboratory and the drafting department on the one hand, and building sites and production floors on the other, became the single most important delineator between men and women in engineering. Wherever women engineers did succeed in gaining employment, they were most likely to be hired in drafting, calculating, or design departments, or laboratories and classrooms. In other words, women engineers joined the rank-and-file of the profession without hope of advancement.

No woman without a family connection ever moved into supervisory positions in family firms. Women like Roebling, Gleason, and Gilbreth, who were steeped in the patriarchal culture of the family business, advanced in engineering through a combination of excellence, perseverance, family connections, and the pooling of resources. But assessing the chances of women's employment in 1940, Olive Dennis (1880–1957), who like Blatch had received her education from Cornell University in civil engineering ('20) in addition to degrees in mathematics, warned that 'unless a woman has a family connection in an engineering firm, or enough capital to go into business herself, her chances of rising to an executive position in structural engineering seem negative.'[49] Employed first as a draftsman in the bridge division of the Baltimore & Ohio Railroad and later transferred to the company's service department for interior coach design, Dennis' response during the 1940s is the most revealing: she was always touted as a woman's success story, both by women engineering advocates and by her employers, especially during the Second World War, when the War Manpower Commission and the Office of War Information launched an intense propaganda campaign to lure women into the technical fields. Dennis's warning is instructive in many ways. It pointed to the split between government rhetoric and women's experience of it; the difference between women with family resources and those trying to make it on their own in the emerging corporate and federal bureaucracies; the gap between women's technical expertise and their ability to move into managerial positions; and the contrast between nineteenth-century ideals and twentieth-century practices.[50]

More explicitly than any woman might have said it, the introduction of the term 'engineering aide' encapsulated the story of women's marginalization as a labor reserve force. With a single linguistic stroke the term placed women with technical ability and training outside the domain of technology.

Facing Male Professionalism

National professional organizations became the most visible, if not the only, institution of the engineering fraternity. Few scholars still regard the nineteenth-century creation of professional identities as a trend towards expertise, knowledge, rational behavior, peer review, and non-ideological values. Most now consider professionalization a form of occupational control by which autonomy for some, with carefully defined jurisdictions and privileges, was guaranteed by a rhetorical mask of political disinterestedness and objectivity. The engineering societies were by no means exceptional; they explored some of the classic strategies pioneered by other professions, looking for new ways to enhance their status and cultural authority. The classic model of professionalism was medicine, which emphasized the work ethic, trust, professional associations, licensing, collegial control, and strong client-practitioner relationships. This was problematic for American engineers, however, as many historians of engineering professionalism have argued. In engineering the classic model therefore also competed with what Peter Meiksins identifies as the business and rank-and-file models of professionalism. The American Institute of Mining and Metallurgical Engineers (AIME) advocated business values, associating itself closely with the culture of family firms, and adhering less strictly to the newly emerging ideology of professionalism: profits were more important than ethics. Its business-oriented policy had an immediate impact on the number of women admitted: while only twenty-five women had majored in mining by 1952 and many states had laws prohibiting women from working underground, the mining engineers admitted more women to their ranks than any of the other major engineering organizations. In 1943, the AIME membership included such daughters of family firms like Jean Horning Marburg, then member of the National Resources Planning Board, Helen A. Antonova, an assayer at the R&F Refining Co., Edith P. Meyer, a development engineer at Brush Beryllium Co., and another 19 women in addition to a large number of female students.[51] For daughters of family firms like Jean Horning business professionalism opened some doors that would have remained closed otherwise. Thus business models of professionalism were more open to women if they were connected to the patriarchal culture of family firms like Gleason, Gilbreth, and Honing. They were closed to women without the proper family connections.

Even if engineers did not succeed in maintaining strict boundaries compared to the other professions, when successful the classic model of professionalism ensured a thoroughly male and middle-class endeavor. The more an organization strove to follow the example of medicine, the more it was inclined to bar women.[52] Here it was not the number of women engineers in either absolute or relative terms that determined the percentage of female membership, but rather the level of professionalism to which the leadership of national organizations laid claim. Significantly more women engineering students opted for civil and electrical than other engineering fields, but this was not reflected in the membership of the American Society of Civil Engineers or the American Institute of Electrical Engineers, whose requirements were far more strict than those of AIME.

The American Society of Civil Engineers (ASCE) and other major societies guarded against any female incursions. Some also granted secondary member-ship without voting rights to unimportant rank-and-file engineers and women who managed to infiltrate them.[53] The American Institute of Electrical Engineers (AIEE), emerging in a field with high aspirations towards a medical model of professionalization, refused to admit Susan B. Leiter, a laboratory assistant at the Testing Bureau in New York, to membership in 1904, when the organization was looking for ways to upgrade the profession.[54] When Elmina Wilson and Nora Blatch applied for membership to the American Society of Civil Engineers, they found the doors closed. Having graduated *cum laude* in civil engineering from Cornell

University, Nora Blatch could claim superior mathematical ability and theoretical engineering knowledge, the kind of credentials advocates of engineering schools thought crucial for any engineer to succeed. But Blatch had more to offer: she also possessed the necessary hands-on experience that advocates of shop floor and field training saw as hallmarks of the true engineer. In 1916, however, the society dropped her from its membership and Nora Stanton Blatch filed a lawsuit against the ASCE. As an experienced engineer, she had accumulated over ten years of experience to meet the Society's requirements. In addition to her four-year education in civil engineering at Cornell, she had taken courses in electricity and mathematics with the famous Michael Pupin at Columbia University. She had practiced as a draftsman for the American Bridge Company and the New York City Board of Water for about two years and as an assistant engineer and chief draftsman at the Radley Steel Construction Company for another three years. Finally she had worked as an assistant engineer at the New York Public Service Commission. Most importantly for the requirements for full membership, she had supervised over thirty draftsmen when working at Radley Steel. Blatch, a feminist, divorcee, and single mother whose income depended on her engineering work at the time, challenged the ASCE when more women and sons of lower-class men were trying to enter the field through the new institutions of higher education and when engineering advocates were busy defining the occupation as a profession by excluding more and more groups of practitioners such as draftsmen and surveyors. Her suit marks one of many contests in which the emerging professions staked out their professional claims by means of border disputes with other competing fields.[55]

In addition to outright exclusion the professional organizations also granted women secondary membership without voting rights. The controversy over Ethel Ricker's Tau Beta Pi membership was yet another drawn-out contest about gender in the fields of civil engineering and architecture, where women were numerous. In 1903, the local chapter of Tau Beta Pi elected Ricker, an architecture student at the University of Illinois, to the engineering honor society. But the national executive board and the society's convention not only overturned the decision to elect her, they went so far as to amend the constitution to specify that henceforth only men would be eligible for membership. During the Depression of the 1930s, when many civil engineers faced unemployment, the Tau Beta Pi honor society introduced a Women's Badge in an attempt to deal with the (small) number of qualified women who had made their presence felt. Women's Badge wearers were neither members nor allowed to pay initiation fees. Disapproving of such 'separate but unequal' recognition, some women refused them. It would take three quarters of a century before women would be accepted as equal partners in the organization: in 1969 the honor society changed its constitution to admit women, in 1973 it cleansed the constitution and bylaws of sexist language, and in 1976 it elected a woman as a national officer for the first time.) By changing its constitution and by designing a 'woman's badge,' the fraternity of young aspiring engineers set up explicit gender barriers around engineering when job markets were particularly tight.[56]

As in the various scientific fields, engineering specializations dealt with the perceived threat of feminization either by excluding women from full membership in professional organizations or by relegating them to a secondary status without voting rights. In the words of Margaret Rossiter, 'the very word *professional* was in some contexts a synonym for an all-masculine and so high-status organization.'[57] Women faced outright exclusion, were relegated to secondary membership, and banished to separate organizations.

Divide and Conquer

Defining separate labor markets for male and female engineering work was another tactic by which to cope with women who started to seek engineering education and employment. The best example comes from the chemical industry, where many women worked as chemists. To deal with these female incursions the American Institute of Chemical Engineering did everything in its power to define the occupation in such a way as to effectively bar women from the field and relegated them to chemistry.

In an effort to distinguish themselves from chemical analysts – a large proportion of whom were women – whose status and pay diminished dramatically around 1900, chemical engineers placed heavy emphasis on the ability to manage other men. In response, production chemists sought to align their occupation with the male world of mechanical engineering rather than with science. Chemical engineers saw themselves running plants, as opposed to laboratories. As one of the most important founders of the American Institute of Chemical Engineers (AIChE), Arthur D. Little (1863–1935), spokesman of engineering professionalism and the nestor of commercial chemical research, introduced a key concept for the development of a distinct chemical engineering identity in 1915. This was the notion of 'unit operations.' Little argued that unit operations involved neither pure chemical science nor mechanical engineering, but distinctly physical, man-made objects in the plant operation rather than chemical reactions in the laboratory. In the same year that Little refined his notion of chemical engineering in a report for the AIChE, his opinion was solicited by the Bureau of Vocational Information, which was preparing a report on employment opportunities for women in chemistry and chemical engineering in the tight engineering labor market after the First World War. Recommending chemical analysis as 'one of the most promising fields of work for women,' Little reserved chemical engineering as an exclusive specialization for men, arguing that 'it is probably the most difficult branch of the profession.' In addition to requiring long hours, extensive travel, and physical endurance, the 'rough and tumble of contests with contractors and labor unions' involved in·the new construction and design of a plant would prohibit women's employment.[58]

Little's rhetorical position was broadly shared by male chemists working in the field. While stressing women's strength in all other lines of work connected with the chemistry lab, the chief chemist of the Calco Chemical Company voiced the general sentiment in 1919. 'It is impossible,' he said, 'to use women chemists on development work which has to be translated into plant practice by actual operation in the plant. This is the only limitation.' Other potential employers of women chemists elaborated on that particular theme by explaining that, 'research men must go into the Plant and manipulate all sorts of plant apparatus, direction [sic] foreign labor of every sort. You can readily see that a woman would be at a great disadvantage in this work …' Or that 'it often involves night work and almost always involves dealing with plant foremen and operators not easy to deal with.' The representative of the Grasselli Chemical Company's research department wrote in 1917 that chemical engineering involved 'large rough mechanical apparatus,· … which work is usually carried on by unintelligent labor, in a good many cases the roughest kind of material.'[59] By establishing such notions as 'fact,' Little and other chemists succeeded in safely associating their work with both the male codes of the machine shop or the plant operation and managerial control. The power struggle in the workplace where matters of class were contested was a matter *between men.* Middle-class chemical engineers thus appropriated the male codes from the struggles of work place.

Despite the way women and men's engineering work was constructed, women did in fact probably do some of this work. Often discrepancies existed between job title and job content. Take the case of Glenola Behling Rose, whose title was chemist but who described her duties in 1920 as follows: 'I left the chemical dept. to go into the Dyestuffs Sales Dept. I have but one man over me and as his assistant, I am the *Executive Office Supervisor* of the Dyestuffs Technical Laboratory and have charge of all dealings with the chemical dept. such as deciding what dyes they shall go ahead to investigate & in what quantities, and keep track of their work in order to see whether they produce the dyes economically enough for us to market them. In a way I am the link between the research, the manufacturing and the selling of dyestuffs … As you will see a good deal of my work is supervisory.' With bachelors' degrees in geology and chemistry, and a Masters in chemistry, the highly qualified Glenola Rose felt she was technically well prepared for such job. In response to the question of what training she thought would be most beneficial to women entering her field, she replied that there was a 'need for a thorough foundation and a training with men,' by which she meant the task of managing men. And Florence Renick wrote that, in fact, 'I have had to deal considerably with labor of all kinds, mostly ignorant and many foriegners [*sic*] among them, and none of them but consider me "boss" so far as the laboratory is concerned.' On this particular point Jessie Elizabeth Minor, chief chemist at the Hamersley Manufacturing Company articulated women's ambivalence: 'There is still much masculine prejudice to combat. Many laboratories are not attractive looking. We come in contact with working men (which may be construed as an asset or liability).'[60] Thus, in these contexts, white women chemists actually did supervisory work and would have qualified as engineers according to the terms Little and other chemical engineering advocates had established for their engineering specialization. In all these instances, technical qualification or experience was less decisive in considerations for job assignments and promotions than the issue of supervision.

Women chemists were thus kept under job titles they actually had outgrown. Significantly, traffic between chemistry and chemical engineering also went in the other direction. Women who trained as chemical engineers ended up in lower-paid positions as chemists. Dorothy Hall (1894–1989) seemed to embody the success story of a woman advancing on the corporate ladder as a research and later chief chemist at GE. But with a Ph.D. in chemical engineering from the University of Michigan ('20) she was overqualified for her job. When asked, most employers said they thought women competent and excellent for research and analysis; few raised objections of a technical nature. But like Arthur D. Little, they all drew the line at work related to the plant operation. Thus in 1948 the Women's Bureau reported that most women who trained in chemical engineering were employed as chemists.[61]

In a limited way, chemical engineering and chemistry offered a niche to women students interested in engineering. Large numbers of women interested in the sciences in the US and elsewhere flocked to the field of chemistry. The same held true for chemical engineering: more women graduated in chemical engineering than in any of the other engineering specialization. In fact, a higher proportion of female engineering students than of male engineering students majored in chemical engineering. But prominent chemical engineering advocates pushed for an explicitly male professional ethic by defining their discipline as an exclusive male domain that required supervisory skills. Tens of thousands of women chemists and chemical engineers were banished to chemical laboratories, where working conditions were dire and the pay was low.[62] In these contexts, the term chemist meant an ill-paid woman, while the title of chemical engineer often denoted a man in command of higher wages and managerial authority.

The male engineers' push for a high-status professional identity was in part a response to the enormous expansion of engineering work, which provided new opportunities to lower-class youths and sons of recent immigrants. The call for clear boundaries regarding class, however, resulted in a reinscription of male middle-class identity.

Census takers colluded in this through their unceasing efforts to find new categories for reliable enumeration; they sought to make classification consistent by excluding more and more groups of skilled workers from the category of engineer. Among those omitted were boat and steam shovel engineers, foremen of radio stations, engineers under thirty-five without a college education, and chemists. According to economists who have worked with the data, however, the statisticians made these adjustments without much attention to uniformity. These statistical and linguistic interventions did little to generate a satisfying set of data.[63] Nevertheless, historians have reproduced many of these figures, thereby accepting the categorizations provided without question. The definition of who would count as a true engineer and the production of statistics to justify this illusion worked together to create an illusory picture of a male, middle-class profession. The example of the chemical industry shows how linguistic construction and social practices made women invisible in engineering.

Organizing at Last

Women engineers responded to such strategies with stoicism but also with collective action. Before the First World War, the early generation of women engineers like Rickter, Wilson, Blatch, and Leiter had tried to gain access to the existing male organizations as individuals. They were rebuffed outright, granted secondary status, relegated to separate-but-unequal organizations, or segregated into different labor markets. A second generation of young women students and recent graduates including Lou Alta Melton, Hazel Quick, Elsie Eaves, Hilda Counts Edgecomb, and Alice Goff, who had found entry-level employment opportunities during the First World War, tried in 1919 to create a separate women's organization. But this failed. The post-suffrage generation championed the cause of women engineers with great enthusiasm. Yet none of the advocates of women's presence in engineering rallied to this feminist cause – even if they grounded their promotion of women engineers as professionals precisely in one of the important principles of modern feminism: as individuals women should be able to develop themselves to their fullest potential. All supported the notion that women had the freedom to choose whichever line of work suited their abilities, without the obligation to appeal to feminine propriety by arguing that such a choice was inspired by higher morals. But all ardently believed the engineering profession's promise of upward mobility. Resisting any direct association with the women's movement, therefore, they claimed instead that they just happened to have a knack for engineering as individuals. Success or failure was down to individual merit. This discourse was particularly prominent in 1943 when the government sought arguments to mobilize women for the war industries.[64] The majority of women engineers had internalized the values of corporate engineering, merit, and self-reliance.

The second generation saw their organizing efforts thwarted partly because they followed a logic of maintaining professional standards similar to that used by the male national organizations. Hence they excluded engineering students and working women engineers without formal education, such as non-collegiate draftsmen, chemists, and testing technicians, from their membership.[65] No doubt they had to do this in an effort to defend against sexism and to garner professional prestige. But emulating high professional standards prevented them from gathering the critical mass necessary for the success of such an organization.

In the same year that American women failed, British women succeeded. They established the Women's Engineering Society (WES), an inclusive organization that brought together women engineers with or without formal collegiate education as well as machinists who were skilled or semi-skilled workers. The British successfully, albeit briefly, united across classes, in part because they did not adhere to the classic medical model of professionalism; instead they combined the tradition of upper-class business professionalism and trade associations with feminist ideals. In the end, as historian Carroll Pursell shows in the next article, the British leaders also abandoned their policy of 'gender solidarity for male privilege and class advantage' and narrowed 'their focus to exclude the great mass of women who had entered the engineering trades during the First World War directly out of the working class.'[66] Between the world wars, when job opportunities virtually disappeared, American women engineers sought temporary shelter with their British colleagues through membership in the British WES and kept in touch through informal networks. Still clinging to the model of medicine, and having failed to create a critical mass of members, American women engineers of the interwar period turned instead trying to shape public opinion. They did this partly through writing biographical sketches of each other according to well-established formulae, which stressed that with hard work and self-reliance women could – to paraphrase the title of Alice Goff's publication during these years – indeed become engineers. Thus American women engineers borrowed male models of meritocracy, but neither questioned the middle-class structure of American engineering nor campaigned for equal rights.

The final push towards organization in the United States did not come from the hundreds of thousands of women working in federal engineering jobs during the Second World War, from the informal networks of academically overqualified women engineers who had learned to be self-effacing during hard times, or from the women urban planners who had been nurtured in the Progressive era. It came once again – as it had in 1919 – from young students and recent graduates eager to enter the job market and yearning for official recognition and respectability.

After the gap between rhetoric and reality had widened once again in the aftermath of the Second World War, American women engineers united at last in 1949. They did so long after other female professionals such as lawyers and doctors. An energetic and ambitious junior student leader on a scholarship, Phyllis Evans, won the support of the dean of women, Dorothy R. Young, and university counsel A. W. Grosvernor to organize the first meetings of women engineers at Drexel University in 1949.(Figure 5) She and her colleagues organized over seventy young women engineering students from nineteen colleges on the East Coast. The goal was to make their 'voices … heard in the technological world' and to address inequities in engineering work. In the greater New York area, a group of women engineers who had been working in war-related industries – students from Cooper Union and City College of New York, coupled with graduates working nearby – also struck up conversations about their plight in the college libraries and Manhattan's coffee shops. Soon the long hidden tensions over leadership and the direction that professionalism should take for women engineers burst into public discussion.[67]

In the founding years, the student group at Drexel University in Philadelphia and a coalition of various groups in the greater New York area were in competition with each other for leadership. The origins of this contest concerned different professional strategies. Phyllis Evans – like so many other young women who had begun studying engineering during the war – was single, about to graduate, and facing unemployment. Echoing the governmental war rhetoric, she cherished high expectations for her future. Explaining her choice for engineering, she told a journalist about how her war experience as a cadet sergeant had inspired her to go into engineering. She wanted her future to be in military research. 'I want to build rockets and I want to go to Mars,' she said with youthful optimism. Establishment engineers such as Lillian G. Murad and Lillian Gilbreth, property owners who were steeped in the ethics of the patriarchal culture of family firms, opted for a more conservative approach, combining supervision and high professional standards. Although Lillian Gilbreth had been supportive of other women engineers, she did not favor a separate women's organization and was disinclined to head the SWE when it first sought a leader. She was most concerned about the bold effrontery of a separate organization and its feminist implications, and warned against blaming men for the difficulties women encountered in entering the field; instead she accused women of lacking credentials. 'The reason for women not being admitted into the National Engineering Society was not because they were women, but rather because they did not yet meet the qualification,' she said. But the elder Gilbreth was somewhat at odds with Dorothy Young, Drexel's dean of women students. Young both pushed for an activist strategy that confronted the inequity between men

Figure 5. Photograph of Phyllis Evans posing in overalls in Rosie-the-Riveter iconography illustrating a newspaper report on the first organizing efforts of women engineering graduates who had been disappointed by their employment opportunities after the peacetime conversion. Reproduced from The Christian Science Monitor *(19 April 1949). Courtesy of Royal Dutch Library, The Hague, The Netherlands.*

and women, but also struck a conciliatory note: women 'need to realize that it is necessary to work cooperatively with men in larger field, planning together to abolish those inconsistencies that mar our democratic society,' she stated.[68]

The strategy of the young organization remained a balancing act: between the impatience of the students of Miller's generation, whose expectations had been raised by the government propaganda of Rosie-the-Riveter, and the cautious, conservative strategy of a previous generation of daughters and of wives, who were wedded to the patriarchal culture of family firms. The SWE tried to inspire

Figure 6. Group portrait of women attending the founding meeting of the American Society of Women Engineers at Green Engineering Camp of Cooper Union, N.J, May 27, 1950, as a response to the government's back-home campaigns just after the war. Courtesy of Labor and Urban Affairs Archives, Wayne State University, Detroit, MI, USA.

younger women to go into engineering by taking individual women and their exemplary careers as models to be emulated. It did this through the establishment of medals and scholarship programs, the writing of biographical narratives, and the dissemination of pictures. In all of these, the organization stressed individual efforts. This was in direct contrast to the new corporate male ideal of team players that was promulgated by companies such as Dupont and General Motors. Thus the SWE highlighted merit and self-reliance, rather than becoming the collective movement for equality for which Nora Blatch had campaigned. (Figure 6)

The SWE never resolved these conflicting goals. It sought to attract more young women to engineering schools, yet necessarily had to deal with the long-entrenched patriarchy of family firms, the exclusionary tactics of the professional organizations, and the discriminatory employment practices of corporations and the federal government. If it openly battled prejudice and sexism in engineering schools and practice, however, it risked frightening off prospective recruits.[69] Though the SWE was established in 1949 in the postwar period of 'adjustment,' when the government and corporations devised policies to push women back into the home, the organization would ride a new wave of ambiguous government encouragement during the Cold War. As in the former communist states like Soviet Union, the German Democratic Republic, and other Eastern European countries, the American military collaborated with corporations in actively recruiting women as technical personnel. 'Woman Power' campaigns were initiated after the Sputnik panic in the West in 1957. Despite its evocative image and name *Woman Power* (the title of a 1953 report issued by the National Manpower Council at Columbia University) had little to do with feminist calls for equal rights: it shied away from controversial issues such as equal pay for equal work and job discrimination. It neither upset gender hierarchies nor helped foster a separate women's culture. (Figure 7) Instead, it provided the conservative part of the women's movement a certain legitimacy, since it stressed that women, if they worked hard, could alleviate national labor shortages.[70] This new society successfully gathered the critical mass necessary for such a separate organization despite its high professional standards because the cold war had created such a high demand for technical personnel where numbers had been lacking three decades earlier. The British women engineers had been so succesful at organizing their sisters because they had been building bridges with both the women's movement and the rank-and-file of their profession. The truth was that in the United States women engineers's individual opportunities were part of America's military-industrial complex and highly depended on it. At times its doors might be open to women, but it almost always reproduced old patriarchal patterns in a new corporate context. (Figure 8)

'Woman Power' and Daughters of Martha

The government's campaign of *Woman Power* was a borrowed model. So was Lillian Gilbreth's appropriation of Rudyard Kipling's 1907 poem 'The Sons of Martha.'[71] At the 1961 opening of the Society of Women Engineers's new headquarters in the United Engineering Building, Gilbreth made the poem relevant to the needs of women engineers by entitling her speech 'The Daughters of Martha.' The modernist United Engineering Building towered high in New York and expressed the coming of age of the engineering professions, but the new headquarters Gilbreth was about to open were tiny and symbolic of women's place in the profession. By invoking Kipling's allegory, Gilbreth sought to empower women in the technical professions through the values of

Figure 7. Photograph taken as Part of the campaign to recruit women for engineering positions for the U.S. military as the Cold War heats up in 1953. In this photograph, gender relations are preserved rather than subverted by the photographer's frog-eyed frame stressing masculine features of the woman in charge and the subordinate position of her reclining colleague. Permission of Schlesinger Library, Radcliffe College, Harvard University, Cambridge, MA, USA.

service, sacrifice, and self-reliance. In so doing, she showed the marginalized place of women in a male technical world and the hardship of women who labored on the lower rungs of the profession as rank-and-file engineers and corporate workers. She thus put forward a model that doubled the burden on women who aspired to be engineers: they were expected to make sacrifices both by virtue of their sex and because of their profession.[72] By the 1960s, Gilbreth's professional model of 'simple service simply given,' or, as she had advised earlier, 'helping others express themselves [as] the truest self expression,' was out of date for women.[73] Her call for the inclusion of women in the profession was based on her own long career, and on the trying experiences of many other women engineers. But principally it rested on a conservative notion

that the best way forward for women professionals generally was to be ever stoical and always overqualified. Despite her attempt to redefine the place of women at the center of engineering, her celebration of service, sacrifice, and self-reliance reinforced some very traditional notions about women's proper place in engineering. Her employment of the Kipling's poem threatened to become a failed allegory. When male engineers used the poem, it could be mobilized to appropriate working-class badges of manliness or to symbolize them as underdogs, but when women mobilized the poem for their cause the figure of Martha turned into an image of subordination, stoicism, and lack of advancement. Men perhaps could pass 'down' but women could rarely pass 'up' the cultural hierarchies as women found here and elsewhere.

Women engineers in the military-industrial complex or in the patriarchal culture of family firms had no appealing role models except the problematic image that Lillian Gilbreth supplied. Outside the military-industrial complex and the patriarchal family firms, however, women were building their own structures. The women of the Progressive era who participated in the women's reform movement also helped to shape an alternative women's technical culture that was nurtured by women's traditions.[74] Throughout the country, from Boston to San Francisco, women reformers helped build the public infrastructure of the civic improvements movement as private citizens rather than as corporate employees. Highly organized in private philanthrophic organizations such as the General Federation of Women's Clubs, these women reformers

Figure 8. General Electric's advertising promoting the company through four of its women engineering employees against a male-coded apparatus. The image countered the working-class image of Rosie the Riveter of the government war propaganda campaigns to show women could be engineers without losing their femininity during the Cold War era. Reproduced from Cincinatti News *(1959).*

campaigned for what they called municipal housekeeping. They conducted surveys, drew up plans for urban infrastructures, pushed for better housing, and helped finance public facilities from streetlights to sewer systems. They forged coalitions with local politicians, architects, civic leaders, and professional women such as Ellen Swallow Richards, Alice Hamilton, and Ruth Carson in public health, science, and social research respectively. The women of the Progressive movement became urban planners of the modern age.[75] There existed, therefore, a rich female heritage of building. But this was not available to women engineers working in family firms or military and corporate industries. In fact, the women of these separate technical cultures never met or sought to bridge the gap between them.

Notes

1. An earlier version of this chapter appeared in Ruth Oldenziel, Making Technology Masculine: Men, Women and Modern Machines in America 1870–1945, Amsterdam University Press, 1999. The author is most grateful to Karin Zachmann for sustained critical readings of earlier drafts.
2. Samuel Florman, 'Engineering and the Female Mind,' *Harper's* (February 1978): 57–64.
3. Florman, 'Engineering and the Female Mind,' 61
4. Peter F. Meiksins, 'Engineers in the United States: A House Divided,' in *Engineering Labour: Technical Workers in a Comparative Perspective*, edited by Peter Meiksins and Smith Chris (London: Verso, 1996), 61–97 and his 'Professionalism and Conflict: The Case of the American Association of Engineers,' *Journal of Social History* 19, no. 3 (Spring 1986): 403–22.
5. Bruce Seeley, 'Research, Engineering, and Science in American Engineering Colleges, 1900–1960,' *Technology and Culture* 34 (1993): 344–86; Peter Lundgreen, 'Engineering Education in Europe and the U.S.A., 1750–1930: The Rise to Dominance of School Culture and the Engineering Professions,' *Annals of Science* 47 (1990): 33–75.
6. Unless otherwise indicated, this article is based on my survey of *all* American accredited engineering schools prior to 1945. A query was sent to 274 schools. This correspondence was followed up by letters to registrar's offices, engineering colleges, university archives, and alumni associations. In total, 175 schools answered: 82 reported on women graduates. Of the 99 schools that did not respond, only 10 schools appear to have awarded a significant number of degrees to women. Based on this survey and other sources, over 600 women in engineering have been identified by name. *American Women Engineering Graduates Papers*, Author's Personal Collection (AWEG Papers, hereafter).
7. Examples of hostility: Vera Jones Mackay in *100 years: A Story of the First Century of Vanderbilt University School of Engineering, 1875–1975* (Nashville: Vanderbilt Engineering Alumni Association, 1975), 116–9; Marion Monet (MIT '43) personal interview March 23, 1990; Author's telephone interview with Dorothy Quiggle November 14, 1989, and *MIT Survey*; example of aloofness: Eleonore D. Allen (Swarthmore '36), 'Lady Auto Engineer: Her Ideas Irreparable,' *New York World-Telegram & Sun* (August 14, 1961). On post-suffrage generation of women professionals, see: Nancy Cott, *The Grounding of Feminism* (New Haven: Yale University Press, 1987), Chapter 1 and Introduction.
8. Rosalyn Rosenberg, *Beyond Separate Spheres. Intellectual Roots of Modern Feminism* (New Haven: Yale University, 1983); Louise Michele Newman, ed., *Men's Ideas/Women's Realities* (New York: Pergamon Press, 1985); John M. Staudenmaier, S. J., *Technology's Storytellers. Reweaving the Human Fabric* (Cambridge, MA: MIT, 1985); Margaret W. Rossiter, *Women Scientists in America: Struggles and Strategies to 1941* (Baltimore: The Hopkins University Press, 1982).
9. Author's interview with Nora Blatch's daughter, Rhoda Barney Jenkins, September 1989; Ellen DuBois, ed., 'Spanning Two Centuries: The Autobiography of Nora Stanton Barney,' *History Workshop Journal* 22 (1986): 131–52, p. 134. Cf. Suzy Fisher, 'Nora Stanton Barney, First US Woman CE, Dies at 87,' *Civil Engineering* 41 (April 1971): 87.
10. Lillian Gilbreth, 'Marriage, a Career and the Curriculum,' typewritten manuscript (probably 1930s), Lillian Gilbreth Collection, NHZ 0830-27, Box 135, Department of Special Collections and Archives,

Purdue University Library, Lafayette, IL (Gilbreth Papers hereafter). See also 'American Women Survey Their Emancipation. Careers Are Found an Aid To Successful Married Life,' *Washington Post* (August 16, 1934); Edna Yost, *Frank and Lillian Gilbreth, Partners for Life* (New Brunswick: Rutgers University Press, 1949) and Yost's description in *American Women of Science* (New York: Frederick A. Stokes, 1943); Ruth Schwartz Cowan, *Dictionary of Notable Women. Modern Period* (Cambridge: Belknap Press, 1980), s.v., 'Gilbreth.' The notion of 'borrowed identity' comes from Margot Fuchs, 'Like Fathers-Like Daughters: Professionalization Strategies of Women Students and Engineers in Germany, 1890s to 1940s,' *History and Technology* 14, 1 (1997): 49–64.

11. See Rossiter's excellent *Women Scientists in America* (1982), 248. The sequel *Women Scientists in America. Before Affirmative Action, 1940–1972* (Baltimore: The Johns Hopkins University Press, 1995) is equally pathbreaking and continues to be an inspiration.

12. *First Century of Vanderbilt University School of Engineering, 1875–1975*, 116–9; Correspondence Vanderbilt University, Special Collections University Archives, Nashville, TN, AWEG Papers.

13. Ovid Eshbach, Technological Institute, Northwestern University, Chicago, IL, February 15, 1945 with Woman's Bureau, RG 86, Box 701, Woman's Bureau, National Archives, Washington, DC, (WB NA hereafter). Cf. Juliane Mikoletzky, 'An Unintended Consequence: Women's Entry into Engineering Education in Austria,' *History and Technology* 14, 1 (January 1997): 31–48. Fuchs, 'Like Fathers-Like Daughters.''

14. Surveying many unusual fields of women's employment, Miriam Simons Leuck reported on a great many widows, some of whom were engineers: 'Women in Odd and Unusual Fields of Work,' *AAAPSS* 143 (1929): 166–79; US Bureau of the Census, *Census* 1890 (Washington, DC: Government Printing Office, 1890).

15. A., 'Women Engineers,' Letter to the Editor, *Professional Engineer* (May 1922): 20; 'Roebling Memorabilia' *The New York Times* (October 4, 1983); Gustave Lindenthal, Letter to the Editor, 'A Woman's Share in the Brooklyn Bridge,' *Civil Engineering* 3, 3 (1933): 473; Alva T. Matthews, 'Emily W. Roebling, One of the Builders of the Bridge,' in *Bridge to the Future: A Centennial Celebration of the Brooklyn Bridge*, edited by Margaret Latimer, Brooke Hindle, and Melvin Kranzberg *Annals of the New York Academy of Sciences* (1984), 63–70; for a biography on Roebling, see Marilyn Weigold's study, *The Silent Builder: Emily Warren Roebling and the Brooklyn Bridge* (Port Washington, NY: Associated Faculty Press, 1984). For an appreciation of her work by the engineering community, see 'Engineers Pay Tribute to the Woman Who Helped Build the Brooklyn Bridge,' (address by David B. Steinman for the Brooklyn Engineers Club, May 24, 1953), reprinted in *The Transit of Chi Epsilon* (Spring-Fall 1954): 1–7.

16. For a treatment of Gilbreth's work, see Martha Moore Trescott 'Lillian Moller Gilbreth and the Founding of Modern Industrial Engineering,' in *Machina Ex Dea: Feminist Perspectives on Technology*, ed. by Joan Rothschild (New York: Pergamon Press, 1983), 23–37. 'Marriage, a Career and the Curriculum'; 'American Women Survey Their Emancipation'; Yost *Frank and Lillian Gilbreth* and *American Women of Science*. Yost was one of the Gilbreths' most important promoters and popularizers. Cowan in *Notable American Women*, s.v., 'Gilbreth.''

17. Correspondence Speical Collecitons Department, University Archives, University of Nevada, Reno; correspondence with New Jersey Institute of Technology, Alumni Association, Newark, NJ, (AWEG Papers); Carolyn Cummings Perrucci, 'Engineering and the Class Structure,' in *The Engineers and The Social System*, eds. Robert Perrucci and Joel Gerstl (New York: John Wiley and Sons, 1969), 284, Table 3.

18 Correspondence Cornell University Library, Department of Manuscripts and University Archives (AWEG Papers); Eve Chappell, 'Kate Gleason's Careers' *Woman's Citizen* (January 1926): 19–20, 37–8; *DAB*, s.v. 'Gleason'; 'A Woman Who Was First,' *The Cornell Alumni News* (January 19, 1933): 179 with excerpts from *The Cleveland Plain Dealer* and *The New York Tribune*; Leuck, 'Women in Odd and Unusual Fields of Work,' 175; *The Gleason Works, 1865–1950* (n.p., 1950); *ASME. Transactions* 56, RI 19 (1934), s.v., 'Gleason.' Cf. Christoper Lindley in *Notable American Women. Supplement* 1 (1934), s.v., 'Gleason.''

19. Judith S. McIlwee and Gregg J. Robinson, *Women in Engineering: Gender, Power and Workplace Culture* (Albany: State University of New York Press, 1992).

20. DuBois, 'Spanning Two Centuries,' 150.

21. Terry Kay Rockefeller in *Notable American Women* (1980), s.v., 'Barney.'

22. ''Barnard Girls Test Wireless' Phones,' *The New York Times* (February 23, 1909):7:3.

23. In *Inventing American Broadcasting, 1899–1922* (Baltimore: The Johns Hopkins University, 1987), 167–7, Susan J. Douglas writes that Nora Blatch's 'contributions to early development of voice

transmission have been either completely ignored or dismissed", (p. 174).

24. ”Warns Wives of 'Careers,' ' *The New York Times* (July 28, 1911) 18:3 and response by Ethel C. Avery, 'Suffrage Leaders and Divorce' (July 31, 1911) 6:5. See also Margaret W. Raven, a graduate from MIT ('39) in General Science and Meteorology, Association of MIT Alumnae, Membership Survey, 1972, MIT, Institute Archives, Cambridge, MA; and McIlwee and Robinson, *Women in Engineering.*

25. David Noble, *A World Without WOmen: The Christian Clerical Culture of Western Science* (New York: Alfred A. Knopf, 1992).

26. Edna May Turner, 'Education of Women for Engineering in the United States, 1885–1952,' (Ph.D. diss., New York University, 1954). This valuable and pioneering dissertation contains little analysis or biographical information beyond the statistics.

27. Correspondence with University Archives, Ohio State University, Ames, OH; Special Collections and Center for Southern History and Culture, University of Alabama, Tuscaloosa, AL; correspondence with Stanford Alumni Association, Stanford, CA (AWEG Papers). Robin found a similar trend for the post-World War era. 'The Female in Engineering,' in *The Engineers and the Social System*, 203–218. Cf. 'Report of the Committee on Statistics of Engineering Education,' *Proceedings of the Society for the Promotion of Engineering Education* 10 (1902): 230–57, p. 238, Table I.

28. Tom S. Gillis, roommate of Henry M. Rollins, Hutson's son, letter to author, October 21, 1989; correspondence Texas A&M University, College Station, TX; Univeristy of Missouri-Rolla, Library and Learning Resources, University of Missouri-Rolla, Rolla, MO, (AWEG Papers); Lawrence O. Christensen and Jack B. Ridley, *UM-Rollo: A History of MSM/UMR* (Missouri: University of Missouri Printing Services, n.d.), 93–4; and his 'Being Special: Women Students at the Missouri School of Mines and Metallurgy,' *Missouri Historical Review* 83, 1 (October, 1988): 17–35; Frances A. Groff, 'A Mistress of Mechanigraphics,' *Sunset Magazine* (October 1911): 415–8.

29. Clipping file, Alumni Office, Swarthmore College, Swarthmore, PA; for further information: *Yearbook* (1942) (AWEG Papers); Edward M. Tuft, 'Women in Electronics,' *National Business Women* (November 1956). See also: Olive W. Dennis to Marguerite Zapoleon, September 3, 1947, Women's Bureau, Bulletins, RG no. 223-225, WB, NA.

30. Sally Hacker, 'The Mathematization of Engineering: Limits on Women and the Field,' in *Machina Ex Dea*, 38–58. After the Second World War, competence in mathematics was the single most common denominator in women's motivation to go into engineering. Martha Moore Trescott, 'Women Engineers in History: Profiles in Holism and Persistence,' in *Women in Scientific and Engineering Professions*, eds. Violet B. Haas and Carolyn C. Perrucci (Ann Arbor: The University of Michigan Press, 1984), 181–205. More work needs to be done on women's education in mathematics at the secondary school level, but see Warren Colburn, 'Teaching of Arithmetic. Address before the American Institute of Instruction in Boston, August 1830,' in *Readings in the History of Mathematics Education* (Washington, DC: National Council of Teachers in Mathematics, 1970), 24–37; Ruth Oldenziel, 'The Classmates of Lizzie Borden' (University of Massachusetts: unpublished manuscript, 1982). See also Robert Fox and Anna Guagnini, 'Classical Values and Useful Knowledge: The Problem of Access to Technical Careers in Modern Europe,' *Daedalus* 116, No. 4 (1987): 153–71. For an excellent discussion on the issue during the 1970s and 1980s, McIlwee and Robinson, *Women in Engineering.*

31. Nora Blatch Photo Album, Courtesy of her daughter Rhoda Barney Jenkins, Greenwich, CT (Barney-Jenkins Papers, hereafter); Records of the Women's Bureau, Women's Bureau Bulletin 223–225; RG, Box 701, WB NA; 'Rice Women Engineering,' *Rice Engineer* (January 1986): 18–23. *Women Engineers and Architects* (March 1938): 1, Box 131 folder 'Women Engineers and Architects, 1938–1940, Society of Women Engineers Collection, Walter Reuther Library, Wayne State University, Detroit, MI (SWE Collection, hereafter); correspondence Woodson Research Center, Rice University, Houston, TX.

32. Karl Drews, 'Women Engineers: The Obstacles in Their Way,' *Scientific American, Supplement* 65 (March 7, 1908): 147–8.

33. Barbara Drygulski Wright in her introduction to *Women, Work, and Technology. Transformations* (Ann Arbor: University of Michigan Press, 1987), 16–7.

34. John W. Upp, 'The American Woman Worker,' *The Woman Engineer* 1 and 'American Women Engineers,' *The Woman Engineer* 1, 11 (June 1922): 156; 186–88.

35. *Report of the Registrar of the University of Michigan, 1926–1941*; correspondence Office of Development, Bentley Historical Library, University of Michigan, Ann Arbor, MI (AWEG Papers).

36. College of Engineering and Applied Science; Special Collections, University of Colorado, Boulder, CO; correspondence University of Nevada, Special Collections Department, Reno, NV; Ohio Northern University, Alumni Office, Ada, OH; *Alumni Directory* (1875–1953), University of Minnesota; *The Minnesota Techno-Log* (May 1925): 11; correspondence University of Minnesota, University Archives, Minneapolis, MN (AWEG Papers).

37. Correspondence with Alumni Records Office, College of Engineering, Bentley Historical Library, University of Michigan, Ann Arbor, MI; ; correspondence with Bertha L. Ihnat; Ohio State University, *Commencement Programs* (1878–1907), Ohio State University, Ames, OH (AWEG Papers) clipping file, General Electric Company, GE Hall of History Collection; Goff, *Women Can Be Engineers*; Elsie Eaves, 'Wanted: Women Engineers,' *Independent Woman* (May 1942): 132–3, 158–9, p. 158.

38. Correspondence with Alumni Association, Iowa State University, Ames, IA (AWEG Papers); Adelaide Handy, 'Calculates Power Transmission for General Electric Company,' *The New York Times* (October 27, 1940).

39. Dennis to Marguerite Zapoleon, September 3, 1947, RG, 223–225, Box 701, WB NA.

40. Nora Stanton Barney letters to the editor, 'Industrial Equality for Women,' *N.Y. Herald-Tribune* (April 21, 1933) and 'Wages and Sex,' *N.Y. Herald-Tribune* (July 21, 1933).

41. Elsie Eaves to Mary Esther Poorman, November 8, 1933; Elsie Eaves to Virginia A. Swaty, March 2, 1936; Elsie Eaves to Jane Hall, November 10, 1938, Box 146, folder 'Earliest Efforts to organize, 1929–1940,' SWE Collection.

42. Box 146, folder 'Earliest efforts to organize, 1920–1940,' SWE Collection.

43. Juliet K. Coyle, 'Evolution of a Species – Engineering Aide,' *US Woman Engineer* (April 1984): 23–4; Robert McMath, Jr. et al., *Engineering the New South. Georgia Tech, 1885–1985* (Athens: The University of Georgia Press, 1985), 212; correspondence John D. Akerman with Curtiss-Wright Corporation, 1942–43, University Archives, University of Minnesota, Minneapolis, MN; Curtiss-Wright Engineering Cadettes Program Papers, Archives of Women in Science and Engineering, Iowa State University, Ames, IA; correspondence George Institute of Technology, Archives and Records Department, Atlanta, GA; 'Engineering Aide, Curtis Wright Program Follow-up of Cadette trained at Rennselaer Polytechnic,' March 22, 1945, Rennselaer Polytechnic Library, NY, (AWEG Papers).

44. C. Wilson Cole, 'Training of Women in Engineering,' *Journal of Engineering Education* 43 (October 1943): 167–84; E. D. Howe, 'Training Women for Engineering Tasks,' *Mechanical Engineering* 65 (October 1943): 742–4; R. H. Baker and Mary L. Reimold, 'What Can Be Done to Train Women for Jobs in Engineering,' *Mechanical Engineering* 64 (December 1942): 853–5; D. J. Bolanovich, 'Selection of Female Engineering Trainees,' *Journal of Educational Psychology*: (1943) 545–53; 'Free-Tuition in Courses Engineering for Women,' *Science* 95, 2455, Suppl. 10 (January 9, 1942): 10; 'Training for Women in Aeronautical Engineering at the University of Cincinnati,' *Science* 97 (June 18, 1943): 548–9.

45. *Training in Business and Technical Careers for Women* (Ohio: The University of Cincinnati, 1944); Harriette Burr, 'Guidance for Girls in Mathematics,' *The Mathematics Teacher* 36 (May 1943): 203–11. Margaret Barnard Pickel, 'A Warning to the Career Women,' *The New York Times Magazine* (July 16, 1944): 19, 32–3 and Malvina Lindsay, 'The Gentler Sex. Young Women in a Hurry,' *Washington Post* (July 20, 1944).

46. Olive W. Dennis to Marguerite Zapoleon, September 3, 1947, WB, NA.

47. Many examples may be found in the Records of the Women's Bureau Bulletins, RG 86, WB NA.

48. Coyle, 'Evolution of a Species"; US Department of Labor, Woman's Bureau, *The Outlook for Women for Women in Architecture and Engineering*, Bulletin 223 no. 5 (Washington, DC: Government Printing Office, 1948); US Department of Labor, Woman's Bureau, 'Employment and Characteristics of Women Engineers,' *Monthly Labor Review* (May 1956): 551–6.

49. Excerpts from Olive W. Dennis, 'So – Your Daughter Wants to be a Civil Engineer,' Box 701, WB NA; excerpts from letter, *Baltimore and Ohio Magazine* (September 1940): 30.

50. Olive W. Dennis, Clipping File, Baltimore and Ohio Railroad Museum, Baltimore, MD.

51. "Women AIME Members Contribute Their Share in Engineering War,' *Mining and Metallurgy* 23 (November 1942): 580–1.

52. For gender differences in the professions: Joan Brumberg, and Nancy Tomes, 'Women in the Professions: A Research Agenda for American Historians,' *Reviews of American History* 10, 2 (June 1982): 275–96; Barbara F. Reskin and Polly A. Phipps, 'Women in Male-Dominated Professional and Managerial Occupations,' in *Women Working: Theories and Facts in Perspective*, eds. Ann H. Stromberg and Shirley Harkness (Mountain View, CA: Mayfield, 1988); Barbara F. Reskin

and Partricia A. Roos, *Job Queues, Gender Queues: Explaining Women's Inroads into Male Occupations* (Philadelphia: Temple University Press, 1990). Se also, Andrew Abbott, *The Systems of Profession: An Essay on the Division of Expert Labor* (Chicago: Chicago University Press, 1988), 98–111.

53. The number of women in the professional organizations was reported in Woman's Bureau, *The Outlook for Women*, 22. In 1946, women accounted for six out of every thousand members in the ASCE and AIEE. Nine participated in the AIChE, 16 in the ASME, and 21 in the AIME. These figures roughly resemble the data gathered in the survey. New research should focus on local engineering organizations, however.

54. A. Michal McMahon, *The Making of a Profession: A Century of Electrical Engineering in America* (New York: The Institute of Electrical and Electronics Engineers Press, 1984), 58–9.

55. Anson Marston to Hilda Counts, May 6, 1919, Society of Women Engineers Papers, Box 146, folder 'Earliest Efforts to organize, 1918–1920,' SWE Collection. On her suit and efforts to rally support for her case, see W. W. Pearse to Nora S. Blatch, January 20, 1915 and another from Ernest W. Schroder to Nora S. Blatch, January 20, 1915, Barney-Jenkins Papers. 'Mrs. De Forest Loses Suit,' *The New York Times* (January 22, 1916): 13; Reports on the case appeared also on Saturday January 1, 1916: 18, 'Mrs. De Forest Files Suit"; January 12, 1916: 7; 'Old Men Bar Miss Blatch"; and in *The New York Sun* January 1, 1916: 7, 'Mrs. De Forest, Suing, Tells her Real Age.' Blatch's employment history may be found in *Notable American Women* (1980), s.v., 'Barney"; DuBois, 'Spanning Two Centuries,' 148; Speech 'Petticoats and Slide Rules", 6, Box 187, folder 'Miscellaneous Correspondence, Elsie Eaves", SWE Collection. The speech was published under the same title in a slightly altered form in *The Midwest Engineer* (1952).

56. "Women Engineers-Yesterday and Today,' *The Bent of Tau Beta Pi* (Summer 1971): 10–2; correspondence University Library, University Achives, University of Illinois at Urbana-Champaign, Urbana, IL, (AWEG Papers).

57. Rossiter, *Women Scientists in America* (1982), 77 and Chapter 4. For a sample of the literature on women in the professions see Barbara Melosh, *The Physicians' Hand: Work Culture and Conflict in American Nursing* (Philadelphia: Temple University Press, 1982); Brumberg and Tomes, 'Women in the Professions."

58. *Dictionary of Occupational Titles* (1939); Arthur D. Little to Beatrice Doerschuk, February 6, 1922, Bureau of Vocational Information, Schlesinger Library, Radcliffe College, Cambridge, MA, microfilm [BVI, hereafter], reel 12. His opinion was extensively quoted in a section on opportunities in chemical engineering in the Bureau of Vocational Information, *Women in Chemistry: A Study of Professional Opportunities* (New York: Bureau of Vocational Information, 1922), 60–1. Terry S. Reynolds, 'Defining Professional Boundaries: Chemical Engineering in the Early 20th Century,' *Technology and Culture* 27 (1986): 694–716, p. 709. On Little, see also David F. Noble, *America By Design* (New York: Oxford University Press, 1977), 124–5.

59. Calco Chemical Company, M. L. Crossley to Emma P. Hirth, December 24, 1919; National Aniline & Chemical Company, C. G. Denck to Emma P. Hirth, December 22, 1917; A. P. Tanberg, Dupont to Emma P. Hirth, August 30, 1921; L. C. Drefahl Grasselli Chemical Company to Emma P. Hirth, December 20, 1917; all BVI reels 12 and 13.

60. Mrs. Glenola Behling Rose, chemist at Dupont Company, questionnaire, February 1920, BVI reel 12; Florence Renick, questionnaire, February 1920, BVI reel 13; Jessie Elizabeth Minor, questionnaire, January 12, BVI reel 12.

61. US Department of Labor, Women's Bureau, *The Occupational Progress of Women, 1910 to 1930*, Bulletin 104 (Washington, DC: Government Printing Office, 1933); *The Outlook for Women*, 46.

62. US Department of Interior, Office of Education, *Land-Grant Colleges and Universities* (Washington, DC: Printing Office, 1930), 805; Ruth Oldenziel, 'Gender and the Meanings of Technology: Engineering in the US, 1880–1945,' (Ph.D. diss., Yale University, 1992), Fig. 6.

63. David M. Blank and George J. Stigler, *The Demand and Supply of Scientific Personnel*, General Series, 62 (New York: National Bureau of Economic Research, 1957) 4, 8–9, 10–2, 87, 192.

64. Cott, *The Grounding of Feminism*, Chapter 1 and Introduction.

65. Box 146, folder 'Earliest Efforts to Organize, 1918–1920,' SWE Collection.

66. Carroll Pursell, this volume. See also Crystal Eastman, 'Caroline Haslett and the Women Engineers,' *Equal Rights* 11/12 (10 April 1929): 69–70.

67. "Origins of the Society by Phyllis Evans Miller,' Box 147, SWE Collection; 'Girls Studying Engineering See Future for Women in These Fields,' *The Christian Science Monitor* (Saturday, April

16, 1949): 4; 'New Members of the Women's Engineering Society' *Women's Engineering Society* (1950), 315.

68. "Girls Studying Engineering See Future"; Mart Navia Kindya with Cynthia Knox Lang, *Four Decades of the Society of Women Engineers* (Society of Women Engineers, n.d.), 12; Lillian G. Murad to Katherine (Stinson), June 8, 1952, Box 146, folder 'SWE history 1951–1957,' SWE Collection.

69. See also Carroll W. Pursell, *The Machine in America: A Social History of Technology* (Baltimore: The Johns Hopkins University Press, 1995), 310, for a passing, but insightful, remark on this issue.

70. Rossiter, *Women Scientists in America* (1995), 28, 59, and Chapter 2.

71. Lillian M. Gilbreth, 'The Daughters of Martha,' speech before the Society of Women Engineers at the opening of the headquarters in the United Engineering Building, SWE celebration banquet, 1961. Box 24, Gilbreth Papers.

72. Today the SWE's office is overcrowded and understaffed, occupying a tiny space in an otherwise imposing building where all engineering societies reside together near the United Nations Headquarters in New York city.

73. "Marriage, a Career and the Curriculum,' Box 135, Gilbreth Papers.

74. For a provocative inquiry, see: Pamela Mack 'What Difference has Feminism Made to Engineering in the 20th Century?' in *Science, Medicine, Technology: The Difference Feminism Has Made* eds. Londa Schiebinger, Elizabeth Lunbeck, and Angela N.H. Creager (Chicago: University of Chicago Press, Forthcoming).

75. Mary Ritter Beard, *Woman's Work in Municipalities* (New York: Arno Press, 1972 [1915]).

Carroll Pursell

2. 'Am I a Lady or an Engineer?' The Origins of the Women's Engineering Society in Britain, 1918–1940

In 1915, *The Englishwoman's Year Book and Directory* carried a brief essay on 'Engineering', simply signed 'C. Griff.'"engineering as a profession,' writes Ms. Griff, 'has not many attractions to most women, but owing to the ever increasing use of machinery in this the twentieth century, there is an equally growing need and place in the professions for the woman engineer.'[2] Russia, Switzerland, the United States and Canada, she continues, were the most progressive on the issue, but 'in England there is not – to the writer's knowledge – any woman but herself practicing as a consulting engineer, and exceedingly few training as qualified engineers in any branch beyond automobilism.' Griff goes on to point out that there are essentially two ways in which engineers in Great Britain can become qualified: either by pursuing theoretical studies at an institution of higher education, or through 'work in the "shops" of engineering works.' She further mentions that the 'next best method is private tuition from a practical working engineer,' in which the two other, more feasible routes are combined. This type of dual qualification, academic as well as practical, which prevailed in Great Britain at the time, made it doubly difficult for women to enter the profession.

It is important to note that the three institutions blocking women's access to careers in engineering – universities, industries, and professional societies all participated in and defended a culture that powerfully supported male privilege in British society. Engineering was just part of a larger system attributing essential characteristics to men and women alike. Complex behavioral systems defined what was both natural and proper for each gender. Notions of masculinity and femininity were never simple nor uncontested, however, and over time certain activities that had formerly met with disapproval could move into the 'approved behavior' slot. Still, the fundamental belief in sexual difference itself continued to be fiercely defended. Unless women were essentially different from men, there was no way of knowing for men that they were, in fact, manly. Moreover, without some agreement on essential differences between the sexes, there would be no firm ground upon which to sustain a hierarchical system of sexual privilege, to ensure men's sole right to power, agency and financial control. This persistence of a fundamental gender hierarchy urged Ms. Griff to declare, in 1915: 'For a long time yet, in England, there will be opposition to expect from the opposite sex, for there is some jealousy in a profession which has so far been considered "safe" from an "invasion of women".'

While the Women's Engineering Society struggled to make itself a resource for women trying to find a place in what was perhaps the most masculine of professions, women themselves had to renegotiate their own gendered behavior and expectations generally. For generations, middle- and upper-class women in Britain had been moving into public roles by effectively using the essentialist notion that, as women, they possessed a moral authority over men. However, as historian Angela Woollacott has recently pointed out, it was around the time of World War I that some women began to combine this claim with another, i.e., that their professional qualifications were equal to those of men.[3] The claim to moral authority continued to be employed by those controlling the working-classes, especially women and children, such as police officers, factory inspectors, welfare supervisors, and the like. An essential female moral authority was less obviously relevant to the field of engineering, which seemed, at least superficially, to be concerned with mastering the natural world, rather than society. Whereas the presumed moral authority of women lay outside the reach of male privilege, the right to professional qualification, however, remained firmly lodged in male-dominated organizations – universities, industries, and professional societies – which privilege its gatekeepers were loath to share with 'the other sex.' This was a problem the members of the Women's Engineering Society continued to struggle with in the years following the Great War.

Early in 1919, a group of well-educated, ambitious British women, who had been recently engaged in munitions work, founded the Women's Engineering Society (WES), in order to protect themselves against risk of

Figure 9. Members of the British Women Engineering Society (founded 1919) – Caroline Haslett second from the right at the back – at its conference at the Birmingham University, April 1923. The organization, with ties to the British women's movement, united upper-class daughters of engineering firms and working-class women of the engineering trades. Between the wars, women engineers elsewhere in Europe and the United States looked to the British for inspiration. Permission and courtesy of Women Engineering Society Papers, The Institution of Electrical Engineers Archives, London, U.K.

Figure 10. The leaders of the Women Engineering's Society gathering in a relaxed atmosphere at their annual conference at Crosby Hall, 21 September 1935. From the right to the left are Mrs Willis, Caroline Haslett, Amy Mollison, Douglas, and Kennedy. Permission and courtesy of Women Engineering Society Papers, The Institution of Electrical Engineers Archives, London, U.K.

being forced out of the engineering business. Its founders immediately saw themselves confronted with the acute and abiding problem of their marginality. First, they had to establish a new professional organization. As a new generation of professional female technicians, they were forced to rely upon, but also resist being dominated by the older, more affluent women, still acting upon a fading style of moral authority and class deference. Second, they had to carve out an identity for themselves somewhere between the leisured gentry and their working-class sisters, while simultaneously attempting to claim a space for their professional expertise. Third, they were similarly caught between a powerful, patriarchal workers' union and an equally patriarchal, male-dominated professional field, both of which they both wanted to re-engender. And finally, they necessarily had to find a proper balance between their femininity and their professionalism, between the gendered roles of woman and engineer. Their struggle with these various issues helped shape the enduring, though problematic social role of British female engineers throughout the twentieth century.

The carnage beginning in August 1914 used up men and munitions at an unprecedented rate, and the need for greater supplies of both at the front created a dilemma back home. By early 1915, the situation had become critical. David Lloyd George, the British Minister of Munitions, was quoted as saying, This is 'an engineer's war, and it will be won or lost owing to the efforts or shortcomings of engineers. We need men, but we need arms more than men.'[4] The solution was to attract women into the munition plants to take the place of the men needed for combat. Tommy's Sister enabled her brother to go to war.

The war came during growing feminist agitation over women's rights, especially over the vote. Suffragettes split in their views about the wisdom and morality of the war. Some, such as Sylvia Pankhurst, fiercely opposed the war and women's participation in it. Pankhurst's mother Emmeline and her sister Christabel, however, ardently supported the war effort, and postponed the fight for the franchise in the face of this larger crisis. Still others, Millicent Garrett Fawcett, for instance, were highly ambivalent about the whole issue, not wishing to support a war in which women had very little part, but not willing, either, to oppose the government at such a critical moment in time. At the end of the war, women over thirty were granted the vote (1918). In 1919, the Sex Disqualification (Removal) Act was passed by the Houses of Parliament, removing barriers to women's participation in the professions, especially in realm of the law. Whatever their prewar expectations or wartime allegiances, British women's activities between 1914 and 1918 created strong new expectations, as well as opportunities.

After the war, the Women's Industrial League circulated a questionnaire among 5,000 firms, asking them about their wartime employment of women. A total of 1,400 companies responded. They reported that, while employing 43,200 women before the war, the total number of women employed during the war averaged 245,300. The general engineering firms had expanded at even a greater rate, growing from 14,100 female employees to 134,600. By the end of the war, probably 90 per cent of munition workers were women.[5] An estimated total of 800,000 women took up munitions work during the war, working in shipyards, aircraft factories, foundries, machine shops, shell-packing plants, explosives plants, as well as a host of other factory sites. Their duties ranged from the turning of shells on single-purpose lathes, to the manufacturing of tools, jigs, and fixtures, and the setting up, and reading of drawings.

The range of wartime work done by women can be gauged by the recollections of several of them who later acquired prominence in the WES. Verena Holmes, for example, entered a propeller factory in July 1916, and was put to work in the gluing department, laminating wood for blades. As she recalled, 'this was not a skilled job and I started, while working there, to attend evening classes in machine work and fitting at the Shoreditch Technical Institute. When I had completed the course I got a job in Willesden as a turner on a centre lathe making small aircraft parts.' She later went on to do more responsible work.[6] Margaret, Lady Moir, organized a 'relief munition movement,' which trained 'so-called leisured women' to take the place of regular munition workers on weekends. The volunteers were put through a three-week training course.[7] Margaret M. Partridge worked in the testing department of an electrical firm, which made instruments for all the branches of the service. N. M. Jeans had been an art student before the war, but was recruited into a mechanical drawing training program to wind up in the drawing room of a munition works.[8]

The hiring of women in a form of trade that was not only considered to lie outside their traditional realm, but also one dominated by the powerful male engineers' union, was executed in a cautiously regulated and limited manner. The Statutory Order, Circular L.2 of February 1916, for example, was headed 'Directions relating to the employment and remuneration of women on munition work of a class which prior to the war was not recognized as women's work in districts where such work was customarily carried on.' An additonal note further revealed that 'these Directions are on the basis of the setting up of the Machines being otherwise provided for. They are strictly confined to the War period....'[9]

The most important policy agreement concerning women's wartime work, however, was that between the government and the Amalgamated Society of Engineers, which asserted that, at the end of the war, women would be forced to quit their jobs to make way for male engineers returning from military service. This so-called Treasury Agreement of March 1915, along with the layoffs starting in early 1918, after the cancellation of Russian armaments orders, urged many female munitions workers to cast about for some sort of job security.

Lady Katherine Parsons played a leading public role in these efforts. Through her marriage to Sir Charles Parsons she had some familiarity with the engineering trades, but it appears that it was her daughter who urged her into taking up the cause of female engineers. Lady Parson's only son had been killed in the war in 1918, while her daughter, Rachael Mary (1885–1956), had started to develop an interest in engineering from an early age onwards.[10] Her father's favorite, Rachael worked with Sir Charles in his workshop at home, and later attended Roedean school and Newnham College at Cambridge, where she took the Mechanical Science Tripos. With the onset of war, she joined her father's Heaton factory, a 'small manufacturing works for the development of steam turbines...and of high-speed electrical machinery.' When her brother was sent to the front, she became the plant's Director.[11]

Lady Parsons may have credited her daughter with taking the initiative for the setting up of a society, but it was she herself who maintained the position and social standing to make it possible. In 1918, she was President of the Newcastle branch of the National Union of Women Workers of Great Britain and Ireland, whose primary concerns were women's security, reform, and employment. In March 1918, she proposed a resolution to the Union's Executive Committee, urging for the appointment of a woman on the government's recently established Reconstruction Committee 'in connection with the engineering trades.' She pointed out that 'if the present proposals were carried, this would involve the displacement of a large number of women engaged in Newcastle in skilled engineering.' The committee was sympathetic and

Figure 11. Advertisement for C.A. Parsons and Co. Ltd., the family firm of Rachel Parsons, one of the founder members of the British Women's Engineering Society, in their house quarterly publication, which was printed by Women's Printing Society. Young women building up micanite insulating sheets. The Woman Engineer *1, 9 (December 1921). Permission and courtesy of Women Engineering Society Papers, The Institution of Electrical Engineers Archives, London, U.K.*

recommended bringing the resolution before the full Council.[12] When Lady Parson's proposal (along with a list of suitable candidates) had finally been sent to the government, the response was that Mary Macarthur was on the Labour Panel of the Reconstruction Committee, and that the ministry was reluctant to add more names. The group considered the reply 'most unsatisfactory' and sent a copy to 'Mrs. Fawcett in case the National Union of Women's Suffrage Societies thought it desirable to take further action.'[13]

The annual council meeting of the National Union of Women Workers, taking place in Harrogate in October 1918, paid special attention to 'The Position of Women in Agriculture and Engineering,' with a panel on engineering in which Florence M. Campbell of the Women's Trade Union League and Rachael Parsons took part. Campbell claimed to speaking on behalf of the minority of women munitions workers who 'wish to study, and learn from a scientific point of view.'[14] Early in December, Rachael Parsons revealed that the (renamed) National Council of Women had 'appointed a Special Engineering Committee, whose objects are to work for equal opportunities for training and employment for women in engineering and allied industries.'[15] A notice appearing in *The Times* announced that 'all women employed in engineering and allied industries wishing to join an association for safeguarding their interests are invited to write, giving their name, address, and descriptions of work to the Hon. Secretary of the Engineering Committee, National Council of Women.'[16]

Whilst the Engineering Committee continued to do its work, early in 1919, the National Council of Women went a step further and orchestrated the birth of an independent Women's Engineering Society.[17] Their advertisement in the journal *Engineering* was brought to the attention of Caroline Haslett who applied for, and acquired, the position of organizing secretary of the fledgling group.[18] In February 1919, she began setting up an organization.

Caroline Haslett (1895–1957) was one of five children from a respectable Victorian family. Her father was an engineer and her mother a founding member of the Worth Women's Institution, as well as president of the Crawley Co-operative Women's Guild. Haslett was reported to have shown an early antipathy to housework, preferring her father's workshop instead, in which he taught his children the proper use of tools. She grew close to her mother when, apparently in spite of her father's disapproval, both became active in the suffragist movement. At the age of 18, Haslett went to London to attend a commercial college, and later joined the Cochrane Boiler Company which ran a plant in Scotland. A friend of her mother's, who had been imprisoned for Sinn Fein activities, was the sister of Cochrane's Managing Director. During the war, Haslett went to work at the plant in Scotland, and by 1918, she was back in London, managing the firm's office in the capital.[19] Starting as its organizing secretary, Caroline Haslett continued to keep close connections with the WES until her death.

Records of interviews given by Haslett during the first days of the society's existence make abundantly clear that she not only set out to find outside allies to support the new group, but also immediately began to receive pleas for help. She called upon and distributed literature among the War Service Women's Legion, the Women's Freedom League, the National Federation of Women Workers, the YWCA, the Women's Auxiliary Force, the Central Bureau for the Employment of Women, and similar groups. A Mr. Philip Bellows of the Friends Institute in Devonshire called in, 'wishing to give lectures.' Haslett noted that Bellows had 'advanced ideas on [the] emancipation of women.' A more typical case was that coming in on February 20th: a 'Miss…Scott, Black Heath, called here had considerable training. Wishes eventually to become Lady Supt. in works. Now taking classes at Woolwich Poly.' On the same day, 'Mrs Robinson and Mrs Knaffelt called here with Miss Selly. Mrs R capstan lathe work for 2 1/2 years & finished on turning machines. Mrs K tool setter and finished on turning machine. Both widows keen to continue their work.'[20] The trouble these women took to seek out the new society, the accounts of their wartime experience, and their clearly expressed desire to continue working, all pointed to the most immediate and gravest problem facing the society: the Restoration of Pre-War Practices Bill.[21]

Within the post-war industrial world, opposition to women remaining in the engineering trades was widespread. A. P. M. Fleming, manager of the education research department of the British Westinghouse Company, expressed the prevailing cultural taboo against women engineers: 'The average woman,' he told *The Daily News*, 'does not possess the same engineering instinct as the average man. For repetition work, yes, but for originality and research, well, there is something lacking – perhaps it is the survival of the Cave days, when the woman stayed at home and the right to live depended upon the man's wits. The great factor that tells against a woman taking up engineering – and I refer to professional engineering as distinct from the engineering workwoman – is marriage.' Not surprisingly, Fleming found it impossible to 'suggest any remedy. It's just human nature.'[22]

The members of the Amalgamated Society of Engineers vented much more specific complaints. 'Most of us,' one of them wrote, 'admit that the woman engineer met an emergency and that the nation has reason to be grateful to her for meeting it as well as she has done. That does not imply that she has been the unqualified success the Government and the Press have represented her to be; nor will all the talk in the world make her what she is not – the skilled man's equivalent as a producer.' 'Do the women,' he went on to ask, 'propose to remain to the permanent exclusion of these men [who had served in the forces], or are those of us who have the right conferred by probation and skilled qualification to take the streets while our wives and daughters effect their economic emancipation at our expense?' Ostensibly, the main fear was that women were not as productive as men are, would therefore be paid lower wages, and thus also drive down male wages. 'If she insists upon equal pay for equal work,' he concluded, 'the employer will not have her; if she offers to compete with men upon lower wage terms, we will not have her.'[23] The wage competition was real, of course, but the threat to the patriarchal order and to masculinity as such is equally palpable. Though founded as early as 1851, the ASE refused to admit women as their members up until 1943, in the middle of World War II. It therewith set a record for the longest period of the exclusion of women among all the craft unions.[24]

Put simply, the proposed Restoration of Pre-War Practices Bill was largely designed to fulfill the government's promise to the engineers' union as it had been set down in the March 1915 Treasury Agreement, and ensure a return to the exclusionary practices pursued before 1915.[25] Rachael Parsons tried to put the question in terms least likely to offend her powerful opponents in both the government and the unions. This, she wrote, 'is no class question; it is not, one may say with thankfulness, a disagreement between capital and labour; for in the ranks of the demobilised women every class is included – the girl who must earn her own living or who wishes to follow the example of her brothers and take up a profession; the widow working to support her children – all are represented.' She went on to scold women themselves for having failed to organize during the war, 'till today they are dismissed with scarcely a word of protest and with little power to make their voices heard.' The WES, she concluded, 'invites all who are interested in this subject to join its ranks and thus help to safeguard the interests of women in engineering.'[26]

The National Council of Women passed a resolution at its annual meeting in 1919, urging the government to 'see that the Restoration of Pre-War Practices Bill imposes no restrictions upon the employment of women beyond those definitely promised in the Treasury Agreement, and in particular begs that the right of women to work in new trades may not be taken from them.'[27] Despite such appeals, the act was passed and, as one scholar comments, 'employers were quick to co-operate with skilled workers in expelling women from their jobs so that production during the post-war boom was not interrupted.'[28] The first issue of the society's journal *Woman Engineer*, which appeared in December 1919, reported that 'the Women's Engineering Society was established nearly a year ago in the interests of women engaged in engineering and allied trades. Since that time the outlook for women in the engineering world has become increasingly gloomy, and with the passing of the Restoration of Pre-War Practices Bill, the position seemed almost hopeless.'[29]

The society nonetheless persevered. In a pamphlet printed by the Women's Printing Society, its leaders declared that they were determined to 'press for the setting up of new peace industries; for the best and most modern machinery; for the most wholesome conditions of work and for equal pay for work that is genuinely equal.'[30] The society set up an office and a technical library, opened four branches outside London, and collected information on training programs that were open to women. Although, they maintained, 'the Society is not essentially an Employment Bureau, we are doing all in our power to find suitable posts for our members.'[31]

If job openings were slow in coming to its attention, the society itself took active steps to increase their number. Lady Parsons reported that she had offered to help finance the new Swainson Pump Company on condition that it hire female factory workers, as well as take on a woman as Assistant Manager.[32] More importantly, Lady Parsons was instrumental in organizing a firm, named the Association of Atalanta, serving with Haslett on its board of directors.[33] Three other women also sat on the board (including one designated Works Manager), in addition to three male engineers from Loughborough Technical College.[34] Haslett reported to the society that the firm was a 'scheme which some of the members had in hand to commence a small factory for the manufacture of bicycles, etc., to be run entirely by women.'[35] *Woman Engineer* explained that 'seeing no scope for their activities, and having the natural road of success barred to them, these women have decided to risk their all and to establish an engineering works where there will be absolute freedom for them to use the ability and skill which they possess.'[36]

Although Atalanta ended eventually in failure and recriminations, the basic idea underlying it was tried out more than once. In 1927, the Electrical Enterprise, Ltd. was established, with Haslett once again on its board of directors, along with her friend Margaret Partridge. Set up to take advantage of the opportunities in rural electrification provided by the Electricity (Supply) Act of 1926, the firm also hoped to 'provide openings for women in the business of electricity supply, and by that means to encourage women to enter a calling which is most important from the point of view of public service, and is essentially one of equal interest to men and women.'[37]

Even more directly in the same line, was the initiative taken by the WES in 1924, to found a similar organization, the Electrical Association for Women, which was also presided over by Caroline Haslett. The membership of this new association showed significant overlap with that of the older one, but it concentrated on bringing the benefits of electricity to women at home, and on creating professional opportunities for women within various branches of the growing field of electric power supply.[38] In 1986, the EAW was ultimately disbanded, when it was clear that its most important tasks had been successfully achieved.

Throughout the first decade of its existence, the WES found that many of its major activities were closely entangled with the problems of class. In wartime analysis for *The New Statesman*, Beatrice and Sidney Webb shrewdly noted that professional associations 'are closely connected with the directors and managers of industry, and with

Employers' Associations; and there are usually among their members some Capitalists who combine professional qualifications with the ownership of the instruments of production and the employment of subordinate labour in pursuit of personal profit.'[39] Within a few months after the foundation of the WES, *The Masses* warned its readers against its activities, claiming that 'the working woman does not show much interest in the society itself.' The journal continued that it was rather 'Mr. Richardson [the editor of the journal *Engineering*], Lady Parsons – the wife of Sir Herbert [*sic*], of Parsons' turbine fame – Rachael, their daughter, Mrs. Ormsby, their Welfare Supervisor, and Mr. Harry Dubery, of the National Alliance of Employers and Employed, who demand for their women the right to re-enter engineering.'[40]

The society's original purpose appears to have been the protection of working rights for all the women who had gained experience in the engineering trades during the war. The invitation to join, appearing in *The Times* in December 1918, was addressed to 'all women employed in engineering and allied industries.' A leaflet that was further issued asserted that it was the 'aim of this Society to unite all classes of women in the Engineering trades.' Its attempts to act as an informal clearing house for jobs forms additional evidence that the society's concerns reached beyond the interests of highly-trained women only.[41] At the first meeting of its executive committee, discussion arose 'regarding the desirability of forming a Trade Union for women who had attained some degree of skill in engineering.'[42] At the same meeting, the society's political determination came to the test through the case of a Miss Wright, a member of the Manchester branch. The minutes reveal that Miss Wright 'had been suspended by the Labour Exchange because she would not accept a post as third housemaid. She had never done domestic work before and had no wish to do it. Miss Warren asked whether the Society had taken any action with regard to this case. Lady Moir pointed out that we were hardly strong enough at the present time to take action, & also that we were not yet incorporated.'[43] The failure to take up such a case, so evidently and directly relevant to the society's proclaimed purposes, bade ill for the future.

Within the year, the society was backing away from its original definition of their core-constituency on the basis of sex, in favor of one more shaped by class interests. From the beginning, Rachael Parsons had sometimes identified her true target group as those who 'wish to study, and learn from a scientific point of view.'[44] Provoked by charges in *The Engineer*, the society asserted that its members 'fully realise the differences between the scientifically-trained engineer and the mechanic, and we agree that an endeavour should be made to effect a distinction between the two.' The suggestion that women only fit within the lower-level, semi-skilled branches of the industry, however, they rejected on the grounds that 'we do not agree that women should be "content to call themselves mechanics, machinists, fillers, and so on."' For one thing, they were quick to point out, it was that route to skilled labor in particular that was cut off by male unions. 'The present attitude of the skilled Trades Unions,' they protested, 'prevents a girl from taking up an apprenticeship in the shops.'[45]

It would not be surprising to find that engineering societies and schools appeared more susceptible to the idea of admitting women than the trades unions. After all, the society's leaders all belonged to the same class as those in charge of these male organizations. A concomitant shift in address can be seen to have occurred within a period of two years. When, in 1921, it was suggested that representatives of the trades unions and the Employers Federation might be asked to speak on 'alternate nights to give their views on women's position in Engineering,' the idea was turned down. 'It was felt that it would be undesirable to ask representatives from Trades Unions at the present time,' the minutes recorded. The next entry, however, conveys that 'Mr. Letter, of the Engineering Employers' Federation, was suggested as a possible lecturer.'[46] In 1922, the society revised its original pamphlet to insert the words 'profession' and 'industries' into its original statement of purpose, which now read: 'The Women's Engineering Society is established in the interests of Women engaged in the Engineering Profession and Allied Trades and Industries.' The word 'experience' was further substituted with 'degree of skill,' and the phrase 'women of engineering and scientific attainment' became 'technical women.'[47] By this time, the Society began to concentrate almost exclusively on the needs of professional women engineers, and largely abandoned the vast number of women with practical experience only.

Around the mid-1920s, it furthermore became painfully clear that class-issues could create deep divisions even within the society's leadership, and, indeed, might destroy the spirit of gender solidarity that had been so evident in the early days, thus endangering the continued existence of the society as such. Lady Parsons and her daughter Rachael had been the society's initiators, its formal patrons, and, more often than not, its public voice. On a range of issues, including her own continued financial support of the organization, Lady Parsons found herself increasingly at odds with both Caroline Haslett's professional style and needs and those of the society's other members. At some point, Haslett actually wrote to a friend, 'when I tell you that I have been harried and worried by my President to the extent of a solicitor's letter, you will I am sure forgive me. She is behaving in a perfectly stupid way but it is wonderful how much damage an obstructionist can do even without brains.'[48]

The quarrel between Lady Parsons and Haslett in fact signaled a more general division among middle-class women operating in the 'public' sphere. Parsons represented an older style of public involvement through specific social issues, one that largely rested on class deference and moral authority. Haslett, a single woman without any apparent financial resources over and above her salary, derived her authority from her technical expertise. The general shift occurring in society in the early twentieth century, its change from a structure based on the moral principles of the Protestant faith to one in which the authority of science and technology began to prevail, thus played a critical role in legitimating Haslett's and her cohort's claims to power.[49]

In her battle for control over the society with Haslett, Lady Parsons enlisted friends among her own social ranks, with whom she attended society meetings in order to sway their proceedings. Two months after Haslett's outburst to her friend, the latter wrote back to her: 'I don't know what upset me most yesterday – the meeting was quite revolting, or the fact that I had to leave before anything was settled.' She cautioned Haslett to 'remember that the Policy of Lady P will be to rake up hosts of new members – or old ones resuscitated to out vote the engineer members. It was quite obvious yesterday that she would far rather kill the society than let anyone else run it. I do wish I could have stayed yesterday – but there was a £3000 tender in question – so I just had to come off. Worst luck – that is rather where the Society ladies get an extra pull over working members. I'm Bolshi!!!!'[50] Margaret Partridge may not have been a Bolshevik, but she had identified a real danger. In drawing back from the needs and concerns of ordinary working women, the society had associated itself with the gentry, with which it had, in fact, little in common. A deep rift between wealthy patrons and volunteers, such as Lady Parsons, and the women actually pursuing their careers in engineering had become apparent.

Lady Parsons had considered closing down the society as early as 1922, confronted as she was with a small membership, an even smaller treasury, and a generally hostile environment for women engineers. A potential alternative, explored by Haslett, was for the society to join with another organization. The question immediately arising was whether a men's or women's group would be best for this purpose. The society's secretary reflected: 'If we can manage it I think we would be better to amalgamate with a man's organisation, as that would give us status. As you know there are several schemes on foot at the moment to draft the smaller engineering Societies into one large group. If we amalgamate with a woman's organisation,' she concluded, 'it seems to me that we should lose our status, although we might retain the propaganda side of our work.'[51] It was a controversial issue, but at least for the moment, gender solidarity lost out to the power combined male privilege and class advantage.

The WES in the end maintained its independence and survived the withdrawal of Lady Parsons' personal and financial support. Parsons had played a critical role in opening doors and providing financial backing in the early years, but by the time the rupture within the society's leadership came fully to the surface, the numbers and concomitant power of its technical members, headed by Haslett, proved enough to sustain it without her. The timely spin-off of its sister organization, the Electrical Association for Women, strengthened the society's public visibility, and furthermore provided financial support by sharing expenses and facilities, as well as membership.[52]

Class divisions among female engineers were a serious problem, but gender differences were seen as more fundamental in defining their overall marginality. British engineering societies had emerged in the Victorian era. They took their example from gentlemen's clubs which, as one critic points out, were based on 'two ancient British ideas – the segregation of classes, and the segregation of sexes.'[53] In a power structure set up along the lines of class and gender, engineers had, since the eighteenth century, insisted on differentiating themselves from, on the one hand, mere mechanics and artisans, and from those involved in less 'manly' occupations, on the other. The two groups were, of course, closely entwined. The engineers rational understanding of principles and processes set them apart from the common worker, but also shifted understanding as such to a more masculine discourse of abstract reasoning. The well-guarded chambers of engineering societies merely formed the outward manifestations of modes of segregation carefully designed, and in due course, painfully achieved in order to claim and preserve the power of privilege.[54]

The WES inquired with all the major engineering organizations whether they admitted women into their membership. The responses ranged from downright negative to unenthusiastic confirmation. In the spring of 1919, the Institute of Naval Architects gave in to the pressure, and changed its rules against female membership. The very next day, three women were elected as Associates, including Rachael Parsons.[55] The Institute of Electrical Engineers had elected its first full female member in 1899; it was not until 1916 that the next would be admitted. From that year onwards, up to 1938, 'about 35 ladies were admitted to various classes of membership.'[56] In 1922, the WES could report that 'most of the Institutions are now prepared to admit women, with the exception of the Mining Institutions.'[57]

A major victory came in 1922, when Caroline Haslett was invited to a meeting of representatives of engineering societies at the Engineer's Club in September. 'At the last moment,' she reported to Lady Parsons, 'Miss Griff turned up and I was able to get her a ticket, which was rather nice as otherwise I should have been the only woman there.' There was 'much fluttering,' she continued, 'among the hall porters when we arrived, as they assured us that no ladies were ever admitted to the Club and that it must be a mistake. However we finally convinced them that we had a genuine invitation.' The subject to be discussed was the proposed formation of an Association of British Engineering Societies, intended to amalgamate a number, if not all, of the societies in existence. After two hours of discussion ('I had no idea that men could talk so much' Haslett remarked afterwards), but a committee of seven representatives was appointed to draft a proposal for a constitution. The Secretary of the WES was among this group.[58] In the end, nothing came of this potential joining of forces with male-dominated organizations, however, partly because the drive for general consolidation among the engineering societies lost its momentum.[59]

The admission records of women taking courses in engineering schools were equally erratic. In a speech before the Scottish Council of Women Citizens' Association, Lady Parsons related the story – which may, in fact, have involved her daughter Rachael – 'that about 1910 three girls, daughters of engineers, who were going to Newnham, persuaded the late Professor Hopkinson to admit them to his course in engineering. He had broken the fact to his class, after standing in the door for a few moments, with the words, 'It has come!'[60] At the beginning of the war, some special training courses were provided for women. In 1916, Loughborough College received a sub-contract for the manufacturing of shell cases, and immediately recruited a group of women who were to acquire engineering skills while actually working the machines. By the end of the war, some 2,305 women had been formally educated at Loughborough through this 'concept of training on production.'[61]

After the war, progress was slow. Haslett was instructed by her executive council to write a letter of reproach to the Oldham Technical School, which was reported to have turned away a woman on the grounds 'that she would be a disturbing influence amongst the boy students and also that her general technical education would in any case be inferior to theirs.'[62] In 1922, the WES was informed by the Board of Education that the Whitworth Scholarships and Exhibitions would not be open to women, 'owing to the present state of unemployment in the engineering world.' The council decided that 'this was not a sufficiently good reason for excluding women,' and decided to send a delegation to the Minister of Education.[63]

In 1932, members of the King's College London Engineering Society heard 'rumors that women students in engineering might enter the College next autumn. This caused uproar amongst the gatherings in the Common Room; and it was decided to send a delegation from the Society to make representations to the Professional Board. A saying of "the Little Professor" was quoted: "there is no such thing as a woman engineer." A petition against the entry of women Students into the Engineering Faculty was signed by over 100 past students and by 110 present Students. Result: No Women!'[64]

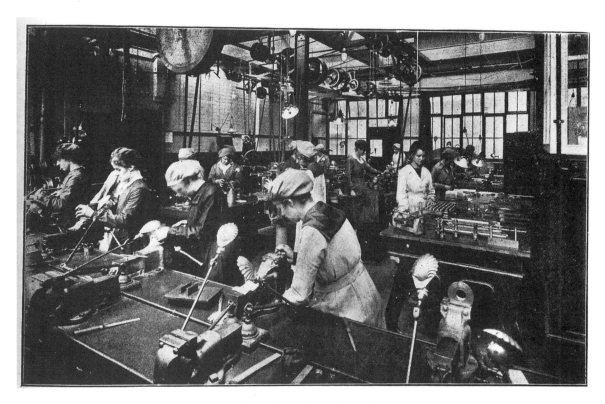

Figure 12. Gauge apprentice shop, Loughborough College. In the UK, the engineering profession also contained lower professional ranks such as metalworkers. Cover of The Woman Engineer *1, 9 (December 1921). Permission and courtesy of Women Engineering Society Papers, The Institution of Electrical Engineers Archives, London, U.K.*

Faced with such male awkwardness and misogyny, creating a new social role for women was not an easy task. What, for example, was expected of female engineers at professional conferences, and what should women's response be? In the spring of 1925, Margaret Partridge wrote to Caroline Haslett about a meeting they were both to attend: 'Please tell me – am I a lady or an engineer – what are you? Do you notice that at each of the papers there is a special side stunt for ladies.... I am entirely in your hands,' she concluded, 'to be organised as you think best – either to join in discussion of the papers – to talk… to the wives – or merely charm in silence. What ho! I've made a new hat for that – don't be angry.'[65]

Haslett's reply was clear and to the point. 'One thing I am quite clear about,' she wrote, 'and that is that we are both going as engineers or semi engineers and not as wives, also I am quite clear about the fact that if I desire to take part in any of the more serious discussions I shall most certainly do so.' About the hat, however, she was more hesitant: 'I am most interested in the news that you have made a new hat for the occasion, I do hope that it is the kind of hat which engineers and not wives would wear. I have vaguely thought about clothes for the occasion but so far have done nothing definite about it.'[66] Partridge was quick to respond: 'No doubt you will be relieved to hear', she reassured her friend, 'that it is strictly a man's hat – It is my Father's top hat to be precise – with a bit of red scarf round it & a tassel at the side to give it the hybrid semi-feminine air which seems suitable to the occasion.' She further enclosed a quickly drawn sketch of 'my new hat.' Yet Haslett was not so easily placated. 'The description and sketch of your hat,' she wrote in return, 'if anything, increases my alarm and anxiety. I do not want to appear sarcastic about your artistic efforts but it seems to me there may be some danger of our being taken for the new pierrot troupe! However, I have no doubt the members at the Convention will welcome a little light diversion of this sort.'[67]

From his epistolary exchange, we may reasonably infer that the affectionate banter between the two close friends served to mask genuine tension about the ways in which they might be expected to act, and their own desires and political conscience in the face of marginality. A hat, of course, was by no way a trivial issue for them to be anxious about. As any other article of clothing, a hat was deeply imbued with specific gender and professional significance.

Haslett later also related a telling anecdote concerning Emmeline Pankhurst. Finding herself next to her at a dinner party, the leader of the suffragettes asked her what she did for a living. Pankhurst met the requested information with a exclaimation: 'But surely, that's a very unsuitable occupation for a lady, isn't it?' Although the story appears to render Haslett more solidly feminist than her companion, Pankhurst went on to inquire whether she ever met with discrimination – which Haslett flatly denied! The territory indeed appeared as yet to be unmapped.[68]

It should not come as a surprise that, at a time when the New Woman of the late nineteenth century, as well as the more recently emerging Bachelor Girl, were social types figuring prominently in the public mind, that the question of marriage in relation to women engineers should provoke much comment. There always had been a general concern that their inappropriate 'male' activities could only be pursued at the expense of traditional 'female' duties and obligations. Margaret Partridge was offended by the existing legal regulations on this issue when she set up her own electrical contracting firm. 'You'll be amused,' she wrote to Haslett,' – re the floatation of my own Co. – I'm having altercations about my absolute right to be married as many times as I please without altering any document or agreement in any way! Don't you support me in this? I *won't* see why my status 'Married' and 'Spinster' should have any more bearing on the case than it would with a man.'[69]

Many WES members were married, some having, in fact, first come to engineering on account of their husbands' interests. Lady Margaret Moir, for instance, who was married to Sir Ernest, a prominent civil engineer, was very active in WES, but still called herself an 'engineer by marriage.'[70] Social expectations were as yet sufficiently powerful for the *Woman Engineer*, upon congratulating three newly married members, proudly to add that 'the fact that all three women are retaining their

Forgive these long screeds — At the moment I am living alone & working alone & see no one at all to let off steam to — So as a little gossipy relaxation I worry you.

Re I.M.E.A. I had filled up the same cards you did. till I got stuck over the dinner & Roedene

I shall do Robinsons paper instead of Roedene — but otherwise I shall do the same as you — cut the dinner & save 23/- ie one night at hotel & ~~th~~ 8/6. for dinner.

My new hat.

Your knowledge of the ways of devils seems pretty accurate — where did you learn it all — Has F.S. helped to your education.

Yours as ever
Margaret.

The woman is not quite as mad as this letter would indicate — only about one third gone yet — M.P.

Figure 13. Margaret Partridge's portrait of herself with "My New Hat" as drawn in a personal letter to her colleague Caroline Haslett, May 29, 1925. For these pioneer women engineers, it was not clear how "feminine" they dare be in what was perceived as a "masculine" profession. Permission and courtesy of Women's Engineering Society Papers, Institution of Electrical Engineers Archives, London, U.K.

membership of the WES, and intend to continue their interest in engineering work might provide, in some measure, an answer to the cynics' point of view, that women lose all interest in their career after marriage!'[71]

The issue of marriage was a moot point for Caroline Haslett who remained single throughout her life – as did, in effect, Rachael Parsons and several of her associates.[72] Indeed, under the headline 'A COSY FLAT FOR ONE,' a 1925 newspaper report dubbed Haslett 'Miss All Alone, the self-sufficing bachelor girl of 1925…working hard in trade or profession, she finds time to equip and run a little home. Everything in it is the expression of her independent personality.' After describing Haslett's electrical appliances (she and Margaret Partridge had wired the flat themselves), the reporter asked rhetorically: 'Lonely? Not a bit of it! "There's a gramophone there",' he quotes Haslett as saying, 'a home-made one. Friends come in. We just take this out of the way, and this, and dance to it on the stained floor. I have friends coming to dinner to-night.' The report concluded that 'Miss All Alone is now the head of a new engineering organisation for women, with all the world before her and the friendship of women who are doing things in the trades and professions.' At least in this reporter's view, Haslett was not a figure of pity, shackled down by a career in a male profession, but a bright new social type, an independent 'Bachelor Girl.'

The 'Bachelor Girl', however, was still 'Miss All Alone,' whether by preference or necessity. The French play 'L'Enfant' by Brieux, staged in Paris in 1923, in contrast suggested that such a choice was both a deliberate and a political one. The play's protagonist, a female engineer, was not only shown to fight her own battle for independence, but also doing so in opposition to the combined forces of family and society. She finds satisfaction in her professional work (building a factory that will employ women engineers), but is lonely in her personal life. Seeing marriage as 'a thralldom and a surrender of her personality,' she puts all her emotional energy in the education of her sister's child. Jealous of the child's devotion to its aunt, and worried about her husband's growing interest in her sister, the child's mother cuts her off from the family. The female engineer then decides to have a child of her own and on her own terms – i.e., out of wedlock – and gets pregnant by her cousin. Her mother unexpectedly agrees to accompany her to Switzerland for the lying-in. When the natural father turns up, he persuades her to marry him. 'But her last words are flung out as a challenge even at the moment of her surrender… "Je cede…je suis bien sur d'avoir raison, cependant. J'ai raison…trop tot".' ["I yield…I am certain I am right in doing so. I am right…but premature."][73] In its review, *Woman Engineer* commented that this not only 'surely [was] the first time in history that a woman engineer has appeared as the heroine of a play,' but also that 'she is taken as the extreme type of emancipation rather than as a technical example, but the subject is treated from so many serious aspects, and illustrates so many problems that face the woman engineer in this country, that it seems justifiable to give a brief notice of it.'[74]

The political choices of the play's French heroine may have appeared premature even to herself, but her professional choice clearly was not. In 1939, with Great Britain plunging into war for the second time within one generation, Caroline Haslett, by then Honorary Secretary of the WES, was elected as its president. 'Miss Haslett,' it was reported, 'would not disown the title of feminist but she would define the term exactly. She believes in women, in their capacities and powers, in their right to contribute the best of which they are capable in every sphere of activity, in their right to opportunities and rewards, but she would not admit any reason for rivalry between men and women…. She has always collaborated harmoniously with her men colleagues.' For two decades, the new president had been the shaping force behind, and in a very real sense become the epitome of the society. 'As a feminist,' the article continued, 'Miss Haslett is pleased that the formation of the Women's Engineering Society has improved the opportunities of women engineers.'[75] The engineering profession, however, had been socially established as a masculine preserve for more than a century, and the growing participation of women in the field did little to re-engender the enterprise as such.

Shortly before World War I, a survey reported 12,271 'Engineers and Surveyors' in the country, none of them women. It was the only profession, other than 'Barristers, Solicitors,' to have no female participants at all.[76] In 1933, Great Britain counted 46 women working in engineering. The fact that this number rose to 200 within a year, not only gives some indication of the problems concerning professional definition, but also attests to the success with which some women managed to find a place for themselves in a traditionally, perhaps *the* most traditionally by male profession.[77]

In the first decades after being founded, the WES moved from precarious existence to permanent establishment. Its leaders, in achieving this success, had found it necessary both to fight patriarchal rule within the profession, to discover their own roles within the field and, more problematically still, to narrow their focus to the exclusion of the large mass of working-class women who had entered the engineering trades during World War I, and who subsequently found it impossible to keep their jobs – and thus their recent gain in wages, hours, and conditions. By restricting their attention to professional goals only, the WES leaders sought to beat the male engineers by joining them, a strategy that in the end ensured their survival. The fact that at the end of the century, engineering is still largely a male preserve, in many ways extremely hostile to aspiring female participants, says more about the difficulties involved in trying to re-engender the field, than that it shows any failure on the part of the WES.

Notes

1. The author thanks especially his colleague Dr. Angela Woollacott for a thorough and critical reading of this manuscript and for help in understanding the role of gender in shaping the history of women.
2. G. E. Mitton, ed., *The Englishwoman's Year Book and Directory* (London: A. & C. Black, Ltd., 1915), 105.
3. Angela Woollacott, 'From Moral to Professional Authority: Secularism, Social Work, and Middle-Class Women's Self-Construction in World War I Britain,' *Journal of Women's History* 10 (Summer 1998): 85–111.
4. Lewis R. Freeman, 'Lloyd George: Minister of 'What-Most-Needs-Doing',' *Review of Reviews* 52 (November 1915): 569.
5. Women's Industrial League, *Report on Women in Engineering and Other Trades* (London, c. 1919), 6. Iris A. Cummins, 'The Woman Engineer,' *The Englishwoman* 46 (April 1920): 38.
6. Quoted in 'Thinking Back to the Last War,' *The Woman Engineer* 4 (Autumn 1939): 308.
7. Ibid.
8. Ibid., 308–309.
9. Quoted in Barbara Drake, *Women in the Engineering Trades*, Trade Union Series No. 3 (London, 1917), 134.
10. Obituary of Lady Parsons, *Daily Telegraph* (10 October 1933).
11. Rollo Appleyard, *Charles Parsons: His Life and Work* (London, 1933), 280. Rachael's brother was killed in the war in 1918.
12. NUWW Executive Committee Minutes, Dec, 1916-Nov. 1920, meeting of March 22, 1918, Archives of the National Council of Women of Great Britain, London.
13. NUWW Executive Committee Meeting of May 10, 1918, ibid. Mary Macarthur was head of the National Federation of Women Workers and the most prominent advocate of women workers. Millicent Garrett Fawcett was arguably the most prominent suffragist in Britain.
14. NUWW, *Occasional Papers, National Union of Women Workers of Great Britain and Ireland, annual Council Meeting in Harrogate, 8th, 9th & 10th October 1918* (n.p., n.d.), 18.
15. *Morning Post* (7 December 1918).
16. *The Times* (3 December 1918).
17. Rachael Parsons wrote about the 'inauguration of the WES by the NCW.' *Daily Mail* (7 February 1917).
18. J. S. to Carrie [Haslett], January 12, 1919, Caroline Haslett papers, Archives of the Institution of Electrical Engineers, London, Great Britain.
19. Rosalind Messenger, *The Doors of Opportunity* (London, 1967), *passim.* The author, Haslett's sister, said that one of her 'Bibles' was Olive Schreiner's book *Woman and Labour.*
20. Booklet headed Interviews, February 5, 1919, Women's Engineering Society papers, IEE archives, London, Great Britain.
21. See Ray Strachey, 'The Restoration of Pre-War Practices,' *Common Cause* 11 (4 April 1919): 628–629.
22. *Daily News and Leader* (3 June 1919). For a subtle and powerful discussion of a closely related question, see Evelyn Fox Keller, *Reflections on Gender and Science* (New Haven: Yale University Press, 1985).
23. *Morning Post* (7 December 1918).
24. Sylvia Walby, *Patriarchy at Work: Patriarchal and Capitalist Relations in Employment* (Cambridge: Cambridge University Press, 1986), 250.
25. J. R. Clynes, 'The Pre-War Practices Bill,' *The Englishwoman* 43 (August 1919): 81–88.
26. Rachael M. Parsons, 'Women in Engineering,' *Queen* 145 (15 March 1919): 287.
27. *Handbook and Report of the National Council of Women in Great Britain and Ireland, 1919–1920* (London, 1920), 90. In fact 'new' trades, such as work on aircraft and electrical appliances, remained open to women. See Miriam Glucksmann, *Women Assemble: Women Workers and the New Industries in Inter-war Britain* (London: Routledge, 1990).
28. Norbert C. Soldon, *Women in British Trade Unions, 1877–1976* (Dublin, 1978), 100.
29. *Woman Engineer* 1 (December, 1919): 2.
30. Undated pamphlet entitled *The Women's Engineering Society*, copy in WES papers and correspondence, IEE archive.
31. *Woman Engineer* 1 (December 1919): 2.

32. WES, Minute Book No. 2, Council meeting, June 11, 1920, > IEE archives.

33. Atalanta was a mythical Greek heroine who was exposed as an infant by her father, who wanted only boys. Having been raised by hunters, excelling at sports, and wishing to remain unmarried, she forced all her suitors to run a foot race with her. The rules were that one could marry her if he won, but she would kill those that lost.

34. Articles of Association of the Atalanta, Limited, October 25, 1920, WES papers and correspondence, 1915–17, IEE archives.

35. WES, Minute Book, No. 2, Council meeting of June 11, 1920. WES papers, IEE archives.

36. *Woman Engineer* 1 (September 1920): 33.

37. *Woman Engineer* 2 (September 1927): 248.

38. Taken from a leaflet *Highlights in History of the EAW* (n.p., n.d.) put out by the Association.

39. 'Special Supplement on Professional Associations,' *The New Statesman* 9 (21 April 1917): 4.

40. Quoted in *The Labour Situation* 68 (14 April 1919): 7.

41. *Times* (3 December 1918) and undated leaflet, *The Women's Engineering Society*, WES papers, IEE Archives. In part this inclusiveness was encouraged by the British use of the word 'engineer' to encompass both those professionals with formal collegiate education and those machinists who were skilled or semi-skilled workers. To complicate matters further, male professional engineers in Britain were slow to require collegiate training and hung on to the tradition of practical work experience for their members. R. A. Buchanan, *The Engineers: A History of the Engineering Profession in Britain, 1750–1914* (London, 1989), 210.

42. WES Minute Book, 1919, minutes of meeting of executive committee, May 15, 1919, WES papers, IEE archives.

43. Ibid.

44. See footnote 12.

45. *Woman Engineer* 1 (March 1920): 10.

46. WES, Minute Book, No. 2, minutes of Council meeting, June 20, 1921, WES papers, IEE archives.

47. WES, Minute Book, No. 2, minutes of Council meeting, June 21, 1922, WES papers, IEE archives.

48. Haslett to Margaret Partridge, May 9, 1925, Correspondence with M. Partridge & Co., 1925–26, WES papers, IEE archives.

49. See note 3. As recent scholars have pointed out, 'the [female] engineer is in a more contradictory position than the writer, since one cannot function privately as an engineer. It is an occupation *which only takes place publically.*' Ruth Carter and Gill Kirkup, *Women in Engineering: A Good Place to Be?* (London: Macmillan, 1990), 76.

50. Margaret Partridge to C.H., July 19, 1925, Correspondence with M. Partridge & Co., 1925–26, WES papers, IEE archives.

51. [Caroline Haslett] to Miss Griff, September 5, 1922, Correspondence with Lady Parsons, 1922–23, WES papers, IEE archives.

52. The popular linking of the WES and the EAW is clearly shown in Vera Brittain, *Women's Work in Modern England* (London, 1928), 41–45.

53. Quoted in R. A. Buchanan, 'Gentlemen Engineers: The Making of a Profession,' *Victorian Studies* 26 (Summer 1983): 413.

54. Buchanan, *The Engineers*, 111.

55. *The Engineer* 127 (2 May 1919): 420.

56. Rollo Appleyard, *The History of the Institution of Electrical Engineers (1871–1931)* (London, 1939), 167–168.

57. WES, Minute Book, No. 2, Council meeting of February 9, 1921, WES papers, IEE archives.

58. [Caroline Haslett] to Lady Parsons, September 30, 1922, Correspondence with Lady Parsons; 1922–23, and WES, Minute Book No. 2, minutes of Council meeting, November 6, 1922, WES papers, IEE Archives.

59. On the worry over the proliferation of engineering societies, and the need to consolidate, see Buchanan, *The Engineers*, 202.

60. Reported in *Fife Herald* (25 May 1927), copy in Correspondence and pamphlets for Conference, c. 1925, WES papers, IEE archives. T. J. N. Hilken, *Engineering at Cambridge University, 1783–1965* (Cambridge: Cambridge University Press, 1967), however, does not list 'women' in its index and says nothing of them in its chapter on WWI.

61. *The History of Loughborough College, 1915–1957* (Published by the Past Student's Association of Loughborough College, 1957), 18.

62. Minute Book, No. 2, Minutes of Executive Council meeting, December 3, 1928, WES papers, IEE archives.

63. Minute Book, no. 2, Minutes of Council Meeting, November 6, 1922, WES papers, IEE Archives.

64. W. O. Skeat, *King's College London Engineering Society, 1847–1957* (London 1957), 16. One is reminded of the related engineering witticism: 'I've heard of radio engineers, aircraft engineers, aeronautical engineers, but how do you engineer a woman?' Carter and Kirkup, 94.

65. M. Partridge to Haslett, May 28, 1925, Correspondence with M. Partridge & Co., 1925–26, WES papers, IEE archives.

66. Haslett to M. Partridge, May 29, 1925, Correspondence with M. Partridge & Co., 1925–26, WES papers, IEE Archives.

67. M. Partridge to Haslett, May 29, 1925 and Haslett to M. Partridge, June 3, 1925, Correspondence with M. Partridge & Co., 1925–26, WES papers, IEE archives.

68. Messenger, *Doors of Opportunity*, 26–27. Pankhurst, of course, could have been using the term 'lady' to denote either class or sex, or, more likely, both at once.

69. M. Partridge to Haslett, n.d., Correspondence with M. Partridge & Co., 1925–26, WES papers, IEE Archives.

70. *The Times* (19 October 1942).

71. *Woman Engineer* 2 (March 1925): 18.

72. Rachel Parsons met a tragic end in 1956 when she was murdered, apparently by 'a local stableman' *The Times* (3 July 1956). She was known 'in the district as an eccentric' and 'a woman of very independent nature.' *Daily Telegraph* (3 July 1956).

73. Review in *Woman Engineer* 1 (December 1923): 292–293.

74. Ibid.

75. 'The President,' *The Woman Engineer* 4 (Autumn 1939): 307.

76. Reported in *Humanity* 1 (February 1914): 6.

77. *The Women's Who's Who 1933. An Annual Record of the Careers and Activities of the Leading Women of the Day* (London, 1933), 11, and *The Women's Who's Who 1934–5* (London, 1935), 21.

Boel Berner

3. Educating Men: Women and the Swedish Royal Institute of Technology, 1880–1930

The institutionalization of engineering as a specialized and socially valued profession occurred in Sweden, as it did in many other industrializing European countries, by the end of the nineteenth century. Engineers acquired prestige and power because of their mastery of modern technology and socio-technical systems.[1] The profession organized itself as a masculine domain, monopolized by men and invested with masculine ideals and ambitions. My purpose in this chapter is to show that such genderization of professional engineering should not be regarded as a self-evident fact, but rather as an ongoing social process, as a form of 'history-in-the-making' within the realm of education and work. We shall see how identities were defined and redefined through social practices, in which the distinction between male and female positions in social hierarchies and the boundaries between masculine and feminine areas of technical expertise were at once being contested, defended, and sometimes, redrawn.

In what follows, I will focus on Sweden's most prestigious school of engineering, the Royal Institute of Technology in Stockholm, to analyze prevailing educational goals and everyday practices at the turn of the century. I will discuss the ways in which the institute prepared selected young men for positions as technical leaders within industry and public affairs. I will look at educational practices in a broad sense, including the school's curriculum, its pedagogical methods, and the homo-social relations and rituals of its everyday life. What kind of masculinity was forwarded to the pupils? What sort of man was being formed in this educational context? And what happened when women attempted, occasionally with success, to gain entry? Did their presence result in boundaries being redrawn, and was the school's homo-social culture challenged?

Previous Studies

Other scholars have addressed questions concerning the masculine character of the field of engineering and the response to women who wished to acquire technical expertise in different ways. There is a growing international body of empirically oriented historical research, some of which appearing in this volume.[2] These studies provide, as we shall see, valuable points of comparison and contrast. Their focus, however, is on the exceptions – i.e. the small number of women engineers – whereas my focus here will be on the rule, that is to say, on *men*. Historians of ideas, such as Brian Easlea, Carolyn Merchant, and David Noble, have addressed the relationship between masculinity and technical expertise.[3] Such studies are broad in scope, and link up the development of science and technology, from the Middle Ages onwards, to ideologies of destruction, misogyny, or control, as well as to the creation of all-male institutions of knowledge, such as monasteries, universities, polytechnics, research institutes, and so forth. While interesting in themselves, both these studies and their arguments appear too straightforward for my purposes here. I regard the relations between masculinity and technology as more ambiguous, more changeable, as well as more contradictory than such studies appear to allow. In order to understand the significance of these relations, we must study everyday practices and modes of daily interaction at a particular time and place.[4]

Instruments for this kind of analysis are provided by the sociological and anthropological work of, for instance, Cynthia Cockburn, Sally Hacker, and Sharon Traweek.[5] Cockburn has shown in which ways manual technical skills became an integral part of the masculine self-image and a source of pride for the typographical workers she interviewed. Hacker has exposed the identification of mathematics and technical problem solving with male power at an elite technical university in the US. Traweek studied the career patterns and the various kinds of sociability that are rewarded in the highly competitive research field of physics, patterns which almost by definition exclude women and benefit a certain type of man.

Whilst these are important studies, enabling us to understand specific expressions of masculinity in different sociohistorical contexts, they do not help us with a detailed analysis of everyday *educational* relations and practices. I therefore also use theoretical tools provided by the sociology of education, despite the lack of gender awareness in this field. Studies by Pierre Bourdieu, and by Bourdieu and Jean-Claude Passeron, have not only enlarged my understanding of the *specificities* of educational institutions in relation to material and symbolic power, but also enabled me to focus on the particular cognitive and social '*habitus*' inculcated in specific educational contexts.[6] Basil Bernstein has discussed in which ways power is organized within *educational practices*.[7] His insights have had considerable influence on my own analysis below. Finally, a by now classic article by Ralph Turner has offered me tools for understanding the ways in which different regimes of *social mobility* produce a (masculine) form of sociability.[8] Turner's concept of 'sponsored mobility' has helped me to analyze the situation in the early twentieth century, when young men from an early age onwards were selected and ultimately incorporated into a societal elite, via grammar school (*Gymnasium*), and – in the present case – through higher technical education. How this was done will shortly become clear.[9]

Figure 14.
The Chalmers
Co-educational School,
the future of engineering
education as seen
through the eyes of the
well-known Swedish
artist Carl Larsson.
The drawing was
published in the college
engineering newspaper
in 1903. Reproduced
from Rasp (1903).

An Education For A New Elite

The period I focus on here, 1880–1930, was one of expansion and con-tradiction. Sweden was transforming itself from a backward, rural, and agricultural community into an industrial and increasingly urban society. As late as 1910, 70 per cent of the population resided outside the towns and boroughs. The income per capita was low, about half the average American wage at the time.[10] Only twenty years later, however, there were more people working in industry than in agriculture and forestry, the urban population had outgrown the rural one, and the standard of living was among the highest in Europe.[11]

The transformation of Swedish society took place in the context of fast economic growth. Early industrialization had been a predominantly rural phenomenon, based as it was on the demand for iron, wood, and agricul-tural products from abroad, but also on an internal demand for agricul-tural machinery and plain consumption goods. Small- and medium-sized companies dominated industrial production, which was closely con-nected to traditional mining and ironworks firms. By the end of the nineteenth century, this mode of production gradually linked up with a new industrial structure. From the 1890s onwards, industry grew rapidly, by the turn of the century, the new industries had replaced agriculture as the main contribution to the Gross National Product. More finished products, such as high-grade steel, engineering products, as well as paper and pulp, became the main products of exportation. Swedish engineers were instrumental in bringing about these developments. Large-scale attempts at exploiting Sweden's natural resources, notably iron ore and water power, were undertaken to benefit an expanding industry. 'Taming the waterfalls' was a major preoccupation of many among the most prominent engineers in the early twentieth century.[12]

Another central feature of this period was that it saw the creation of several important engineering firms, which, to this very day, form the backbone of Sweden's industrial and export sector. These industries – such as SKF in spherical ball bearings, Alfa Laval in cream separators and steam turbines, ASEA in the generation and transmission of electrical power, and Ericsson in telecommunications – enjoy almost mythical importance in the Swedish public mind. Most of these firms found their basis in Swedish innovations, and were built on mainly Swedish capital. They soon set up subsidiary companies abroad, and became successful competitors on the foreign market.

Not surprisingly, then, professional engineering managed to establish itself during this period as a vocation for a modernizing elite. It was engineers' expert knowledge that lay the foundations for the important new industries of mechanical and electromechanical engineering, and for the new methods of processing in the paper and pulp industry. Engineers constructed the railroads, the gas works and the electricity systems for a modernizing country. They promoted scientific technology and 'scientific management,' and introduced new consumer goods, such as bicycles, telephones, sewing machines, and so forth. To many contemporaries, and indeed many subsequent generations, engineers were the heroes of industrial expansion and the modernization of society. Their ambitions for leader positions in both state and industrial domains were frequently fulfilled.

In order to keep down their numbers, and to avoid creating an 'engineering proletariat' such as the ones thought to exist in Germany or the US, the Swedish engineering community also managed to maintain high entrance qualifications and low student numbers in the most prestigious schools of engineering.[13] To enroll in the Royal Institute of Technology (KTH) in Stockholm, and to a lesser degree, in the Chalmers Technical Institute in Gothenburg, was thus a sure ticket to a position of leadership, whether in industry, state administration, or independent engineering practice. Most contemporaries took it for granted that this was to be an exclusively male domain.

Women's Employment and the Engineering Profession

The years around turn of the century were also those in which women's fight for education, work and social liberation gathered momentum and achieved important results. Many discriminatory practices regarding women's and men's respective social positions hence were not only established and defended, but also challenged and overturned. The universities opened their doors to women in the 1870s. Middle-class women subsequently tried, and in the face of considerable opposition, to acquire teaching positions at the *Gymnasia,* to gain academic recognition, and to take up new occupations related to technical progress, such as telegraphy and telephony.[14]

Sweden, with its low average income and strong rural orientation, knew a long tradition of female employment. Unlike the women in more prosperous countries, they were forced to work for a living. In 1920, 36.3 per cent of Swedish women registered as gainfully employed, as compared to 19.5 per cent of American women of European descent. If farm-wives had also been included in the Swedish total, the figure would have risen to 54 per cent. Outside the rural areas, the pattern of predominantly male breadwinners continued, however, and by the end of the nineteenth century, the notion of separate spheres for the different sexes was widespread. In view of the low per capita income, we can nonetheless safely assume that Swedish women, also those in the lower middle classes, had to supplement the family income through going out to work.

Add to this that Sweden had for a long time known a substantial numerical dominance of women over men in the total population; a difference that became even more marked when, around the turn of the century, many men left for the USA. In 1910, there were 105 women in Sweden to every 100 men. What is more, only 46 per cent of Swedish women were married, as compared to 60 per cent in the USA. All this entailed that many women had to earn their own incomes, not only to maintain themselves, but also to support their dependents. The number of single mothers rose in particular at this time: in 1920, the proportion of illegitimate children reached a maximum of about 15 per cent.[15]

The struggles for female civic, educational, and employment rights gained important successes after World War I. Universal suffrage was established in 1919. In 1921, most of the professional schools that had not done so before, e.g. institutes of engineering, opened up to women. In the late 1920s, it was no longer allowed to exclude women from middle- and higher-level state positions in, for instance the realms of teaching, law, and medicine. In 1924, finally, the major route to the middle classes, i.e., a baccalaureate from the public *Gymnasium,* which in turn led to professional and academic positions, was opened up to women. The only possibilities to gain a baccalaureate for women had earlier been through attending private schools or study at home.

In order to acquire much-needed jobs and to expand their influence in society, women thus tried to open many doors that had previously been closed to them. Especially the male educational and occupational monopolies led to several bitter fights. In the field of engineering, everything remained quiet, however – or almost everything.

In 1892, a young woman tried to enroll in the KTH. She was refused admission on the basis of the school's 1876 statutes, which stated that it was intended for 'young men who wish to pursue a technical occupation.' Several professors considered the statutes outmoded; women were, after all, allowed into the universities and higher schools of law and medicine. A committee was installed to determine whether women could be admitted into KTH as regular students. In its report, issued in 1893, the committee came up with a negative answer, despite its acknowledging women's legitimate interest in studying technology. It also found, however, that the presence of women in the school would lead to important 'inconveniences,' which was the reason why they should not be admitted. Consequently, women would have to wait till 1921 before they were allowed to pursue higher technical education on the same basis as men.

What kinds of 'inconveniences' the committee members had in mind, we will look at in a moment. At this point, we may note that women's entry into higher technical education was not an important issue for an otherwise highly active feminist movement. Sweden did not witness anything like the attempts in Russia and France at establishing a technical institute for women only, as discussed by Dimitri and Irina Gouzévitch and Annie Canel in this volume – an endeavor, incidentally, which appears to have been quite unique in an international perspective.[16]

This lack of interest in engineering among early feminists may have various reasons. First, Swedish upper- and middle-class women had, after all, been able to study chemistry, mathematics, and physics at the universities since the 1870s, and many women indeed did so.[17] Second, women of fewer means were able to acquire some professional skills by attending the many lower-level technical schools that were open to them. And third, a degree in engineering would have been of little use to women at the time, since relevant state positions were closed to them, while openings for university-trained female engineers in industry were few and far between. Normally, a degree from KTH, Chalmers, or a Technical *Gymnasium*, was supposed to lead to managerial or leadership functions, i.e., positions that, by definition, were considered unsuitable for women.

The masculine character of a degree in higher engineering was thus largely taken for granted by both women and men, and much energy was put into 'identity work' in order to authenticate the profession as one exclusively reserved to a *male* elite. At the same time, new ideas and technological developments nonetheless produced a subtle redrawing of gender boundaries in the realm of engineering. This was also noted and supported by the 1893 advisory committee.

The Creation of 'Technical Man'

The cultural association between masculinity and technology was taught and perfected in the Royal Institute of Technology as well as in the less prestigious Chalmers Technical Institute in Gothenburg. Both the educational experience as such, and the knowledge they acquired, united the 'young men' admitted into these schools within a brotherhood of technical and industrial expertise. Through the brotherhood the schools' graduates

PÅ LABBIS!

Figure 15. Translated from Swedish the cartoon reads: "In the Lab". The white respectable coat and gentleman's garb contrasts with the student's frivolous play with the lab equipment. Appearing in the engineering student newspaper in 1913, the cartoon might be a self-mockery of the working codes in the chemical lab and a display of the competence and arrogance of a new elite. Reproduced from: Rasp (1913).

not only distinguished themselves from women but, I would suggest, even more importantly, from other men with different qualifications and destinies in society. Group *solidarity* and *distinction* were the most significant effects of the educational process on the aspiring engineers, characteristics that would subsequently be used in the struggle for positions of male leadership in a rapidly changing society. 'If it is true that a new time is coming, and that the sun is setting on the age of the lawyers, the pedants and the bureaucrats – then it is also certain that a place of honor in this new era will be taken up by the engineers,' a leading figure in the association of engineers cried out in its professional paper *Teknisk Tidskrift* in 1909.

What kind of man was this 'honorable' leader of the new era supposed to be? In what ways was the connection between masculinity and technology expressed in the training and professional practice of engineers? To answer these questions, I will proceed by taking a closer look at the educational processes at work at KTH and Chalmers. Contemporary documents seldom take into account that both schools were exclusively catering to 'young men.' Only occasionally, when a break in the normal pattern occurred, was the schools' masculine orientation made explicit. This happened, for instance, when a woman applied, therewith provoking reactions of either rejection or delight. In combination with other, more implicit, characteristics of both institutions' ambitions and preferred practices, this gives us a fair picture of their gendered character.

In the next section, I will first try to show that the *knowledge* that was considered useful for engineers was coded as more or less suitable for women. It was also a means of distinguishing the university-trained engineers from other, less consequential men. I will go on to discuss *everyday educational practices* to make clear that these were intended to construct a particular versatile, mobile, and industrious man. These practices were simultaneously regarded as too strenuous, hence unsuitable for female students, who were thus excluded from the school. I will finally argue that the young men, through the school's *everyday homosocial relationships*, received the appropriate psychic and social preparation in a 'world without women' to take up their future positions as men of power within an well-educated elite.

Gendered Repertoires of Knowledge

The 1893 committee's report mentioned previously is particularly interesting for our purposes here, in that it actually singles out certain areas of knowledge as, in fact, suitable for women. The criteria for making them so are not exactly spelled out, but they can be inferred from various other documents, as well as from a general understanding of the social context in which engineering was practiced at the time. We may further note that, in a comparative European framework, they reflect a shared international understanding; an ideological position embraced, or so it seems, by women and men alike.

The first area of knowledge considered suitable for women was architecture. This was seen as a technical profession, hence offered at KTH, but also one in which aesthetics and ethical values formed constituent parts. Late nineteenth-century ideology stressed women's particular sense of beauty, as well as their moral duty in building a harmonious home. Many middle-class women took active part in public debates about the critical function of hygiene and cleanliness in both society and in individual homes. In the general crusade against disorder and disease, female architects were expected to take up an important role.[18] As Ellen Key, one of the leading feminist thinkers of the time, put it in 1896: Female architects' essential 'motherliness' will result in harmoniously and hygienically designed homes, and thus complement the more monumental work of their male counterparts.

Ellen Key had significant influence.[19] She moved in the same liberal circles in Stockholm as many well-known engineers and teachers of KTH, including the school's future vice-chancellor. Her suggestions – which were based on the idea of an essential *difference* between women and men – were nonetheless challenged by many of her feminist contemporaries. Do you want us to accept this 'castrated and semi-developed female identity?,' one of them exclaimed in a pamphlet published in 1896:

Shall the man build the front of the house and the woman its back? Shall the man be in charge of the ornaments and the woman take care of the installment of the sewage pipes? What a beautiful entente between the sexes, or – what a terrible struggle for space between the marble-staircase-building man and the sunny-nursery-decorating woman![20]

The very same year in which the pamphlet appeared, a young woman by name of Agnes Magnell applied with the school of architecture at KTH. By Royal decree, and on the supporting advice of the collegiate teaching staff, she was accepted as a special student. Since there had been no change in the statutes, and it was only a few years ago that the advisory committee had recommended women's continued exclusion, the question arises how this could happen.

Various circumstances appear to have worked in Agnes Magnell's favor. First, there were, at least according to KTH, only a limited number of male applicants for the school of architecture that year: Magnell would thus not be taking the place of a man. Second, the ideological climate was, as we have just seen, favorable in pointing out the need for a 'woman's touch' in architecture. Then there was the example of Vivi Lönn, the first female architect to graduate in Finland in 1896; there were many ties uniting the Finnish and Swedish engineering communities at the time. And finally, Agnes Magnell herself was an asset in both social and cultural terms. She held a degree from the Technical School in Stockholm, one of the many lower-level technical schools open to women. This type of schools provided women with training in drawing, design, xylography, and similar skills considered useful for teachers, and qualifying students for lower-level positions in industry, handicraft, and public services. Magnell's former drawing teacher was also a lecturer (and later a professor) at KTH, and he encouraged her first to get a baccalaureate, and then apply with KTH. He may hence be expected to have put in a good word for her with the selection committee. And Magnell's social capital was not insignificant. Her father was a military officer as well as a landowner, and one of her relatives was a prominent engineer who would twelve years later become vice-chancellor at KTH. Taken together, these various specific personal and social circumstances thus worked in Agnes Magnell's favor in trying to gain admission as the first female student at KTH. Such conditions were not to recur for nearly twenty years after, however, when, in 1915, the second female student of architecture made her entrance into KTH.[21]

The second field of expertise singled out by the 1893 committee as 'particularly suitable for women' covered chemistry, chemical technology, chemical metallurgy, physics, and electrotechnics', a quite surprising combination of subjects.[22] We might reasonably assume that these science-based modes of technological training would be classified as 'male.' Indeed, throughout the nineteenth century, only boys were thought to require the intellectual training and social status that came with taking a strictly scientific curriculum. Secondary-school education in science and mathematics was therefore mainly offered to middle-class boys at the public *Gymnasia*, and seldom to the students at the girls' schools. Why, then, should women be allowed to take up such subjects at KTH?

Figure 16. Cartoon in painterly fashion and soft strokes represents Vera Sandberg, the first Swedish woman student at Chalmers Technical Institute, as a walking Cupid striking her male co-students with love. A variation on the theme of women students' distracting effect on male students. She is wearing trousers, the traditional students' cap, and the architect's tool of the trade, a drawing board. Reproduced from: cover, college newspaper Rasp *(1914).'*

The idea that science-based technological subjects were suitable for women first of all appears to have sprung from a genuine divergence of opinion among the reigning elite in Sweden. An influential group of highly educated men in the dominant scientific, literary and business circles of Stockholm strongly opposed the conservative academic culture prevailing at the traditional universities. These men felt that everybody needed scientific training in order to fight superstition and to further social liberation. In fact, the same men had asked the Russian mathematician Sonia Kowalewski to take up a professorship in advanced mathematical analysis at the newly founded private University of Stockholm. Some engineers and teachers at KTH belonged in this liberal milieu, frequenting its social clubs, but they also taught at Stockholm University.

An interesting case in this respect is that of Leonilda Sjöström (1836–1898), who, for many years, taught mathematics and shipbuilding at her father's private school of navigation in Stockholm. She also gave courses in mechanical engineering, and was, for 25 years, a teacher of mathematics at the Stockholm Technical School. The latter establishment was open to both women and men and had – as we have seen in connection with Agnes Magnell – in many ways tied up with KTH. Sjöström was yet another living proof of women liberating themselves through science and mathematics; to my knowledge, however, her case was never mentioned in KTH documents of the time.[23]

Secondly, changes in technology and industry had led to a both a 'scientification of industry' and the 'industrialization of science.' This may have opened up new job opportunities for women. The *laboratory* made its entrance as an important site for technical work, in which quality analysis, development activities, and even research took place.[24] We recall that women in Sweden had to work for a living. The laboratory offered clean, appropriate jobs for middle-class women with some scientific training, acquired either at university or at KTH. Female laboratory workers generally functioned as the patient, careful, and competent assistants to male engineers, scientists, or doctors, for salaries that few men with a similar educational background would be willing to accept. This, I think, provides a viable explanation for the committee's decision to designate precisely these subjects as suitable for women. In practice, however, this particular niche in the female job market in Sweden did not actually develop until much later, primarily in the years following World War II.[25]

Masculine Fields of Technology

Since other technological fields were strongly gender-coded as masculine, women appeared to have found no place in them, even as assistants. Around the turn of the century, some 80 per cent of the approximately four hundred students at KTH studied mechanical, civil, and mining

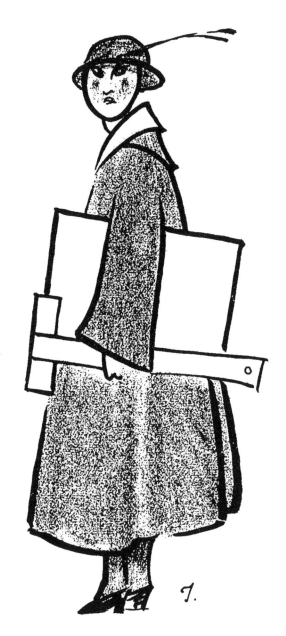

Figure 17. Set in strong modernist strokes, cartoon of Anna Lous-Moor, the first woman student to earn a degree in architecture from the Swedish Royal Institute of Technology, in college newspaper in 1915. Here portrayed as a driven, singular woman without sex appeal and as the polar opposite of the sexual/flapper woman, the only alternative gender script available at this time. Compare this with the illustrations from Austria. Reproduced from: Gåsblandaren, Stockholm (1915).

engineering. The latter formed the core areas of engineering expertise and were considered crucial to Sweden's industrial growth. Mechanical engineering found its roots in metalworking, a traditionally all-male field of competence. Civil engineering had previously been taught at the military academy, and its graduates were part of a state elite corps that was by definition restricted to men. Mining engineering had been a highly important, and exclusively male, area of concern for both state and private industries from the seventeenth century onwards. Apart from traditionally being seen as 'masculine,' these areas of knowledge led to jobs with specific characteristics that appeared very inappropriate for women to the contemporary eye.

Since the mid-nineteenth century, Swedish university-trained engineers had increasingly begun to define themselves as the carriers of 'scientific technology.' They had fought for the setting of high entrance requirements at KTH. To enter, the student needed a baccalaureate, something which was obtainable only for the small percentage of the male population, who was destined to assume positions in an educated elite, or, what the Germans call, a '*Bildungsbürgertum*'.[26] Training at KTH was geared towards 'scientific' technology, with a curriculum based mainly on mathematics, physics, and descriptive geometry, in addition to various subjects in applied technology. Such theoretical training was considered necessary for those in leading positions in industry and in public life, and would place the school's graduates several steps above such engineers as had a mere background in lower technical education. The latter were seen as lacking in both theoretical knowledge, as well as the kind of general edification provided by the *Gymnasia*. The graduate engineers' 'capacity to calculate,' i.e., their analytical skills, would, it was argued, place them way above those who were only able use their professional experience and 'common sense.'

Knowledge of mathematics and theoretical science was thus used as an argument in maintaining and thus reinforcing a *public hierarchy between men*. The notion of the 'scientific' engineer's superior character gained perhaps most importance in traditional technical fields such as metal, mining, and civil engineering. In these fields, academically trained engineers had to compete for the favors of potential employers with men having longstanding practical experience and lower salary demands.

And now we reach the major argument used in the 'scientific' engineers' favor, a characteristic that largely explains why the 'masculine' areas of technology were considered unsuitable for women. What the university-trained engineer supposedly had, was a *unique combination of theory and practice*. A successful engineer could never be a pure theorist, the kind usually produced by universities. He needed to know how to apply his theoretical knowledge in practice in order to solve problems in machine design, tunnel construction, iron smelting, and so on. He should further be able to combine calculation with leadership, and be adept at handling mathematics, as well as men and machines. Only by knowing how to get their boots and their hands dirty would academically trained engineers be able to aspire to technical leadership. Future civil, mechanical, and mining engineers thus spent many hours on field trips and excursions, in drawing rooms, mines, and on construction sites, learning to speak both the workers' and their superiors' language. The purpose of such practical experience on the shop floor or 'in the field' was, as Ruth Oldenziel has remarked with respect to the situation in the US, for engineers to acquire *managerial* skills. Being able to give orders, manage large numbers of men with different skills and functions, move around and get to know all the details of a complex work process: these were class-specific skills that only men should master.[27]

The 1893 advisory committee saw in these everyday educational experiences – often taking place late in the evening, in primitive conditions in the field, and far removed from the controlling eyes of their superiors – yet another argument to exclude women from higher technical education. Women could simply not participate in such activities under the same conditions as, and together with, young men without causing 'serious inconveniences.' Without such practical training, however, women could naturally not become proper engineers. It would be much better, then, not to let women in at all.

So far, we have seen that the very notion of scientific technology in Sweden at the turn of the century was specifically gendered. Technological education was designated for a certain kind of man only, a man in the new mechanical age who was superior to other men. In other countries of Europe, engineering was similarly 'academized.' This gave women who managed to acquire a baccalaureate in science, at least in theory, the chance to apply with institutes of higher learning that were formally open to them.[28] In practice, however, the particular technologies that dominated the professional field at the time – mechanical, mining, and civil engineering – were not considered part of 'feminine' expertise. The specific combinations of theory and practice, and of technological skills and management they required were strongly gender-coded as 'masculine.'

Other forms of technology, however, were seen as more closely related to the home environment, hence to what were considered specifically female tasks and talents. Architecture is here the prime example. Despite its otherwise masculine image, i.e., common perceptions of builders and architects as heroic visionaries, there were certain niches in the professional field that were apparently regarded to suit women's 'natural' abilities and concerns. Various new technical fields, those that had not yet entered the heartland of industrial engineering, were also considered suitable for women. Advanced training in chemistry or electrotechnics, i.e., professional domains that led to white-collar, non-careerist jobs, in the quiet, home-like environments of laboratories and drawing offices, were among these. In her analysis of similar debates taking place in Austria at the time, Juliane Mikoletzky quite rightly remarks that it were, in effect, the more dreary and less prestigious jobs, such as those of laboratory assistants and drafts(wo)men, or jobs in less prestigious fields, such as interior decoration, that came to be defined as 'feminine.' She additionally comments that the line seems to have been drawn between 'indoor' jobs, considered to be more suitable for women, and the more

Figure 18. Five women students from the Swedish Royal Institute of Technology, all of whom were members of the Association of Female Engineering Students *which lobbied for improving their working conditions. Photographed and interviewed for the Swedish women's magazine* The Housewife *in 1927, the women pose in a collegial and convivial atmosphere. Laughing straight into the camera to their female audience and lined up to suggest a succession of women, they serve as potential role models for their female audience. Reproduced from:* Husmodern, Sweden (1927).

'manly' professions that were performed 'out of doors'. This was meant both in a literal sense in that they required engineers to work out in the field, and in a more general sense, in that they entailed public visibility, were positions of social authority.[29] It is further interesting to note that the subjects taught at the unique Russian Polytechnic for women around the turn of the century, were nearly the same as those considered suitable for women elsewhere: architecture, building, electromechanics, and chemistry.[30]

The message the Swedish school (reluctantly) sent out consequently entailed that, should women eventually be admitted, they could only enroll in certain fields. The 1901 version of a popular book by a leading engineer, who also worked as vocational counselor for young people, voiced the same opinion, by stating that: 'Women may be employed in technical work but only in certain limited capacities, that is, as draftsmen or chemists.'[31] In its previous editions, the book had not even mentioned the possibility of women in technology, so times were definitely changing.

In 1901 the discriminatory paragraph about 'young men' was removed from the KTH statutes; primarily, however, to allow slightly *older men* in. Women were admitted as (paying) students with 'extra' or 'temporary' status, but only 'if space and other favorable circumstances so permit.' Early twentieth-century Sweden thus presents us with a similar picture as, for instance, Austria, Germany, and the Netherlands, countries in which women also were admitted step-by-step into higher technical education as 'paying guests,' as 'extraordinary students,' and on individual basis, before finally gaining entry under the same conditions as men.[32] The Swedish school's major concern was the education of *men*.

We will proceed by taking a closer look at everyday interactions at KTH. In what manner did the school prepare young men for their future positions in society? And how did this affect women's integration into or exclusion from professional engineering? Two aspects are of particular interest in this context: the pedagogical organization of everyday study activities, and their homo-social organization of school culture.

A Gendered Pedagogy

The explicit reasons provided by the advisory committee in 1893 for the continued exclusion of women offer illuminating insight into KTH's pedagogical methods. First, there were the field trips and other forms of practical work, which its members considered inappropriate for women. Secondly, they perceived serious obstacles even in the areas thought suitable for women, obstacles which were related to the kind of education offered by the school. Women's 'more delicate constitution' might suffer from the kind of hard work that 'even the most gifted of students' had to undertake in order to keep up with the educational requirements. Why should women be allowed in, if, as a result, their health would deteriorate?

To what extent was the argument of female frailty actually relevant? Many women were working long, arduous days in mines, factories, on farms, and in their homes at the time. They had to be both physically strong and determined to survive. These women were probably not the ones the committee had in mind, however. Middle-class women with the appropriate educational background were, it appears, frail and more delicate creatures. Their health had to be protected from the hard work at KTH. That many of them already were taking university degrees was irrelevant, the committee ordained; the kind of study required at KTH was much more demanding. A similar argument was – as Karin Zachmann reminds us – eloquently put forward around the same time by the (in Sweden also) influential German engineer and Professor Franz Reuleaux. In 1897 Reuleaux argued that women would encounter physical rather than intellectual difficulties in studying mechanical engineering. They furthermore would – horrible thought – in many respects have to give up their 'pure femininity.'[33]

What especially interests me here is whether the educational regime actually was so physically demanding as these various advisers made it out to be, or whether this was just a pretext for sustained discrimination. Study activities at KTH do appear to have been more strenuous than what was required at the universities. Everyday life at school was strictly organized. Students were divided up into annual cohorts. They all took the same courses within a fixed schedule. Each day was tightly controlled by professors or teaching assistants, and days were filled with many hours of lectures, drawing lessons, practical exercises, laboratory work, and the like.

As it turns out, many young men found the pace and organization of every life at school rather hard to take. Adverse psychosomatic responses, in the form of headaches and neurasthenia, were not uncommon, causing great concern among the teaching staff. Each year several students had to re-sit exams, or take long breaks from their studies. Despite student protests and demands for a lighter course load and less teacher control, the school did not budge. The high number of courses and the tight work-schedules were, or so teachers maintained, necessary for helping students pass their exams on time, keep them away from the 'temptations of life in the big city,' and to satisfy external demands.

While claiming to offer university-level scientific training, KTH did *not* really teach its students to think independently and scientifically. In this respect, it did not follow the university model, nor did it resemble the highly theoretical French *Ecole Polytechnique* – otherwise greatly admired by Swedish educators. Instead, the school's pedagogical aims were modelled after the Swiss and Austrian *Technische Hochschulen*, which also prepared engineers for careers in public as well as private industries by combining theoretical and practical training. Still, everyday educational practices at KTH appear to be have been based primarily on the example of the Swedish *Gymnasium,* the type of secondary school all of its teachers as well as the students had at some point attended. The Military Academy at Marieberg may have set another example. The Academy formerly used to educate a small, elite corps of civil engineers, among whom, at the end of the nineteenth century, the KTH vice-chancellor.[34] The educational principles prevailing at both institutions involved a disciplined schedule and strict control of students' time and morale.

In the light of the school's endeavor to create a *certain kind of man,* adherence to such a rigid organization of work bears further interpretation. The strict demands set on the boys prepared them for a life of tight work schedules, and helped them internalize a positive attitude to assigned work. Lying behind pedagogical practices was a certain image of the successful engineer. We discern a disciplined and adaptable person, oriented towards getting things done on time, and used to working under orders, but also one ready to take on a variety of technical problems in very different environments, and to solve them effectively. The successful engineer could rely upon his broad range of skills, which prepared him for work in any circumstances, from 'shipyards and workshops in America, to railway buildings in Russia and laboratories in Zurich, to drawing offices across the world,' as a student newspaper put it in 1906.[35]

KTH may not always have been successful in producing such men, but the image of the tough and versatile engineer appears to have been what it wished to present to the outside world. The 'habitus' that the school desired to create was focused on mobility and responsibility, with an ambition to take up leadership positions in the public world of men. Why, indeed, should women risk their health in trying to reach a goal that was impossible and unsuitable for them to begin with?

Homo-social Culture and the Creation of Community

Both teaching staff and school administrators at KTH prided themselves on providing an education that kept the students working arduously. Most students obtained their degrees within the allocated three- or four-year period. In this, the school maintained, its record compared favorably to other, more liberal institutes of technology in, for example, Germany.

An alternative way of seeing such a 'care taking' mission is what has been defined by the American sociologist Ralph Turner, as an educational system of 'sponsored mobility.'[36] In effect, engineering studies formed part of a larger, coherent system of public education that was created in the nineteenth century, a system that took the *Gymnasium* as its base and provided a mobility machine for men. Just as in many other educational systems in Europe at the time, it selected a limited number of male eleven-year olds, in order to train them for their future membership of a growing industrial and state-employed middle class. Therefore, the student entering higher education had already been socialized in seeing himself as a member of a select social group. KTH and Chalmers subsequently provided a final all-male environment in which students were not only trained as engineers, but which also would assist the young men in making their way in a competitive and demanding working life.

One aspect to take into consideration in this respect is students' relations with their professors. Initially, student-teacher interaction would be rather stiff and formal, with strict and demanding attitudes on the part of the students' superiors, who were duly depicted as quite awesome and distant. Over the years, when the young men began to approach professional maturity, such formality would give way to a more informal mode of exchange. This, at least, is the impression we get from students' memoirs written about their lives at KTH. A particularly striking story in this context is that of the mining students, whose period of practical work took place at the end of three years of study. For several weeks, they would share their everyday lives in the company of their professors, performing calculations and experiments in iron works and mines. Students and professors also interacted after work, however. It is hence not surprising to find the final exams (set in the following year) depicted as something of a family affair: taking place at the professor's home, him with his dog curled up on his lap, the exam being rather casually dealt with.

This particular story further informs us about the ways in which the school helped students, within the context of its 'manager's course,' in learning to establish informal social connections with their future employers and colleagues. Students were introduced into the homes of prospective employers in order to familiarize themselves with the social norms of interaction among their future bosses, colleagues, and subordinates, as well as learning practically to apply their knowledge of mining geology and steel production. They would drink tea with the wives of the mine owners, play tennis or dance with their daughters, and learn the tricks of the trade from the resident mining engineers over copious intakes of schnapps and punch.

The school itself and student life offered many occasions for such 'anticipatory socialization' into the industrial and technical elite to which the young men would soon belong. Many of the students came from lower middle-class backgrounds.[37] Their fellow students would teach them the norms and mores of middle- and upper class sociability. Students were additionally introduced to their superiors in the male hierarchy of engineers within the student organization, which was part of the professional engineering organization. The latter would help them fraternize with captains of industry, famous entrepreneurs and inventors, government ministers, and higher civil servants. Ambitious students were given the chance to distinguish themselves during debates, and to co-organize lectures and parties. The transitional step from being a student to becoming an adult engineer was thus partly taken while they were still at KTH.

The most important experiences of socialization, nevertheless, were those shared with the other young men at KTH. The student community was very much a homo-social world. Social interaction was almost exclusively among men, which led to the forging of strong bonds of camaraderie and friendship. About half of the student population came from outside Stockholm, hence lived in rented rooms, either by themselves or with other students. They spent long hours together in lecture halls, drawing rooms, and on field trips and excursions. Interaction with other students was thus intense. As one student wrote in 1915:

It is not, as it is in other schools, that you only meet in classes and labs under teacher supervision: all the unscheduled time, you see each other in the drawing rooms. There you draw, whistle, and hum as soon as the attendant teacher has gone out the door, and everyone shares the hardships of work. Good comradeship is created there, better than anywhere in the world; there is no jealousy, no envy of the successes of a fellow student, and only rarely some fretting over an unexpectedly low grade… no fawning about giving assistance, and never is a request for help turned down.[38]

Both formal and informal activities outside the lecture halls and drawing rooms normally took place in a largely homo-social environment as well. Students at Technical Institutes were supposed to be *different* – more boisterous, hard drinking, and jointly united than other university students, who did not have the same kind of everyday intensive community experience. Student culture at KTH and Chalmers comprised a great many formal rituals of belonging and brotherhood: elaborate initiation rites for those who had just entered school, ritualistic pranks and carnivals, community symbols, such as the student cap and ring, elaborate insider jokes in student newspapers, and so on. Students' associations flourished; pub going and parties were favorite all-male pastimes. Alcohol was an important aspect of student culture. A 'real engineer' knew how to drink and sing with his peers, i.e., to drink quite a lot, but in such a way as to be still able to attend 8 a.m. lectures, or give early-morning orders to miners or railway workers. Student culture in many ways was highly organized and regulated, and what the young men learnt – the happy-go-lucky ideology notwithstanding – was, in fact, to deal with an excessive alcohol-intake in a socially acceptable and responsible manner.

The social world of engineering studies thus combined seriousness and irresponsibility. As one contemporary observer noted, KTH was 'both a primary school and a university.' The occasionally silly, occasionally serious relationships and rituals served, as far as I can see, two major purposes. First, they offered *individual* students a last refuge of organized irresponsibility before they would enter the world outside and assume professional and family responsibilities. Second, important social networks and friendships were created on the basis of intellectual similarity and mutual interests. They constituted a *collective* power base to be used to further individual careers and raise the profession's social standing.

Women in a Men's World

Both the social culture and the pedagogical regimes at KTH and Chalmers in many ways served, as we have seen, to unite men. However, both were also largely maintained to keep women out.

Around the turn of the century, both schools and society at large, were characterized by a strong sense of *gender dualism*. To many men, women were alien and distant creatures. They nonetheless entered the future engineer's (all-male) world in two major guises: as servants and as dream figures. The former belonged to another social class. They were the students' landladies, janitor's wives, and waitresses, i.e., the kind of women with whom he most frequently interacted. These women were often gratefully remembered in students' memoirs as surrogate mothers in their lonely school lives.

The second form in which women appeared on male students' horizon was that of 'Woman,' the love object of their dreams. This fantasy figure

Figure 19. Hollywood-styled cartoon of the movie star and the engineer as seen in the 1906 college newspaper. Even the most angular engineering student will attract beautiful women seems to be the somewhat ironic message, suggesting the high male sex appeal of the profession to women. Reproduced from: Rasp (1906).

was envisaged as the loving and soft alternative to the disciplined and rational world of engineers. She was everything that technology and engineering was not. She belonged to a private realm, not to the public world of men, and was, indeed, as one engineer retrospectively remarked, seen as an 'individual means of intoxication.'

It was generally considered an anomaly that women might like to enter the realm of engineering, occasionally even seen as a threat. Some men deplored the envisaged loss of their secret knowledge and togetherness if 'Eve' were allowed to enter schools of engineering. Such an intrusion would disrupt the all-male camaraderie of student life, and spoil the collective pastimes of drinking, singing, debating, and playing games. As late as 1927, students at Chalmers voted for the exclusion of women from their 'Valpurgis Night' celebrations. Others found the idea that women would acquire the same kind of knowledge as men simply 'intolerable:' this would turn them into future competitors for what they considered to be all-male jobs. Such hostility, however, did not express the predominant mood. Even if the late nineteenth-, early twentieth-century engineering community was uniformly male, it was neither misogynist, not even uniformly dualistic in its views on questions of gender.

We recall that late nineteenth-century liberal scientists and engineers actively encouraged the recruitment of women students and scholars. The idea of a few 'token' women appealed to many of the students as well.[39] Several excited young men, giggling and curious, actually attended Agnes Magnell's public entrance examinations in 1897. The few women who, in the early twentieth century, completed courses in chemistry as temporary students, or the later (full-time) students of architecture, were the focus of much adoring male attention. The female students were treated as fascinating others, even if they had to struggle to be regarded as 'one of the boys.'

In this respect, two poems, written in 1915 by the first woman student at Chalmers, Vera Sandberg, are of special interest.[40] They accurately capture the sense of being 'outsiders inside' that many female engineers continue to experience to this very day. In the first poem, Sandberg deplores the loss of her 'female logic' and 'female point of view,' having learned to think and talk in a rational manner. There is 'sorrow in her heart': 'I am not a woman and I am not a man.' In the second poem, however, her tone of voice has changed. Work in the 'lab' is fun, and everything is going well. Sandberg is making sulfate now, and can no longer be bothered with cooking or dish washing. Professors bow to her and the 'lads' cheer her on! She ends on a note of optimism: 'The world will be ours when women are free.'

The prevailing, contradictory mix of gallantry and lack of understanding of women's needs was later described by the first cohorts of female students that began to be admitted on a formally equal basis in Sweden during the 1920s.[41] It can further be discerned in the experiences of early generations of female engineering students in Germany during the years between the two World Wars, as well as in the apparently even tougher atmosphere described by the first women students at the Case Institute of Technology in the US in the 1960s.[42] These early female students were, or so it seems, unexceptionally seen as 'guests' in the world of engineering, having been invited in on account of the goodwill of those who really felt 'at home' in it.[43] Not surprisingly, many of these women tried to 'normalize' their situation by forging emotional alliances with their fellow students. They married engineers, from whom they derived emotional and professional support both during their time at KTH and later on, in their working lives.

Figure 20. The 1930s cover of engineering student newspaper of women engineers on the march in socialist-modernist poster style. Displaying boyish masculinity and wearing trousers, the women may only be identified by their make-up. Reproduced from: Rasp (1931).

Moving in from the Margins as 'outsiders inside'

In 1921, women were formally allowed to enter KTH on the same basis as men. Thus ended a period when their admission had been a question of negotiation by individual case. From the late nineteenth century onwards, the occasional woman with exceptional qualities and/or connections had been allowed to enroll in KTH. But engineering – and especially university-level forms of training – was strongly gender-coded as male. The field's masculine orientation implied, as we have seen, a focus on technical expertise that was closely linked up with social and economic power in a rapidly industrializing Sweden. Studying engineering involved a pedagogical approach and a homo-social mode of interaction that served to prepare men for professional mobility and sociability, skills considered unsuitable for women. Both the manifest and hidden aim of the higher technical schools was to forge bonds of solidarity among a select group of men, and to enable them to distinguish themselves from their less worthy competitors. This was a male game in a public world of industry and technology, of leadership and careerism; a world in which women were thought not to belong.

Even when the formal barriers that precluded female participation had been removed, the schools' informal, cultural message was still exclusionary male-oriented: they offered higher education for 'young men' destined to take up their positions in a societal elite. A great many Swedish women, more than in many other countries, were gainfully employed, mostly out of economic necessity. 'Women's work,' however, as, for instance, in the textile industry, did not lead to positions requiring particular expertise; in fact, no formal form of education of textile engineers existed in the country. Relatively few women found manual or office work in the engineering, mining, or paper and pulp industries, and thus did not get any firsthand experience in fields of technical expertise. Another important factor was that Sweden – in contrast to most of the other countries discussed in this volume – was not actively involved in either of the two world wars. Swedish women were therefore never called upon to do the 'men's work' in engineering, heavy industry, or armaments production. They hence had no opportunity to find their entry into, or acquire the required skills for a career in engineering. Thus, the majority of women who did find their way into engineering had been obliged to take the formal educational route.[44]

A trickling stream of quite exceptional young women began to find their way to the technical universities in the decades following 1921. Still, it would take until the 1950s for women to make up more than 5 per cent of the overall student body. As was the case in many other countries in Europe, most female students chose to study architecture and chemistry, both relatively small and marginal fields within engineering generally. Forty-three out of the sixty women graduating from the Royal Institute of Technology between 1924 and 1962, were chemists. Another seventy-eight earned their degrees in architecture.[45] The major professional fields, i.e., mechanical, civil, and mining engineering continued, until very recently, to remain specifically male preserves. Despite earlier predictions, electromechanical engineering also remained a strongly male-dominated area in Sweden, perhaps because of its early and lasting association with heavy engineering, as well as with the heroic, 'manly' taming of the waterfalls in the inaccessible far North.

The greater significance of these large, male-dominated domains has contributed substantively to the preservation of a strong homo-social culture and a male-oriented educational tradition at Swedish technical universities up until the present day. Engineering studies have offered a sense

Figure 21. I attend Chalmers Technical Institute because my chemistry teacher was so handsome," reads the caption of the 1941 college newspaper cartoon. Caricature of a woman chemical engineering student, looking like a sleepwalker with an extra large students' cap and puffy eyes with no knowledge or interest in her studies.
Reproduced from: Rasp *(1941).*

of identity and community to generations of eligible 'young men,' and helped prepare them for careers in a public world in which men take the decisions. Female engineering students – latter-day successors of Agnes Magnell and Very Sandberg – make up larger numbers now than in the past, however, and many have successfully entered various formerly exclusively male professional domains. Their competence is highly appreciated and actually even in demand in the current technical labor market. Still, most are still made to feel that, doing what they do, they are 'not a woman, and not a man.' Many feel treated as 'guests' in engineering, or as 'outsiders inside,' by their male colleagues and superiors. The 'inconveniences' caused by women engineers in a man's world may be of a different nature today than a hundred years ago; they are still so much present as to influence many a woman's life.[46]

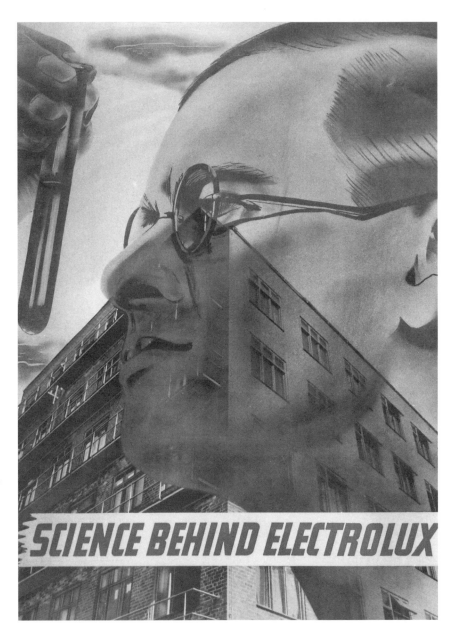

Figure 22. In characteristic modernist collage style, this advertisement for the Swedish company Electrolux geared to its English-speaking market mobilizes the symbol of the male inventor for corporate use, suggesting male creative power of science. Reproduced from 'Science behind Electrolux' Folder (Stockholm: Electrolux, 1939).

SCIENCE BEHIND ELECTROLUX

Notes

1. Boel Berner, 'Professional or Wage Worker? Engineers and Economic Transformation in Sweden' and other contributions in Peter Meiksins and Chris Smith, eds., *Engineering Labour. Technical Workers in Comparative Perspective* (London: Verso, 1996).
2. Special Issue: 'Gaining Access, Crossing Boundaries: Women in Engineering in a Comparative Perspective.' *History and Technology* 14, no. 1 (1997).
3. Brian Easlea, *Fathering the Unthinkable: Masculinity, Scientists and the Nuclear Arms Race* (London: Pluto Press, 1983); Caroline Merchant, *The Death of Nature. Women, Ecology and the Scientific Revolution* (New York: Harper & Row, 1980); David Noble, *A World without Women* (Oxford: Oxford University Press, 1992).

4. This is not to say that cultural analyses of the metaphors and rhetorics linking technology to masculinity are not important; see, for example, Carroll Pursell, 'The Construction of Masculinity and Technology,' *Polhem* 11, 3 (1993): 206–219; Boel Berner, *Perpetuum Mobile? Teknikens utmaniningar och historiens gång* (Lund: Arkiv, 1999), chapter 5.

5. Cynthia Cockburn, *Brothers. Male Dominance and Technological Change* (London: Pluto Press, 1983); Sally Hacker, *Pleasure, Power and Technology* (Boston: Unwin Hyman, 1989); Sally Hacker, '*Doing it the Hard Way'. Investigations of Gender and Technology* (Boston: Unwin Hyman, 1990); Sharon Traweek, *Beamtimes and Lifetimes. The World of High Energy Physicists* (Cambridge, MA: Harvard University Press, 1988).

6. Pierre Bourdieu, *La noblesse d'état. Grandes écoles et esprit de corps* (Paris: Editions de Minuit, 1989); Pierre Bourdieu & Jean-Claude Passeron, *La reproduction* (Paris: Editions de Minuit, 1970).

7. Basil Bernstein, *Class, Codes and Control. Volume 3. Towards a Theory of Educational Transmission* (New York: Routledge & Kegan Paul, 1977 [2nd ed.]).

8. Ralph H. Turner, 'Modes of Social Ascent through Education: Sponsored and Contest Mobility' in A.H. Halsey, J. Floud, and C.A. Andersson, eds., *Economy and Society. A Reader in the Sociology of Education* (Publisher's place: The Free Press of Glencoe, 1961).

9. The article builds upon the research reported in Boel Berner, *Sakernas tillstånd. Kön, klass, teknisk expertis* (Stockholm: Carlssson, 1996). See also Boel Berner, 'Explaining Exclusion: Women and Swedish Engineering Education from the 1890s to the 1920s,' *History and Technology* 14, no. 1 (January 1997): 7–29; and Boel Berner and Ulf Mellström, 'Looking for Mister Engineer: Understanding Masculinity and Technology at two Fins de Siècles,' in Boel Berner, ed., *Gendered Practices. Feminist Studies of Technology and Society* (Stockholm: Almqvist & Wiksell International, 1997), 39–68.

10. Lena Sommestad, 'Human Reproduction and the Rise of Welfare States: An Economic-demographic Approach to Welfare State Formation in the United States and Sweden,' *Scandinavian Economic History Review* XLVI, 2 (1998), 107.

11. Berner, 'Professional or Wage Worker?', 173ff.

12. Bosse Sundin, *Ingenjörsvetenskapens tidevarv* (Stockholm: Almqvist & Wiksell International, 1981).

13. Boel Berner, *Teknikens värld. Teknisk förändring och ingenjörsarbete i svensk industri* (Lund: Arkiv, 1981) for contemporary comparisons with Germany and the USA.

14. Christina Florin and Ulla Johansson, '*Där de härliga lagrarna gro...' Kultur, klass och kön i det svenska läroverket 1850–1914* (Stockholm: Tiden, 1993).

15. Discussion and figures from Sommestad, 'Human reproduction', 107ff.

16. Gouzévitch and Gouzévitch, and Canel this volume. For the similar disinterest among German bourgeois feminists, see Margot Fuchs 'Like Fathers – Like Daughters. Professionalization Strategies of Women Students and Engineers in Germany, 1890s to 1940s,' *History and Technology* 14, no. 1 (1997), 53.

17. Thus there was no need for separate women's courses in mathematics such as the ones organized in the Austrian Technical University in the late 19th century. See Juliane Mikoletzky, 'An Unintended Consequence: Women's Entry into Engineering Education in Austria', *History and Technology* 14, no. 1 (1997), 33.

18. Boel Berner, 'The Meaning of Cleaning: The Creation of Harmony and Hygiene in the Home,' *History and Technology* 14, no. 4 (1998): 313–352.

19. For recent discussions about the feminist movement in Sweden around the turn of the century, see Ulla Wikander, ed., *Det evigt kvinnliga. En historia om förändring* (Stockholm: Tidens, 1994).

20. Quoted in Berner, *Sakernas tillstånd*, 62.

21. This was Anna Lous-Mohr from Norway, a country which already knew some female architects. She was, in fact, the first woman to gain a degree from Swedish Royal Technical Institute, since Agnes Magnell merely fulfilled three of the required four years at the school and then left to marry a fellow student in civil engineering. Magnell may, however, have practiced her profession in collaboration with her husband. The second female architect in Sweden, who graduated in 1922, was Ingegerd Waern-Bugge, who came from a radical family, and knew Ellen Key. She was later to play an important role in the housing projects associated with the social-democratic construction of the Swedish welfare state. For sources and further discussion of this development, see Berner, *Sakernas tillstånd*, chapters 1 and 4.

22. Note that the field of 'General studies,' which attracted many female students at the German technical universities, did not (and does not) exist in Sweden. Fuchs, 'Like Fathers', 54.

23. Berner, *Sakernas tillstånd*, 317.

24. Berner, *Teknikens värld*, 110ff; Göran Ahlström, 'Industrial Research and Technical Proficiency. Swedish Industry in the Early 20th Century,' *Polhem* 12, 3 (1994): 272–310; Sundin, *Ingenjörsvetenskapens tidevarv*.

25. In 1914, both Swedish Royal Technical Institute and Chalmers admitted their first women students of chemistry. The one at Swedish Royal Technical Institute disappeared after one year, but the one at Chalmers, Vera Sandberg, stayed on to graduate and to work for many years as an industrial engineer.

26. A concept used by the German historian Jürgen Kocka to denote the educated and professional part of the upper and middle classes, in contradistinction to the part which derives its position from economic and industrial activity. The professional engineer, in effect, often belonged to both. Jürgen Kocka, ed., *Bürger und Bürgerlichkeit im 19. Jahrhundert* (Göttingen: Vandenhoeck & Ruprecht, 1987).

27. Ruth Oldenziel, 'Decoding the Silence: Women Engineers and Male Culture in the U.S., 1878–1951, *History and Technology* 14, no. 1 (1997), 84.

28. Notably Austria and Germany where most areas of higher education, including engineering, were opened to women in the early decades of the century. Mikoletzky, 'An Unintended Consequence', 35; Fuchs, 'Like Fathers,' 51.

29. Mikoletzky, 'An Unintended Consequence', 39. A similar attitude was expressed in the US, according to Oldenziel, 'Decoding the Silence', 87.

30. Gouzévitch and Gouzévitch, this volume.

31. Quoted in Berner 'Professional or Wage Worker?' 183.

32. See the references in note 28 and for the Netherlands: Frida de Jong "Sisters in Engineering. Women Engineers in the Netherlands", paper presented at conference at Frauen(t)raum Technik, Berlin, January 1999.

33. Karen Zachmann, this volume.

34. It should be noted, that Swedish university-trained engineers (with the exception of architects) were all called 'civil engineers,' irrespective of their area of expertise. Originally, this title was reserved for graduates from the Marieberg Military Academy, who, around the mid-nineteenth century, had enrolled in the Academy in order to assume non-military state employment. 'Civil engineer' became the official title for Swedish Royal Technical Institute graduates in 1915. There hence never was any Swedish state corps of engineers, like that in France, or in Russia at the turn of the century.

35. Quoted in Berner, *Sakernas tillstånd*, 43.

36. Turner, 'Modes of Social Ascent.'

37. Between 1881 and 1910, the fathers of 75 per cent to 80 per cent of the students at Swedish Royal Technical Institute and Chalmers were either business leaders or white-collar employees. See Rolf Torstendahl, *Dispersion of Engineers in a Transitional Society* (Stockholm: Almqvist & Wiksell International, 1975).

38. Quoted in Berner, *Sakernas tillstånd*, 91.

39. Rosabeth Moss Kanter, *Men and Women of the Corporation* (New York: Basic Books, 1977) is the classical study of 'tokenism'.

40. Published in the student journal *Rasp*, quoted in Berner, *Sakernas tillstånd*, 116.

41. Anna Karlqvist, *Från eftersatt till eftersökt. Om kvinnliga studeranden på Kungl. Tekniska Högskolan* (KTH: Avdelningen för teknik- och vetenskapshistoria, 1997).

42. Fuchs, 'Like Father…' for Germany; for USA, see Stephanie Smith-Divita, "Skirts and Slide Rules", paper presented at conference on Frauen(t)raum Technik, Berlin, January 1999.

43. For the concept of 'guest,' see Silvia Gherardi, *Gender, Symbolism and Organizational Cultures* (London: Sage, 1995).

44. A small number of women entered lower-level draftsmen or laboratory assistant jobs from the inter-war period onwards. For a study of engineers taking 'the long way' via the shop floor and evening courses from the 1930s to the 1960s, see Boel Berner, 'The Worker's Dream of Becoming an Engineer,' *History and Technology* 15 (1999).

45. The same numbers apply in Germany, Austria, and Greece; see elsewhere in this volume, and *History and Technology* 14, no. 1 (1997). In contrast, many women trained in civil engineering and aeronautical engineering; in the USA see Oldenziel 'Decoding the Silence', 79.

46. Ulf Mellström, *Engineering lives: Technology, Time and Space in a Male-Centered World* (Linköping: Department of Technology and Social Change, 1996); Boel Berner 'Women Engineers and the Transformation of the Engineering Profession in Sweden Today' *Knowledge and Society* 12 (2000).

Dmitri Gouzévitch and Irina Gouzévitch

4. A Woman's Challenge: The Petersburg Polytechnic Institute for Women, 1905–1918

'Since it happened so long ago, perhaps very few people know that the first higher technical school for women in the world was founded in Russia in 1906. Although this school functioned for 12 years before the revolution and had more than 1000 students, it is difficult to find any information about it in the guides.'

So wrote Boris Krutikov, one of the first professors at the Petersburg Polytechnic Institute for Women, in his unpublished memoirs.[1] This higher technical school, initially called the Petersburg Polytechnic Courses for Women, has been ignored for nearly eighty years. There is little information on it available in scholarly works.[2] Fortunately, Krutikov's manuscript still exists, and through it we can begin to retrace the early history of this unique institution.[3]

It was very unusual for women to have their own single-sex higher technical schools in the early twentieth century.[4] For this reason, the Petersburg institute has acquired a symbolic meaning. Its history is the story of a half-century of battle for women to have free access to higher education and the professions. The school also helped shape education and the paths by which women could gain technical skills. Its history was tied to two key moments in twentieth-century Russia; generated by the Revolution of 1905, it lost its autonomy after the second Revolution of 1917. Why and how did this institution come into being in Russia? How did it differ from similar institutions in Western Europe, and what was its place in the male-dominated world of engineering?

In this study we look at the history of the Petersburg institute, and the political and social context that allowed it, during the period of Russian state reform. We pay special attention to the shape of state-run higher engineering education, which was strictly hierarchical and regimented according to earlier, eighteenth-century models. We also stress the strength of the women's movement which played a major role in opening gates to women into higher engineering education during the second half of the nineteenth century. Finally, we would like to show why the history of Russian women engineers may be considered as part of the nineteenth and twentieth centuries' European history; references to the French and Swedish cases are used to highlight this point.

The Male World of Engineering: Industry as a Niche for Women

From the very beginning, engineering in Russia has been strongly influenced by the military. At the turn of the twentieth century when industry was booming, the private industrial sector, which lacked skilled staff, became a niche for women.

Engineering as a profession was institutionalized in Russia in the early eighteenth century as part of Peter the Great's reforms which aimed to put an end to the century old political, cultural and economic 'backwardness' of the Moscow State. It led to the establishment of special technical administrations, the so-called engineering corps, in the main engineering fields: artillery, ship-building, mining, surveying, and military engineering. Some of the main technical schools emerged earlier, at the turn of the eighteenth century (Gunner School and Navigation School in Moscow founded 1701, Navy Academy in Moscow founded 1715, Artillery School in Saint-Petersburg founded 1721); others were created later together with the relevant corps. Like in the French engineering system, they all supplied the commissioned staff of technical divisions. The only way to achieve the rank of engineer was to enter the service of the state, where one's professional status was categorized according to the Table of Ranks (*Tabel' o rangah*).[5] In addition – state departments had their own professional scale; this resulted in different opportunities to progress careers. Some departments, such as Mining offered more attractive professional scales than others. Later in the nineteenth century, in order to upgrade and ennoble the engineering status, engineer graduates were awarded both engineering and military ranks as in the *Engineering Corps of Ways of Communication* (1810).

Technical education, whose importance grew during the eighteenth century, was given a substantial boost by Alexander I (1801–1825). He was inspired by European examples, especially France, in his drive to modernize pre-industrial Russia. Why did he look to France? First, the Russian monarchy had a French-oriented foreign policy before Napoleon tried to invade Russia in 1812. Second, Alexander recognized the urgency of the need for better transportation and communications, required if Russia was to recover from its economic 'backwardness.' There was a clear lack of competent people able to build and maintain roads, and no specialized technical schools which could become high-level engineering schools in this field. Finally, this was facilitated by the absence of a language barrier, since the Russian nobility from which the different engineering corps drew their members spoke French not only in salons but also as the language of science.

The first high-level school he established was the Institute of the *Engineering Corps of Ways of Communication* in Saint Petersburg in 1809/10. This was a unique blend of two French models, the theoretical-oriented curriculum of the Ecole polytechnique and the Ecole des ponts et chaussées specialized in civil engineering.[6] During the following half-century, Alexander's reforming drive gained momentum and gave rise to a network of special high-level schools. Similar to the French 'grandes écoles', these trained engineers for state service in the main engineering fields.

All engineering staff, including graduates from the technical institutions, were enrolled by the corresponding technical corps following a military or militarized pattern. Thus engineers could either advance their careers along a double ladder (both military and civil) or only along a military one. The only exception were the so-called civil engineers: engineer-architects who graduated from the School of Civil Engineering were granted civil ranks. All military, militarized, and even civil, engineering administrations were ruled by both a very rigid system of internal regulations and a clear state ranking. Identified by its corporatist mentality and conceived without any consideration for the world of women, this system tended to remain monolithic and thus closed to women, even if, miraculously, they were professionally well-trained. This was mainly because such a situation was just not foreseen by the state regulations. However, in 1863–64 there was a pragmatic decision to allow women's professional employment as technical staff of the telegraphic and railway agencies. These opportunities came, however, without any right to enter the system of ranks, to gain professional advancement, or to receive pensions or rewards for long service.

At the same time, there were few opportunities for state engineers to operate as free entrepreneurs. Yet private industry suffered from an acute shortage of skilled engineers. Consequently, a few institutions trained technical experts for the private sector. Among these were the *Saint Petersburg Practical Technological Institute* (founded 1828), the Moscow Manufacturing School (established 1830), and, to some extent, the Moscow School of Commerce (opened 1804). However, all of these were secondary institutions. Their graduates could call themselves technicians or masters, but not engineers. Even if they performed engineering tasks, they could not gain official recognition because they did not have the required higher education degrees.

In the late 1860s, the state engineering corps were de-militarized. (The 1867 Imperial decree was crucial in this regard.) This allowed skilled engineering staff to move between state administration and private industry, which had not been possible before. By this time two of the secondary technical institutions already mentioned, the Moscow Manufacturing School (which was now called the Moscow Technical School) and the Saint Petersburg Technological Institute, became higher-level schools and thus gained the right to award engineering diplomas. Henceforth, their graduates were allowed to enter state administration, obviously much more attractive because it brought a high social status and financial security.

At the end of the nineteenth century, new higher technical schools emerged in Russia. Following the example of German technical universities, these Polytechnic institutes did not have the traditional conservatism and corporatist links of most Russian technical institutions, and were therefore especially attractive to women students. These schools mainly offered opportunities in the industrial sector. However, at the turn of the twentieth century, the private industrial sector which could not enjoy the same prestige or offer equivalent guarantees in spite of its growing importance, still lacked skilled staff. This became a niche that women engineers could fill.

Women Struggles for Education: Opening a Gate for Women to Engineering

Important political events such as the defeat of the Crimean war, the death of Nicolas I (1855), the abolishment of serfdom (1861), and also the push for industrialization deeply perturbed the Russian society in the middle of the nineteenth century. This period was a turning point in the social situation of Russian women. More and more numerous were those who had to earn their living as shown by the increasing proportion of women in the urban population. This encouraged the creation of female gymnasia with a classical curriculum all over the country from 1858.[7]

In the early 1860s, following these major changes and allowed by the 1835 University Regulations which did not state any limitation to the entry of women in higher education, the influx of women into higher institutions dramatically increased. It was also encouraged during the early reform period by the 1857–1858 statement which canceled restrictions in numbers of admissions previously limited to 300. Public lectures were allowed as well. Indeed, the most important factor was the exclusion of military training from the university curriculum.

As the very first female students proved to be very active politically, they immediately integrated the student revolutionary movement. Equal educational rights for men and women became a major claim which also dominated the entire women's movement.[8] According to Sophie Satina, this point differentiates the Russian feminist movement from those in the Western world. For those movements, the fight for general education was complementary to the main goal of gaining civil, political and economic rights for women. However, their active participation in the revolutionary movement contributed to their further exclusion from universities. Minister V. N. Tchicherin stressed in his Memoirs: 'To admit young women to the university whereas we don't know how to manage the young men was the height of madness.'[9]

Struggles of Russian women took many radical forms. One cultural expression was the provocative Nihilist movement of the 1860s. Nihilists

Figure 23. "Let's smash men's power like this empty glass!" reads the translation of the Russian caricature drawing for a Russian popular magazine. Women's issues such as equal educational rights were strong among nineteenth-century revolutionary groups and linked to extreme radicalism in the public mind. The women's rights activists played a key role in Russia's revolutionary movements and paved the way for the women's education in engineering in the early twentieth century. Reproduced from Iskra *37 (25 September 1959), 365.*

rejected the religious and moral principles of the day in favor of personal freedom. Both men and women were part of this movement, but it became associated most clearly in popular culture with women's emancipation. Women were instantly caricatured in cartoons and novels. Wearing short hair and nothing but black clothes, this ill-mannered and short-haired monster smoked, swore, and declaimed loudly on economics, politics, and science.[10] For many Russian women, adopting Nihilist attitudes was a protest against unequal education and the popular notion among men that higher education would destroy women's femininity. These women therefore rejected the old, restrictive kind of femininity altogether. But in doing so, they inadvertently helped to create the stereotype of the un-sexed woman – disturbing and dangerous to society. This imagery was a significant factor in slowing women's access to equal rights and higher education.

As a consequence, 1863 official University Regulations were drawn up providing a way to restrict the access of women's students de facto. In addition, several internal ministry statements were circulated within all universities, closing the doors for women into higher education. This resulted in a dramatic exclusion of women from Russian higher education. In response, many left Russia to study abroad. The women from the Russian Empire constituted for nearly half a century the most numerous population of female students in the main European universities, especially in Berlin, Bern, Königsberg, Liege, Paris and Zurich. Those who could not or did not want to leave intensified their struggle in Russia. This resulted in different experimentations, especially the home universities, called Universities-on-the-Wind, a very mobile form of private higher institutions which attracted the best men professors, like I. Sechenov, A. Borodin, D. Mendeleev, I. Mechnikov, V. Kovalevskij.

Feminist movements further pushed for creating higher courses for women, which soon became a path toward a new social status and an opening toward scientific and professional education and the professions. Some of the first educational institutions were created almost simultaneously in Moscow (Lubjan Courses, 1869, and Vladimir Courses, 1870) and in Saint Petersburg (Alarchin Courses, 1869–1870). They were secondary-level schools; however, some of them in fact adopted the university curriculum. This initiative was followed in several important cities like Kiev (1871–72/78), Kazan (1872/76), Odessa, Varsow, Kharkov.[11]

As these first initiatives were a success, further questions arose on how to face this push. Creating a university for women still seemed a dangerous idea. The government preferred more moderate proposals. One of the most important female institutions created in this period was the Higher (Bestuzhev) Courses for Women. Opened in 1878, this institution was conceived de facto as a university since its beginning. Its graduates progressively gained the right to work as teachers in gymnasia, and from 1913 on their diploma was considered as being equivalent to university degrees. With its curriculum based on natural sciences, physics, mathematics, and astronomy, it helped women open a gate into the male professional world by proving women's ability to cope with such scientific matters. This came together with another breach made in popular opinion on the lesser performances of women in mathematics thanks to the striking success of some Russian women trained abroad, like Sophia Kovalevskaja, famous Russian mathematician and mechanician. Remarkably, the Bestuzhev Courses was the only institution to survive the years of reaction (1886–1889). In 1919, it was integrated into the University of Petrograd.

There is a fascinating contradiction between the male corporatist and military mentality which characterized engineering in Russia and the

112 Etudiante. 1883   Курсистка. 1883

*Figure 24. Nikolaj Jaroshenko's (1846–1898) emphatic portrait of a 1870s Russian female student on her way to class, symbolizing the long and lonely road for aspiring academic women. Jaroshenko's 'The Woman Student' (*Kursistka*) of 1883 belonged to a generation of women students many of whom emigrated to Western Europe after the Czarist clamp down on revolutionary groups to study at European universities like the University of Zurich, Switzerland long before their European sisters. Oil on canvas. Reproduced from:* Les Ambulants. La Société des Expositions Artistiques Ambulantes, 1870–1923 *(Lenigrad: Edition d'Art Aurore, 1974).*

practice of establishing single-sex institutions for women. Boel Berner has also stressed this paradox in the case of Sweden. Male state engineers strongly encouraged a small group of women activists in their educational goals, and some of the famous senior men in the profession became prominent advocates of technical education for women. It was only with their assistance that the Saint Petersburg Institute achieved success when other, similar experiments failed.

The Russian case illustrates the internal logic that can impede, or stimulate, a dialogue in a professional context between two traditionally opposed worlds: the male corporate world and the female emancipating world. Engineers, as individuals, have always been less rigid and hostile towards training for women than many other professional groups, especially medical men. It is important to stress such divisions when talking about women's access to the labor market. Unlike more liberal professions, the state engineering world, with its well-defined fields and its well-established staff hierarchy, remained closed to women for many years. In our research we could find no record of negotiations for access to the state engineering schools in the historical literature.[12] For this reason, the state engineers, especially those who held important senior positions, did not have to fear competition from women. Possibly this allowed them to take an altruistic interest in women's education. At the same time, the increasing industrialization of Russia in the late nineteenth century opened many new opportunities and areas, while the shortage of well-trained industrial engineers continued. Senior state engineers rapidly became aware of this need, as many of them worked closely with private industrial companies. Knowledge of this problem must also have led state engineers to give their crucial support to the Petersburg Polytechnic Institute for Women.

From Drawing School to the Petersburg Polytechnic Institute for Women[13]

As Nikolaj Shchukin, Director of the Polytechnic Institute for Women, stressed, a polytechnic institute for women was a natural step on a path taken long ago. Women first entered teaching professions, they further gained access into the medical world and later entered the scientific world mainly as astronomers. The Polytechnic Institute for Women was indeed the next step which allowed them into the engineering world.

In 1898, a graduate of the Bestuzhev Courses put forward a proposal to the Russian Women's Philanthropic Society for a drawing school. This was Praskov'ja N. Arian, who had studied physics and mathematics, and whose husband was an engineer. She was one of the most vocal advocates of women's right to enter the professions.[14] Her project was not supported by the Society, but one of the members, Argamakova adapted the idea by introducing a special drawing course at her private professional school.[15] Four years later Arian submitted the proposal again, helped by two professors in civil engineering, Nikolaj Belelûbskij and Valerian Kurdûmov. They sponsored a presentation of her ideas to a session of the Permanent Committee for Technical Education, which was connected to the Russian Technical Society (RTO), in 1902.[16] The RTO approved of this proposed 'techno-graphical institute' and its three-year program of study, but there was a delay of several years before it was finally established. Praskovj'a Arian remained its guiding spirit, though she succeeded in gaining some very influential supporters from within the engineering profession and fashionable society. The regulations of the Society for Providing Funding for Women's Technical Education established in 1904, were signed by several professors from the Institute of Ways of Communication such as V. Kurdûmov, N. Belelûbskij and A. Nebolsin. But the signatories also included A. Filosofova, P. Stasova, E. Konradi, V. Taranovskaja, Countess Panina, and Baron Gincburg, who were figures of considerable social standing.[17] On 20 March 1905 the ministry of public education approved the regulations. At the beginning of May the society's members held their first meeting, chaired by Nikolaj Shchukin, who was a leading railway engineer and a professor at the Technological Institute and Military Engineering Academy. Donations of 12,000 rubles were collected the same day.

The social and economic crisis that resulted from Russia's defeat by Japan, and the disastrous political situation, actually helped this educational project. The government was aware of the serious lack of industrial knowledge and the need to move away from outdated forms of technology. Hence it was sympathetic to the schemes promoted, and agreed to consider women engineers as a reserve labor pool of technical specialists. On 22 August 1905, the society, headed by Nikolaj Belelûbskij, was allowed to establish the Petersburg Polytechnic Courses for Women (PPCW). This institution opened its doors on 15 January the following year. The more neutral-sounding courses which were created instead of the suggested institute were cautiously pushed by the Ministry in order to avoid allusion to the higher scientific and technical school for men.

The organizers had to accept this compromise in order to proceed, thus abandoning official recognition for their graduates as professionals. They accepted this, indeed, to gain time and to push ahead with their project whose major goal was clearly specified in the regulations: 'promoting higher technical education for women in those branches of technology where the use of female labor seemed to be the most suitable.'

The founders were very ambitious. The new institution had four engineering departments: architecture, building engineering, electromechanics and chemical engineering. Earliest proposals for drawing school were three years. In order to upgrade the female technical education, a five-year curriculum was proposed and soon extended to seven years. This extension was needed to broaden the curriculum which, 'should not go below the equivalent male curriculum.'[18] To fulfill this requirement, the Polytechnic Courses for Women were quite successful in recruiting the best professors in Saint Petersburg, starting with its first permanent director, Professor Nikolaj Shchukin. Other faculty were eminent teachers at higher technical institutes, such as the famous bridge builders N. Belelûbskij, G. Perederij, I. Aleksandrov, S. Kareisha, N. Puzyrevskij, and V. Lâhnitckij from the Institute of Ways of Communication; L. Serk and R. Gabe from the Institute of Civil Engineering; B. Rosing from the Technological Institute; and Professors L. Benua, V. Pokrovskij, I. Fomin, I. Pâsetckij, S. Beljaev, A. Jakovlev, and V. Jastrzhembskij, who were architects and members of the Academy of Arts. By 1916, there were almost one hundred faculty.

The early years were difficult. The time of the first Russian Revolution was a period of extreme social tension and general scarcity. And the government was not terribly supportive of the new project. Although official recognition was granted to the institute by the Russian state which controlled its affairs, the state refused to promote its graduates or provide any funding. The institute's founders therefore decided not to apply for financial assistance until they had achieved clear results.[19] They depended on other income such as tuition fees, annual membership fees from the professional societies, donations, and proceeds from benefit lectures and concerts. The funds raised from these sources covered classroom space, equipment, and wages, plus small allowances for students. While the PPCW was autonomous, it rented a few private apartments on the upper floors of the Mussin-Pushkin house (68/2 Zagorodnyj Avenue). Two flats served as the physics and chemistry laboratories in the first academic year. About fifteen other laboratories were added subsequently, for organic chemistry and electro-mechanics among other subjects. The school also had its own geodetic study. In addition, students had access to the best laboratories in the higher technical institutes of Saint Petersburg, including the famous mechanical laboratory of the Institute of Ways of Communication. This was headed by Belelûbskij, an early supporter of the PPCW and a signatory to the regulations of its founding society.

Students were generally selected on the basis of their secondary education, in order to meet the same requirements that men's higher schools insisted upon.[20] This meant that prospective students passed through a full course at a girls' gymnasium and achieved high scores in mathematics. But special complementary training in mathematics and physics was provided because male gymnasia were more advanced in these disciplines.

During the first two years all students followed the same curriculum. In the third year they chose a specialized program. In 1908 the common faculty was divided into the architecture and building departments. At the same time, the Board of Professors redefined the four specialized programs. The curriculum included lectures, practical and laboratory

Figure 25. Building of the Women's Polytechnic Institute (1905–1918) at the Mussin-Pushkin House, 68 Zagorodnyi Avenue in Saint Petersburg, Russia, where the school was housed since 1906. Established right after the 1905 revolution, the engineering school for women had been an initiative of a coalition of women with ties to the radical Russian women's movement and members of the male engineering elite, who were closely associated to the czarist militarist state. Photograph by Evgenij Kraev (1984), personal Collection I.Gouzevitch, Paris, France.

work, site visits (to plants, factories, laboratories, electric power stations, and building works), an undergraduate thesis, summer training in industry, and a final project. In each department the Board of Professors organized the curriculum for its engineering specialty. For example, the students in electro-mechanics were allowed to pass their exams at any time in the academic year.

Unfortunately, information on the number of students and faculty in each department is incomplete. According to Kurbatov, an eminent professor in chemistry, in 1906 architecture attracted the most students. This was due, among other factors, to the presence of Professor Pokrovskij, a famous painter-architect who was a member of the Academy of Arts. Initially there were three (later four) sections within this department, each of which was headed by a well-known master-architect (Pokrovskij, Lâlevitch, Lidval, and Benua). The dean of architecture, V. Pjaseckij, and the departmental secretary, the engineer and painter Ruvim Gabe, were both admirers of the arts and famous in their fields. The practice of holding 'architectural Wednesdays' started within this department and was enlivened by notable painters and architects, such as the constructivist architect V. Shchuko. These attracted many students to hear presentations and discussions on various architectural themes by the best practitioners in the field. Students also recalled the 'sculptural evenings' conducted by the eminent artist, L. Shervud. In addition they practiced water-color painting and modeling. Many of the student's architectural projects reached a high artistic level, though they also showed the influence of the building department in their attention to structure. According to Belelûbskij, some of his students' projects made a favorable impression in 1910 on Rabut et Ocagne, who taught at the French École des Ponts et Chaussées.[21]

In addition to their study of architecture, students had to do elaborate projects on water supply and sewage, heating and ventilation, road construction, and building structures. At the end of their studies, the students had to complete an architectural project chosen from such themes as 'building for a history museum' and 'building for the state Duma'. In 1916, an unusual proposal came from a senior woman who came to the institute from a town in the countryside to find a woman engineer able to direct building works in the religious institution she headed. She proposed a project on 'Women's Monasteries.' But no student wanted to specialize in this domain.[22]

The construction department trained specialists in a broad range of fields: ways of communication (bridges and roads), industrial building, uses of water power, sanitation, and organization of public services and amenities. Besides the theoretical disciplines, each student had to undertake to defend several undergraduate theses. The topics proposed for graduation projects included large span bridges, railroads and railways stations, weirs, hydro-electric power stations, irrigation systems, dockyards and ports. Vassilij Starostin, dean of the building department from 1913, was in charge of the foundation course. His colleague Nestor Puzyrevskij, the Russian expert in hydrography, headed the chair of waterways. Beginning in 1913 practical works in the field were conducted by Ljahnickij, who had participated in the building of the Panama Canal. Professor G. Perederij, one of the greatest Russian bridge builders, and a theoretician in elasticity, delivered lectures on bridge building and directed final projects.

The chemistry department was also favored by the students. In spite of insufficient and inadequate pre-requirements, its curriculum, which was inspired by that of the Technological Institute, was much appreciated thanks to its dean, Yulij Zalkind, an eminent professor. He was also in charge of the organic chemistry course. Inorganic chemistry was taught by two professors, A. Bajkov and V. Kurbatov. This department had the first woman chemist, N. Vrevsklaja, as a member of its teaching staff from the beginning. A graduate of the Bestuzhev Courses, she was recommended by Bajkov to be an assistant in the laboratory. A number of specialized laboratories were gradually established in general chemistry, qualitative and quantitative analysis, organic and physical chemistry, and chemical technology. After doing an undergraduate thesis in applied mechanics and architecture (*sic!*), students had to go through two practical training periods, one of them in a factory, during the summer. Later on all had to prepare for a diploma in chemistry, focusing on an original topic of their choice. Forty-five women chemists who specialized in various fields of fundamental and/or industrial chemistry (metallurgy, technology, oil, silicates, and the dye industry), graduated from this department by 1922.

The electro-mechanics department offered an extended and wide-ranging curriculum that qualified its graduates for any branch of applied mechanics, electrical or heat engineering. It covered a variety of topics in theoretical and applied mechanics and electro-technics, but it also dealt with special equipment and construction. It promoted a series of classes on DC and AC machines and transformers, signal and relay protection systems, telegraphy and lighting for electric tramways, high tension lines and turbines, steam and internal combustion engines, machine tools, and hoisting gears. The professors in the various fields of mechanics and electrical engineering more than held their own intellectually with members of the other three departments of the institute. Boris Rosing, dean of the department from 1908, was the inventor of an original electronic system which allowed one to obtain pictures using a cathode-ray tube. He also gave lectures on electric measurements to the students. The course in general electro-technics was taught by Mikhail Shatelen, who had been the first specialist professor of electro-technics in Russia. Osadchij and Vojnarovskij, his colleagues from the electro-technical institute, were in charge of special courses in telegraphy and electric networks. Among the teaching staff were representatives (and even directors) of three other higher engineering institutes: the Polytechnic Institute, the Technological Institute, and the Institute of Ways of Communication.

The response to the very first call for candidates was spectacular. In October 1905, less than two months after the government decision to allow the school to open, 700 applications were registered. This suggests the impact that the women's movement was already having, as emancipated Russian women worked toward a technical education and entrance

Механическая лаборатория в последней трети XIX в.

Figure 26. The women students at the Women's Engineering Institute of St. Petersburg were trained at Nikolaj Beleljubskij's Mechanical Lab at the Institute of Engineers of Ways of Communication *in the city. Belonging to Russia's engineering elite, Beleljuskij (1845–1922) was one of the founders and professors of St. Petersburg's Women's Polytechnic Institute and its first professor.*

to the male-dominated professional world. After the 1905 Revolution, the country had entered a new phase of industrialization. Activists in the women's movement hoped that women might serve as a new labor force as was so evidently needed in a variety of technological fields. And these larger goals sometimes coincided, of course, with the desire of individual young women to escape provincial routine or find self-fulfillment in higher education.

The number of students increased steadily every year. From 224 in 1906, it grew to 415 students in the next academic year. During the decade prior to the 1917 Revolution, between thirty and 200 additional candidates were admitted each year. There were 1500 students in 1916–1917, and a peak of 1750 in 1923–1924, the last year the institute was independent.[23]

The numbers decreased slightly in 1910–1911, probably because of the fact that the institute still couldn't grant full certification. The numbers rose again from 1912–1913, as new regulations were anticipated. These were finally agreed on in 1915. This should have created a dramatic increase in enrollments, but data for the period are difficult to find. Information on the 1920–1921 academic year shows a significant decrease in student admissions. This was doubtless linked to the catastrophic impact of the civil war.

Boris Krutikov's memoirs of his time at the institute offer some insight into the character and interests of the students. He identified three categories: 'enthusiasts who worked with an extreme zeal (they were the majority), the average candidate who just satisfied the requirements, and finally those who went to Petersburg on the pretext of seeking an education, but enjoying all the blessings of the capital.'[24] All the available sources stress the difficult financial problems faced by most students. The annual tuition fees, initially one hundred rubles or less according to the income of the student or her family, rose quickly to 150 rubles. Only a very few small grants were available from the sponsoring society. Rent was large portion of students' budgets, since few had help from their families, especially if they came from other cities to attend the institute.

Figure 27. The drafting room of the Russian Women's Polytechnic Institute in Saint Petersburg around 1906, where women were schooled in drafting, mechanics, and chemistry. Trained separately to fill technical jobs in Russia's nascent industries, the school helped to preserve Russia's core male military engineering culture. Courtesy and permission of the Personal collection of Margot Fuchs, Munich, Germany.

Students, male and female, therefore founded self-help organizations of various kinds. For example, the women's institute had a mutual assistance fund, a students' dining-room, and an employment office. (Available work included giving evening tutorials, copying, statistical work, drawing, and so on.) These data make it likely that most women students came from urban families of middle-class status but having a low income – such as poor gentry and parish clergy.[25] The Women's Polytechnic Institute's sixtieth anniversary book, *The First Women Engineers (Pervye zhenshchiny inzhenery)*, was published in Leningrad in 1967, and it has valuable biographical information on 227 women who entered the Institute between 1906 and 1923. Unfortunately, these are very brief notes and they provide little information on women's professional careers; moreover, the biographies are completely silent about the social background and family lives of the former students. This suggests that, at the time of writing, it was undesirable to admit that one was from a bourgeoisie family.[26]

Another important characteristic of the women students was that they were very politically active. They were true inheritors of the militancy of the earlier women's movement of the 1860s and 1870s, though their activism took a different form. Prior to 1905, women students struggled on two fronts. First, they sought support for their specific educational needs while also demanding equal civil and political rights in solidarity with men. In December 1905 the new Duma granted some of these rights but, despite many promises, (the most famous being in the Czarist manifesto of 17 October) women did not get the vote. The revolutionary movement against the Czarist regime continued, and, according to V. I. Lenin, radical students were 'in the forefront of all democratic forces.'[27] This considerably influenced the way women articulated their particular claims. The revolutionary movement, however, was neither homogenous nor monolithic. Politically active student groups therefore developed in a number of different directions. Women from the PPCW were directly involved. The social-democratic group of E. Smitten, G. Schvarz, A. Kuz'mina, and R. Lianozova, rapidly joined the general student revolutionary movement. According to K. Ivanovskaja, the women at the Polytechnic Institute were all part of the numerous student groups that emerged 'whenever there was an occasion' in most of the higher schools. Some were arrested while demonstrating with other students against the death penalty on 9 and 10 November 1910. Further events showed the truth of Lenin's assertion that 'the small beginnings of minor academic conflicts is a great start, because they will have – if not today then tomorrow, if not tomorrow then the day after-tomorrow – important consequences.'[28]

Despite the political turmoil, teaching went on as planned. The first three engineers graduated six years after the opening of the Women's Polytechnic Courses: Appolinarija Nichipurenko and Aleksandra

Sokolova in electro-mechanics, and Agnija Ivanickaja in chemistry. Forty-seven women graduated over the next three years before the new regulations came into effect. There were three reasons for this low number of graduates. Krutikov commented that not all students were motivated to finish their studies, for example if they got married. There was also a desperate demand for educated, highly skilled workers in the provinces; talented students might be recruited by provincial employers before graduation. The third and the most important reason was the uncertain status of the graduates. The regulations of 1905 did not allow the institute to award professional certification.

If women were now tolerated in laboratories, the presence of 'long-skirted engineers' on building sites was still opposed for various reasons. It was said, for example, that women might find it difficult to climb ladders because of their attire. Shchukin and Belelûbskij were able to smooth the way to practical training in the summers because of their high status and personal contacts in the industrial world. After initial worries, women students proved to be capable workers and were invited to return the following summer as masters and technicians. In fact, extensive demand for women engineers arising from this experience actually led the ministry of ways of communication to ask that they be granted official engineering credentials and rights.

As in many other European countries, especially in France, the place of women in the engineering profession changed during the First World War. From being the reserve labor force many passed directly to the active labor force, and even held top professional positions. As male engineers left for the front, they were replaced by women from the PPCW who proved to be just as efficient. By 1915, women's professional skills were finally recognized by the government, and the Courses were reorganized according to new regulations in September. The school changed its name to the Petrograd Polytechnic Institute for Women, 'with all the rights granted to such an institution.'[29] Soon after this, yet another reorganization took place. This followed the October Revolution, which overthrew the administration of the old regime. The institute was taken over by the People's Commissar for Education by a 1918 decree. It became co-educational, lost its autonomy and was renamed the Second Petrograd Polytechnic Institute.[30] It also moved its premises to Vassiliev Island. It had 250 graduates in 1923. It was mentioned for the last time in the 1924 Address Book. It was then merged with the First Petrograd Polytechnic Institute and so it lost its special status.[31]

As most of the graduates from the Petersburg Polytechnic Institute for Women advanced their careers during the Soviet period, information about their professional activity is very scarce. There are some memoirs which provide an impressive list of projects on which the young women engineers worked. By 1916, they had already participated in numerous construction projects on railways, bridges, plants, power stations and ports all over the country. Although they had to live through very difficult times, many succeeded in integrating themselves into the new society. A few of them reached a high social status and had successful careers. Their progress was based both on having equal educational and professional rights with men, which were guaranteed by the Soviet state from its inception, and on a shortage of engineers due to discrimination against the bourgeois intelligentsia by the new regime. Indeed many engineers, trained under the Czarist system, emigrated to the West to escape the threat of dismissal and incarceration.

Another factor that facilitated the advance of women from the late 1920s on was the lower level of engineering training. This resulted from political emigration of trained engineers, loss during the civil war and lower educational backgrounds of new candidates coming from workers' and peasants' families. Many teachers left their positions, and relatively unqualified workers, soldiers, and peasants entered high schools in large numbers without any preparation for academic training. As a result, the engineering institutes in the twenties and thirties turned out badly trained engineers, which exacerbated the shortage of good ones. And this was a society that needed to mobilize all resources for industrialization. Highly skilled young women benefited from this, as most were offered senior professional opportunities.

To understand the role of the PPCW, we must look at how its students and graduates were eventually employed. Information comes from two main sources: an article by Professor Kurbatov in the late 1940s that mentions the achievements of twenty-five graduates, and the 1967 yearbook already mentioned. Three pioneers graduated in 1912, all of whom had started in 1906. Apolinarija Fedorovna Nichipurenko, who graduated in electrical engineering and received her doctorate in technical sciences, started work as a senior researcher in the All-Union Research Institute for Hydrotechnics, specializing in the study of dams. In 1942 she moved to the Research Institute for Land-Improvement and Hydrotechnics at Tbilissi (Georgia). Author of a series of works on barriers, her professional career lasted thirty-four years. Aleksandra Ivanovna Sokolova-Marenina graduated in mechanical engineering and also did a doctorate in technical sciences. As a physicist, she worked on measuring instrument technologies and electro-physiological methods of investigation in the domain of higher nervous activity. Author of a textbook on the theory and calculus of measuring instruments and other works related to the hypnotic state of the body, she was also member of the Soviet of Leningrad from 1932 to 1938. She worked professionally for forty-nine years. Finally Agnija Petrovna Ivanickaja, who graduated in chemical engineering, worked mainly as a teacher, though she also did research at the Research Institute of Chemistry. Professor and chair of inorganic and analytical chemistry at the State University of Belorussia, she published findings of five projects before 1927. No information about her career is available after this date.[32]

Some of the women engineers who graduated from the Polytechnic Petersburg Institute had remarkable careers in the industrial building sector, especially in projects to build electrical power-stations in the Caucasus and Central Asia, in scientific and industrial research, and as professors and lecturers in prestigious higher institutes such as the Leningrad Polytechnic Institute, the State University of Tbilissi, the Tashkent Institute for Agricultural Irrigation and Mechanization, the Leningrad Textile Institute, and the Leningrad Institute of Civil Engineering. According to the 1967 biographical index, twenty women (or 8.8 per cent) held academic ranks or degrees (as 'candidates,' 'doctors,' 'dozents,' or 'professors'), and thirty-three of them worked as researchers without any academic degree. Nineteen taught as lecturers in higher schools. This was 31.7 per cent of the total number of women graduates.

Five women engineers reported additional higher education in medicine, foreign languages, or the arts. Thirteen, or 5.7 per cent of the total, occupied key engineering and/or administrative positions within industry, as directors, managers, experts and instructors, chief engineers and architects, scientific secretaries and members of ministerial committees. Eight women were members of the Republics' governments, hold positions as senior officials in the Party, and were managers in non-technical fields. In addition, another eight women were elected in the regional and city Soviets. Twenty-nine (or 12.8 per cent) were senior industrialists, administrators, or Party and Soviet officials. It is interesting to note that they had very long careers. Forty-eight had careers lasting up to twenty years, and ninety-nine women had careers that lasted forty years or more. Tat'jana Kolpakova, a mechanical engineer, worked for more than fifty-two years. Such long professional

Table 1. St. Petersburg Polytechnic Institute for Women's graduates and their professional activities at the end of the 1960s. Specializing in civil, electrical, mechanical, and chemical engineering, the women graduates worked as technical personnel in areas that were crucial for Russia's industrial development.

Engineering fields	Number of women engineers	%
Engineers of ways of communication	17	7.5%
Engineer-chemists and engineer-technologists	27	11.9%
Mechanical engineers	27	11.9%
Engineer-architects, civil architects	37	16.3%
Electrical engineers, power engineers	40	17.6%
Building (civil) engineers, hydraulic engineers	59	26.0%
Architects, painter-architects, painter-engineers	15	6.6%
Revolutionary, party workers and administrators	3	1.3%
Non-engineering professions, secondary higher education	2	0.9%
TOTAL	227	100%

careers suggest that the dramatic shortage of male engineers during the Second World War created opportunities for women engineers.

The sources only make mention of careers that were successful, a bias that must be noted. They were anniversary publications, published in 1947 and 1967 respectively, late Stalinist and early Brezhnev periods, and in this sense were not allowed to mention failed careers or politically suspect biographical information. Moreover, nothing was said about emigrants, victims of repression, or displaced individuals. Consider Agnija Ivanickaja, one of the pioneer students, about whom we know nothing after 1927. This lack of information about a professor and chair is disturbing. The same holds for Natalia Kobylina, the first graduate of the building department, who was admitted to the PPCW in 1906. She 'occupied chief engineering positions at the People's Commissariat of Ways of Communication, for the building of the White Sea channel (*Belomorkanal*) and Moscow channel (*kanal im. Moskvy*).' Nothing more is said of her in the anniversary book except that she died in 1939. She must have overseen thousands of working prisoners, but we do not know whether she was a civilian employee, a military officer, or a political prisoner herself. The year of her death, 1939, was also the period when Beria got rid of all the senior Gulag officials who had worked there under his forerunners Jagoda and Ezhov. Other biographies also raise questions. Records show relatively few women with a revolutionary past or with long average periods in the Party. Yet, given the extent of women's political activity, we would expect to find more of them politically involved. We speculate that the most activists of them suffered from repression, so became 'unknown.'

Conclusion

The Petrograd Polytechnic Institute for Women which was founded in Russia in 1906, on the wave of the first Russian Revolution, resulted from a half-a-century struggle of Russian women for emancipation and equal educational and professional rights with men. It emerged from a difficult consensus among numerous actors, governmental and administrative, private and public, associative and corporate. Several converging factors, especially the strengthening revolutionary movement, the industrialization of the country and the various experiences in higher technical schools for women, allowed the institute. The most successful resulted from private or public initiative and financial support, whereas the State educational system, universities as well as higher technical schools, remained closed to women for many years to come. Nevertheless, the State engineers, in spite of their strong corporate mind, proved to be more opened to women's claims for professionalization than their male colleagues in many other professional fields. Aware of increasing industrial needs and protected by their solid status within state administrations, they strongly encouraged women into engineering. The male Polytechnic Institute, created on the German model in 1901 and free of any corporate attachment served as a prototype for testing a new kind of engineering institution for women. Created outside the state system and sponsored by private and public associations, it functioned as an autonomous technical university which could benefit from both the administrative experience of higher women's courses and the intellectual potential of the best state engineering schools. In spite of difficult challenges, the first results proved to be convincing enough to open a path to its graduates towards professional integration. However, by unique historical circumstances, the women engineering students trained in the Russian Empire had to progress their professional careers in the quite different Soviet context.

Abbreviations

Otd. 1   Otdelenie 1, Pravitel'stvennye rasporâzheniâ (in ZhMNP)
Otd. 3   Otdelenie 2, Kritika i bibliografiâ (in ZhMNP)
Otd. 3   Otdelenie 3, po narodnomu obrazovaniû (in ZhMNP)
Otd. 4   Otdelenie 4, Sovremennaâ letopis' (in ZhMNP)

Notes

1. B. A. Krutikov, Vospominaniâ Inzhenera Putej Soobshcheniâ (vypuska 1900 g.) Krutikova Borisa Aleksandrovicha: 14 iûnâ 1877–15 ânvarâ 1968 (Saint Petersburg 1963), 206–208, Manuscript, KP.XII.4, Institut inzhenerov putej soobshcheniâ, biblioteka, Saint Petersburg.

2. Among the reference books we mentioned see for example: Sophie Satina, *Obrazovanie zhenshchin v dorevolûcionnoj Rossii* (New York, 1966); Sophie Satina, *Education of Women in Pre-Revolutionary Russia*, Transl. from Russian by A. F. Poustchine (New York, 1966), esp. Myra M. Sampson, 'Foreword,' ii-iii; Sophie Satina, 'Moskvskie vysshie zhenskie kursy' *Novyj Zhurnal* (N. Y.) 75 (1964): 195–217; and her, 'Obrazovanie zhenshchin v dorevolûcionnoj Rossii' *Novyj Zhurnal* (N. Y.) 76 (1964): 161–179; O. Kajdanova, *Ocherki po istorii narodnogo obrazovaniâ v Rossii i SSSR na osnove lichnogo opyta i nablûdenij* (Bruxelles, 1938–1939), 2 Vols.; D. K. Shedo-Ferotti, *Pis'ma D. K. Shedo-Ferotti o narodnom obrazovanii v Rossii, adresovannye grafu D. A. Tolstomu* Zhurnal Ministerstva narodnogo prosveshchenâ (hereafter ZhMNP) 143, Otd. 3 (1869): 75–94; 'Inostrannaâ uchebnaâ hronika' *ZhMNP* 144, Otd. 4 (1869): 90–102; V. Lemonius, 'Obshcheobrazovatel'nye i special'nye zhenskie uchebnye zavedeniâ v Berline' *ZhMNP* 154, Otd. 3 (1871): 34–70; 'Obshchij ustav Ital'ânskih universitetov' *ZhMNP* 191, Otd. 4 (1877): 53–94; 'Vysshee zhenskoe obrazovanie v Anglii' *ZhMNP* 231, Otd. 3 (January 1884): 1–29; 'O zhenskom vrachebnom obrazovanii v razlichnyh gosudarstvah Evropy i v Severnoj Amerike' *ZhMNP* 236, Otd. 4 (November 1884): 24–56; M. Stefanovich, 'Zhenskie gimnazii v Germanii' *ZhMNP* 254, Otd. 3 (December 1887): 51–70; F. Leontovich, 'Kommercheskie uchilishcha za granicej i v Rossii' *ZhMNP* 154, Otd. 3 (1871): 34–70; 251, Otd. 3 (May 1887): 1–35; 252, Otd. 3 (July-August 1887): 1–15, 49–63; 254, Otd. 3 (November 1887): 1–27 and others; Vl. Farmakovskij, 'Zhenskij trud v nemeckoj narodnoj shkole' *ZhMNP* 2 (Novaâ seriâ) Otd. 3 (April 1906): 130–174; I. Ul'ânov, 'Zhenskoe professional'noe obrazovanie v Bavarii' *ZhMNP* 33 (Novaâ seriâ) 5, Otd. 3 (May 1911): 24–60; I. Krasnopevkov, 'Professional'noe obrazovanie v zapadno-evropejskih gosudarstvah' *ZhMNP* 37 (Novaâ seriâ) 1, Otd. 3 (January 1912): 58–66; Kamilla Tol'mer, 'Nachal'noe i srednee zhenskoe obrazovanie vo Francii' *ZhMNP* 38 (Novaâ seriâ) 4, Otd. 3 (April 1912): 205–221; L-r*, 'Pis'ma iz Parizha' *ZhMNP* 147, Otd. 2 (1870): 190–194; 148, Otd. 2 (1870): 480–489; 225, Otd. 4 (1883): 130–131; 227, Otd. 4 (1883): 42–43; 234, Otd. 4 (July 1884): 48–51; 257, Otd. 4 (July 1888): 57; 28 (Novaâ seriâ) Otd. 4 (August 1910): 85–86; 32 (Novaâ seriâ) 4, Otd. 4 (April 1911): 33–34, 42; 42 (Novaâ seriâ) Otd. 4 (November 1912): 52–53; Ruth Arlene Fluck Dudgeon, 'Women and Higher Education

in Russia, 1855–1905' (The George Washington University, diss., 1975); G. I. Shchetinina, *Universitety v Rossii i ustav 1884 goda* (Moscow: Nauka, 1976); A. E. Ivanov, '[Review]: G. I. Shchetinina. Universitety v Rossii i ustav 1884 goda. (Moscow: Nauka, 1976), 232 s., tirazh 2000…' *Voprosy istorii* 4 (1978): 133–136; A. E. Ivanov, *Vysshaâ shkola Rossii v konce XIX – nachale XX veka* (Moscow: In-t istorii SSSR AN SSSR, 1991); Û. S. Vorob'eva, *Obshchestvennost' i vysshaâ shkola v Rossii v nachale XX veka* (Moscow, 1994); V. M. Budinov-Dybovskaâ, 'Pervye zhenskie politechnicheskie kursy' *Èlektrichestvo* 7 (1970): 91–93; B. A. Kurbatov, 'Zhenskie polytehnicheskie kursy i pervye zhenshchiny-inzhenery' *Priroda* 3 (1947): 76–80; S. L. Èvenchik, 'Vysshie zhenskie kursy v Moskve' in *Opyt podgotovki pedagogicheskih kadrov v dorevolûcionnoj Rossii i v SSSR: Sb. statej* (Moscow: MGPI im. V. I. Lenina, 1972), 4–99; Z. V. Grishina, 'Vysshee obrazovanie zhenshchin v dorevolûcionnoj Rossii i Moskovskij universitet' *Vestnik Moskovskogo universiteta*, Serie, Otd. 8 *Istoriâ* 1 (1984): 52–63; Z. P. Bogomazova, T. D. Kacenelenbogen and T. N. Puzyrevskaâ, 'Iz istorii pervogo zhenskogo vtuza' in *Pervye zhenshchiny-inzhenery* (Leningrad: Lenisdat, 1967), 1–21. We are grateful to our Russian colleague Sergej Snâtkov who provided us with a copy of the last book.

3. An important collection of documents relative to the pre-revolutionary history of this institution unfortuneatley is not available today because the municipal archives of St Petersburg have been closed to the public for several years. F. 871, 5 op., Central'nyj Gosudarstvennyj Istoricheskij arhiv Saint Petersburg (hereafter CGIA Archives Saint Petersburg).

4. Another example was the French Ecole Polytechnique Féminine, founded in 1925. Please see Annie Canel's in this volume. For Europe and the United States, see this volume *passim* and *History and Technology* 14 (1997), special issue.

5. The so-called *The Table of Ranks* law was introduced in January 24, 1722 and regulated state service. It was historically significant because it rationalized the administrative service: hierarchical divisions no longer depended on one's social origin, but on professional merit.

6. For the history of engineering education in Russia – and particularly the origins of its early nineteenth-century French inspiration – see also: D. Gouzévitch and I. Gouzévitch, 'Les contacts franco-russes dans le monde de l'enseignement supérieur technique et de l'art de l'ingénieur' *Cahiers du Monde russe et soviétique* 34, no. 3 (1993): 345–368; I. Gouzévitch, 'La circulation des modèles d'enseignement: problématique, méthodologie, chronologie' Paper presented at the XXth International Congress of History of Science, Liège, July, 20–26, 1997; her 'L'Institut du Corps des ingénieurs des voies de communication de Saint-Pétersbourg: Des modèles étrangers à l'école nationale: 1809–1836' *La formation des ingénieurs en perspective: transfert des modèles et réseaux de sociabilité: XIXe – XXe siècle* (Paris: L'Harmattan, 1999, forthcoming); her 'École Polytechnique comme modèle de référence: L'École polytechnique et l'internationale: bilan et perspectives (22.10. 1998): Bilan du colloque (Palaiseau: École Polytechnique, 1999, in press); S. L. Èvenchik 'Vysshie zhenskie kursy v Moskve' in *Opyt podgotovki pedagogicheskih kadrov v dorevolûcionnoj Rossii i v SSSR: Sb. Statej* (Moscow: MGPI im. V. I. Lenina, 1972), 4–99; Satina, 'Obrazovanie zhenshchin'; Grishina, 'Vysshee obrazovanie zhenshchin.'

7. One of the first female gymnasia, sponsored by Empress Maria Federovna, opened in Saint Petersburg in 1858.

8. Satina, *Education of Women in Pre-Revolutionary Russia*, 1.

9. According to Tishkin G.A. *Zhenskij vopros v Rossii: 50–60-e gody XIXv* (Leningrad.: Izd-vo Leningradskogo un-ta, 1984), 138.

10. The most well-known literary image of such a nihilist woman was portrayed by Ivan Turgenev in his famous novel *Fathers and Children* (Evdoxia Kukshina).

11. For higher education classes for women, including pedagogy and commerce, see: Satina, 'Obrazovanie zhenshchin'; Dudgeon, 'Women and Higher Education'; Ivanov, *Vysshaâ shkola Rossii*, 361–362; Vorob'eva, *Obshchestvennost' i vysshaâ shkola*; K. Shohol' 'K voprosu o razvitii vysshego zhenskogo obrazovaniâ v Rossii' *ZhMNP* 40 (Novaâ seriâ) 8, Otd. 3 (August 1912): 153–195; 44 (Novaâ seriâ) 3, Otd. 4 (March 1913): 1–36; 46 (Novaâ seriâ) 7, Otd. 4 (July 1913): 1–58; 'Obzor deâtel'nosti Ministerstva narodnogo prosveshcheniâ za 1879, 1880 i 1881 goda: Zhenskie gimnazii i progimnazii' *ZhMNP* 254, Otd. 1 (November 1887): 94–99; 'Izvlechenie iz vsepoddannejshego otcheta g. Ministra narodnogo prosveshcheniâ za 1875 god: Zhenskie gimnazii i progimnazii' *ZhMNP* 193, Otd. 1 (1877): 70–74; '… za 1867 god: Zhenskie uchilishcha' *ZhMNP* 142, Otd. 1 (March 1869): 62–66; 'Prilozhenie ko vsepoddannejshemu otchetu Ministra narodnogo prosveshcheniâ za 1867 god' *ZhMNP* 143, Otd. 1 (1869): 52–62; 'Izvestiâ o deâtel'nosti i sostoânii nashih uchebnyh zavedenij: Zhenskie uchilishcha' *ZhMNP* 143, Otd. 4 (1869): 115–125; Ibidem *ZhMNP* 151, Otd. 4 (1870): 164–178; 'O poseshchenii pansiona A. A. Staviskoj Eâ Imperatorskim

Vysochestvom Velikoû Knâgineû Aleksandroû Iosifovnoû' *Ibidem*, 266; 'Ob otkrytii novyh uchebnyh zavedenij' *Ibidem*, 267–268; 'O raspredelenii uchashchihsâ po klassam i kursam v uchebnyh zavediniâh vedomstva Ministerstva narodnogo prosveshcheniâ' *ZhMNP* 190, Otd. 4 (1877): 97–104; 'Izvlechenie iz vsepoddannejshego otcheta Ministra narodnogo prosveshcheniâ za 1868 god: Zhenskie uchilisha, chastnye uchilishcha i domashnee uchenie' *ZhMNP* 147, Otd. 1 (1870): 46–51; *Zhenshchiny revolûcionery i uchenye*, edited by F. P. Dashevskaâ *et al.*, (Moscow: Nauka, 1982); È. P. Fedosova, *Bestuzhevskie kursy-pervyj zhenskij universitet v Rossii: 1878–1918* (Moscow: Pedagogika, 1980); *S.-Peterburgskie vysshie zhenskie kursy za 25 let: 1878–1903: Ocherki i materialy* (Saint Petersburg 1903); 'O deâtel'nosti Vysshih zhenskih kursov v Kazani za 1877–78 uchebnyj god' *ZhMNP* 198, Otd. 4 (1878): 102–107; 'Pravila o pedagogicheskih kursah dlâ prigotovleniâ uchitelej i uchitel'nic nachal'nyh uchilishch: utverzhdeny g. ministrom narodnogo prosveshcheniâ 31-go marta 1900 goda' *ZhMNP* 329, Otd. 1 (June 1900): 83–86; 'Cirkulârnoe predlozhenie Ministerstva narodnogo prosveshcheniâ popechitelâm uchebnyh okrugov o pedagogicheskih kursah dlâ prigotovleniâ uchitelej i uchitel'nic' *Ibidem*, 92–93; 'Odesskie vysshie zhenskie kursy v 1909 godu' *ZhMNP* 300, 11, Otd. 4 (November 1910): 1–18; 'O kursah dlâ podgotovleniâ uchitelej i uchitel'nic srednih uchebnyh zavedenij, May 1, 1912, no 18387, Cirkulâr Ministerstva narodnogo prosveshcheniâ (hereafter Cirkulâr),' *ZhMNP* 40 (Novaâ seriâ) Otd. 1 (August 1912): 155–156; 'Ob otpuske iz gosudarstvennogo kaznachejstva sredstv na vydachu posobiâ vysshim zhenskim kursam v gorode Har'kove: [Uzakonenie, July 4, 1913]' *ZhMNP* 47 (Novaâ seriâ) Otd. 1 (October 1913): 41–42; 'Ob ustanovlenii zvaniâ uchitelâ srednih uchebnyh zavedenij, ob uchrezhdenii odnogodichnyh kursov dlâ podgotovleniâ uchitelej i uchitel'nic srednih uchebnyh zavedenij i kratkosrochnyh kursov dlâ uchitelej i uchitel'nic teh zhe uchebnyh zavedneij, a takzhe o nekotoryh izmeneniâh v ustave sih kursov i v uzakoneniâh o porâdke priobreteniâ zvaniâ uchitelâ (uchitel'nicy) srednih uchebnyh zavedenij: [Uzakonenie, July 3, 1914]' *ZhMNP* 54 (Novaâ seriâ) Otd. 1 (November 1914): 7–12; 'Barac, Semen Moiseevich' in *Evrejskaâ ènciklopediâ* 3 (Saint Petersburg: Ob-vo dlâ nauch. evr. izd-nij; Brokgauz-Efron, s.d.), 790–791.

12. Scholars of institutional history have entirely ignored this issue. Women are simply absent from their accounts, even for the Soviet period.

13. For the history of this institution, see Krutikov, 'Vospominaniâ Inzhenera Putej Soobshcheniâ'; Kurbatov, 'Zhenskie polytehnicheskie kursy'; F. 871, 5 op., CGIA Archives, Saint Petersburg; *Ibidem*, op. 1: *Predislovie k fondu* Otd. 871, p. 1; D. Gouzévitch and I. Gouzévitch, 'Pervyj v mire zhenskij VTUZ' *Nash put'* (Leningrad, LIIZhT) 28 (September 28, 1984): 4; D. Gouzévitch, 'Zhenskij Politehnicheskij institut' in *Sankt-Peterburg, Petrograd, Leningrad: Ènciklopedicheskij spravochnik* (Moscow: BRÈ, 1992), 209; 'Zhenskie Politehnicheskie kursy' in *Ves' Peterburg na 1908 god* (Saint Petersburg, 1908); *Ibidem* '… na 1909 god' (Saint Petersburg, 1909); 'S.-Peterburgskie vysshie zhenskie Politehnicheskie kursy' in *Ves' Peterburg na 1914 god* (Saint Petersburg, 1914): 624; 'Petrogradskie vysshie zhenskie Politehnicheskie kursy' in *Ves' Petrograd na 1915 god* (Saint Petersburg, 1915), 513; 'Petrogradskij zhenskij Politehnicheskij institut' in *Ibidem* 'na 1916' (Saint Petersburg, 1916), 538; *Ibidem* 'na 1917' (Saint Petersburg, 1917), 577; 'Vtoroj Petrogradskij Politehnicheskij institut' in *Ves' Petrograd na 1923 god* (Saint Petersburg 1923), 29; *Ibidem* in *Ves' Leningrad na 1924 god* (Leningrad, 1924), 59; D. Gouzévitch, 'Pervyj zhenskij VTUZ: K 80-letiû Vtorogo Petrogradskogo Politehnicheskogo instituta' Leningrad, manuscript, 1986, authors' Personal Papers; N. Belelûbskij, *Petrogradskij Zhenskij Politehnicheskij institut* (Saint Petersburg, 1915 [Reprint]); V. N. Pâseckij, *Ocherk vozniknoveniâ S.-Peterburgskogo zhenskogo tehnicheskogo instituta* (Saint Petersburg, 1905).

14. During her youth, Pzaskov'â Naumovna Arian belonged to the young people's popular democratic movement of the 1870s ('narodnichestvo'). In 1899, she started publishing an almanac 'First Women's Calendar' ('Pervyj Zhenskij Kalendar'), one of the earliest publications advocating women's equality. She was also member of numerous societies and associations. Bogomazova, 'Iz istorii pervogo zhenskogo vtuza.'

15. B.A. Kurbatov, 'Zhenskie polytechnicheskie kursy i pervye zhenschchiny-inzhenery' 3 *Piroda* (1947): 76–80.

16. According to Pâseckij, an active participant in these events, Arian's 1902 presentation at the RTO was followed by the establishment of women's drawing courses in Moscow and in Saint Petersburg. Pâseckij, *Ocherk vozniknoveniâ…*, 6.

17. Here we are probably dealing with Baron David Goracievic Gincburg (1857 -after 1910), a representative of Jewish family which succeeded in uniquely combining its Jewish faith while at the

same time belonging to the titled Russian nobility. Ivanov, *Vysshaâ shkola Rossii*, 370; *Evrejskaâ ènciklopediâ* 6 (Saint Petersburg: Ob-vo dlâ nauch. evr. izd-nij; Brokgauz-Efron, s.d.), s.v., 'David Goracievic Gincburg.'

18. The male institutions of reference were: Institute of Mining Engineers, Higher Artistic Scool, Imperial Academy of Arts (for Building and Architectural departments), Technological Institute and Institute of Electrical Engineering (for departments of Chemistry and Electromechanics). Kurbatov, 'Zhenskie polytechnicheskie kursy', 77; Belelûbskij, *Petrogradskij Zhenskij Politehnicheskij institut*, 51.

19. Program read as follows: 'not to ask for any financial support from the government nor any professional rights for students until results are obtained which demonstrate the useful activity of graduates in the field of the sciences and technology,' Kurbatov, 'Zhenskie polytechnicheskie kursy,' 77).

20. Ibid.

21. Belelûbskij, *Petrogradskij Zhenskij Politehnicheskij institut*, 58.

22. Z. P. Bogomazova, T.D. Kacenelenbogen, and T.N. Puzyrevskaâ, *Iz Istorii pervogo zhenskogo vtuza Pervye zhenshchiny-inzhenery Lenigrad* (1967), 14–15.

23. Kurbatov, *Petrogradskij Zhenskij Politehnicheskij institut*, 78; Gouzévitch, 'Pervyj zhenskij VTUZ', 4.

24. Krutikov, *Petrogradskij Zhenskij Politehnicheskij institut*, 207–208.

25. Orthodox parish clergymen not only could but were obliged to get married.

26. Bogomazova, Iz Istorii pervogo zhenskogo vtuza in *Pervye zhenshchiny-inzhenery*.

27. Ibid, 18 with reference to: V. I. Lenin, Polnoe sobranie sochinenij, Vol. 11, p. 351.

28. As quoted in Bogomazova, Iz Istorii pervogo zhenskogo vtuza in *Pervye zhenshchiny-inzhenery*, 20.

29. Belelûbskij, *Petrogradskij Zhenskij Politehnicheskij institut*, 43; Predislovie k fondu no. 871, p. 1., F. 871, op. 1, CGIA Archives, Saint Petersburg.

30. For the Complete Collection of Decrees and Orders of the Workers' and Peasants' Government *Sobranie uzakonenij i rasporjazhenij rabochego i krest'ânskogo pravitel'stva* 28 (1918), p. 367, Predislovie k fondu no. 871, p. 1, F. 871, op. 1, CGIA Archives, Saint Petersburg.

31. *Ves' Petrograd na 1923 god* (Saint Petersburg, 1923), s.v., 'Vtoroj Petrogradskij Politehnicheskij institut'. See also for the years following in the series.

32. For the biographies, see: Bogomazova, Iz Istorii pervogo zhenskogo vtuza in *Pervye zhenshchiny-inzhenery*, 201.

Annie Canel*

5. Maintaining the Walls: Women Engineers at the École Polytechnique Féminine and the Grandes Écoles in France

In 1925, a bright young female engineer named Marie-Louise Paris founded an engineering school for women, the Women's Polytechnic School (*École Polytechnique Féminine*). Herself a graduate from the electro-technical institute in Grenoble, this ambitious woman was concerned with the need to provide women with formal scientific and technical training so that they could gain access to professional engineering. She also wanted to help them advance their careers. She achieved her ambition. In 1938, the Commission for the Title of Engineer included the Women's Polytechnic School on the list of institutions that could grant the title of engineer to their graduates. Since the end of the Second World War, all sorts of highly responsible jobs have been offered to the school's graduates, a fact that underlines the impact of this woman and her ideas.

The underlying premise of the Women's Polytechnic School was the belief that women would not succeed in engineering if they go through the male engineering schools. Setting up a separate school of engineering for women that would offer the same education as its traditional male counterparts was regarded as a more promising project than striving for admission to existing institutions. Young women who entered the major established schools for engineering (*grandes écoles*) after the Second World War, in contrast, felt they would have a better chance at professional equality if they went through the same schools as their male colleagues. Both strategies remained in play through the twentieth century, alternately important according to shifts in the political and social landscape.

The history of female engineers in France shows the link between women's access to engineering and industrial performance. It further illustrates the significance of the period after the first World War in the realm of engineering, when women, pushed forward in the belief that education and professional activity were major factors in the struggle for emancipation.

The system of higher scientific and technical education in France is probably one of the most centralized and hierarchical in the industrial world. Its main axis of organization is the differentiation between upper-level and lower-level institutes of engineering. The former focuses on science and mathematics, while the latter is more concerned with practical training in connection with industrial activity. There are four major groups of schools and engineering institutes.[1] The traditional *grandes écoles*, such as the *École des Ponts et Chaussées*, the *École des Mines*, the *École de Génie*

Militaire, and *the École Polytechnique*, were established in the eighteenth century. Their curricula largely consisted of science and mathematics, and their training methods followed a military line. Other *grandes écoles*, including the important schools of arts and crafts (*Écoles des Arts et Métiers*) were founded in the early nineteenth century to produce engineers and technicians for the new industries. National institutes of engineering, associated with the science faculties, emerged at the end of the nineteenth century. As research-oriented institutions aimed at assisting regional industry, they went into rapid decline during the economic crisis of the 1920s and 1930s. The new *grandes écoles*, such as the School of Physics and Chemistry (*École Supérieure de Physique et de Chimie*), the School for Electricity (*École Supérieure d'Électricité*), and the School of Aeronautics (*École Supérieure d'Aéronautique*), which were more or less equivalent to the German *Technische Hochschulen*, were established in the years before the First World War to provide training in applied science and engineering. They were quick to upgrade their curriculum by raising their level of mathematics, by drawing students from the upper social strata, and by elevating their position within the educational hierarchy.[2]

My focus in this chapter is on the traditional *grandes écoles*, the uppermost branch of the educational system, which had a key role defining the boundaries of the world of engineering. They were the last educational establishments to admit women – which happened only in the 1960s. In addition, I will discuss the Women's Polytechnic School, the sole engineering institute for women which reached a higher scientific and technical level. Comparing the founding of the all women's engineering institution and women's entry into the traditional *grandes écoles* illuminates the double strategy by which French women forced their way into engineering. Russia also developed a single-sex institution for women's engineering education.[3] There were some striking similarities between the French and Russian systems of higher scientific and technical education from the beginning, not least because Russia took France as its model in this regard.[4] Both systems developed into highly centralized and hierarchical organizations, and their lack of flexibility forced women to seek to create separate institutions to enter the male world of engineering. The fact that a women's movement emerged earlier in Russia partly helps to explain why the women's school in Saint Petersburg was founded twenty years before the French polytechnic institute.

In such rigid hierarchical systems, establishing separate institutions for women may have been the only possible alternative. Moreover such institutions have been highly supported by influential male professors and eminent industrialists in both countries who viewed those institutions as a means to improve industrial performance. Both schools appeared to address a pressing need, they made a strong connection between training and industry whereas industrial needs were only partly met by the *grandes écoles*.

The manner in which women forced their way into engineering greatly depended on the boundaries that were drawn around and within the world of engineering. Since they were not allowed to enter engineering through the regular main gates, women devised different strategies to find their way in, according to the specific boundaries established in the different countries. When industrial technology became a viable route to social and political legitimacy, women in Germany tended to concentrate their efforts on academic performance. In France, however, given that social and political legitimacy was closely linked to 'anti-utilitarian' science, women advocates of female engineering devised a strategy based on training that was directly linked to industrial needs.[5]

Defining the Engineering World

The first engineers in France were state engineers. From the late seventeenth century on, they were organized into state engineering corps. Most were also part of the military, whose influence on engineering education was considerable. Their emergence on the scene and their designated roles were closely linked to state interventionist policies. Engineers were to regulate and direct the country's entire technical infrastructure, in order to further economic growth and ensure better living standards. They worked in all the sectors in which state intervention and regulation were considered necessary.

Various engineering corps were established at different moments, in line with the nation's needs for technological change and economic growth. Each engineering corps stood in charge of the design and improvement of a specific set of techniques. The state corps of civil engineering (*Corps des Ponts et Chaussées*) was founded along military lines in 1716 by Jean-Baptiste Colbert, the General Inspector of Finance under Louis XIV. (Figure 28)

It was responsible for the assessment of bridges and roads as well as of the navigation of rivers. One of the main tasks of the state engineers was to supervise entrepreneurs, since contractors were considered to lack appropriate professional expertise in civil engineering. Contractors were furthermore expected to be more concerned with their private interest than with general public interests. State engineers, in contrast, saw themselves as serving public interests. Through the improvement of the national infrastructure, state engineers were expected to serve the country's trading needs.[6] (Figure 29)

Another area regarded as crucial for state power was mining. The mining corps (*Corps des Mines*) was established in 1783 to lend expert technical support to the establishment of public policies on mining. State engineers were responsible for developing and controlling the mining sector and the new iron industries. Their main task was to protect what were considered to be scarce, non-renewable resources, resources that were entirely in state

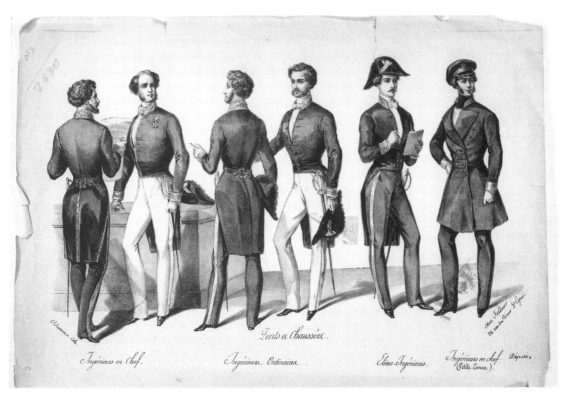

*Figure 28. Drawing of state civil engineers (*Ingénieurs des Ponts et Chaussées*) in uniforms according to their administrative rank in the nineteen century. The depicted uniforms – illustrating French engineering's high status and close ties to the military – did not change until the late nineteenth century. Permission and courtesy of École Nationale des Ponts et Chaussées, Media Library, Paris, France.*

hands at the time. In order to preserve these limited resources for future generations, waste prevention was among the top priorities. Engineers were further expected to encourage innovation and to provide support to those sectors of industry believed to maintain only short-term perspectives, and to be merely concerned with running milling activities. Engineers were to safeguard and guarantee long-term economic policies. Moreover, the state considered private industry to be unscrupulous, to lack any sense of moral integrity. The expertise of state engineers was held to be necessary for ensuring both quality and excellence.[7]

In due course, technical schools were instituted to educate state engineers of particular corps. Among these were the School of Civil Engineering (*École des Ponts et Chaussées*), established in 1747, the School of Army Engineering (*École du Génie Militaire*, founded 1748), the School of Mining (*École des Mines*, founded 1783), and the School of Telegraph Engineering (*École Supérieure de Télégraphie*, founded 1878).[8] Each school was placed under the administrative supervision of a specific ministry.

Figure 29. "La construction d'un grand chemin" oil on canvas by Joseph Vernet (1774). State civil engineers supervising a construction site from their high position on horse back. Unlike engineers elsewhere, French engineers maintained their elite position as high rank civil servants until the present day. Oil painting. Permission and courtesy of The Louvre Museum, Paris, France. Photo RMN-R.G. Ojeda

The French Revolution strengthened the centralization of recruitment and education of state engineers. The *École Polytechnique*, established in 1794, served as a preparatory school for all state corps from its founding day, and rapidly became 'the decisive eye of a needle opening access to the corps of state engineers.'[9] The training it offered focused strongly on mathematics, and was aimed at providing engineers with the basic knowledge and theoretical tools they would need when they would go on to work in the technical departments of the state administration. The technical schools became 'schools of application' (*écoles d'application*), in which state engineers were trained to work in specific fields. The ministries determined the criteria for recruitment, educational requirements, and the promotion of individual engineers.[10] The military status of the state corps was further reinforced under Napoleon's reign, when the *École Polytechnique* was taken over by the Ministry of Army. Henceforth, a period of military training was added to the regular curriculum. Today, different paths beside the *École Polytechnique* lead to the state corps, but the alternatives are few and entrance requirements high. The *École Polytechnique* has maintained its monopoly on state engineer recruitment.[11] This has ensured uniformity in terms of training, but also in the background and values of state engineers.

The first women to attend the *grandes écoles* were 'civil engineers' (*ingénieurs civils*), also known as civilian engineers or those trained to work in private industrial sectors. This specific title was meant to differentiate them from the state engineers, who were educated at the military *École Polytechnique*. The emergence of this new category of engineers mainly resulted from the economic and industrial changes occurring in the nineteenth century. At the turn of the nineteenth century, French industry was confronted with relatively slow industrial progress, at least compared to that in Britain in the same period. The United Kingdom's rapid industrial growth was widely attributed to its civil engineers, who were in charge of new technological developments. The education of the English civil engineers centered on experimental approaches and pragmatic learning situations: direct involvement in engineering projects inside factories was considered the most efficient way to implement new technologies. The new French civilian engineers were expected to become the main collaborators of entrepreneurs. They took up the consulting or managerial positions sorely needed by private industries, and were to provide the 'remedy to the disease' from which French industry was suffering.[12]

The *École Centrale des Arts et Manufactures* (School of Arts and Manufactures), later simply referred to as the *Ecole Centrale*, originally a private initiative of three scientists and industrialists, one of whom, Thomas Olivier, had graduated from the *École Polytechnique*, was founded in 1829. It was one of the first major engineering schools to offer training to civilian engineers. Women began to enter this school as early as 1918, while the first female state engineers were allowed to attend the *École Polytechnique* as late as 1972. Unlike that of the *École Polytechnique*, the curriculum at the *École Centrale* did not concentrate on mathematics and theoretical science. Instead it developed a practical orientation, being largely concerned with technology and applied science. 'Civil engineers' were supposed to forge a new discipline called 'industrial science,' in which science was consistently linked to practical engineering projects, as distinct from the direct application of theoretical insights to industrial problems. Industrial science emphasized specialized knowledge of technical processes and industrial operations[13]. The *École Centrale* sought to differentiate itself from the theoretical approach prevailing at the *École Polytechnique* by stressing engineering over mathematics, and applied over theoretical science. This attempted demarcation can also be discerned in the attempt of the *École Centrale*'s graduates to establish 'civil engineers' as a recognized professional body in 1848, when they founded the *Société des Ingénieurs Civils*. In 1857, the *École Centrale* was taken over by the Ministry of Commerce and became a public institution. In the late nineteenth century, its teaching staff attempted to upgrade their profession by shifting admission requirements and training programs toward the more science-oriented and high-status model of the *École Polytechnique*, shaping the identity of the French engineer.[15]

The decision to admit women to the *École Centrale* in 1917 was preceded by heated debates. Advocates of women's right of admission invoked international examples, such as Hungary and Russia.[16] In 1917, Bodin, a member of the *École Centrale*'s Board of Directors mentioned that he visited an engineering women's school, in Saint Petersburg, in 1906.[17] At the same meeting, Compere stressed that, during the Turin Exposition, Italy, he was struck to see a young Russian woman among the leaders in the electrical area.[18] The idea that a woman could be educated in engineering proved difficult to accept. Advocates argued that the country had an economic stake in allowing women to enter into industry. The Russian example legitimized this as an objective for the school's presidents. In fact, in the years between the wars, nine percent of the women graduating from the *École Centrale* – that is, five of fifty-five – were foreign students, including two from Russia and one from Italy. Women advocates also stressed that some women had developed a particular interest in science and engineering – rather than a masculine concern with professional careers – as a way to contain the threat of women's employment. The first female students graduated from the *Ecole Centrale* in 1921. Their number increased through the following years, but decreased in the period between the wars when the economy shrank considerably. In 1921, there were seven female graduates; in 1922 there were nine. From 1923 onwards, the number gradually decreased, to level off at an average of one to four women until the start of the second World War. (Figure 30)

About eighty percent of the women who graduated from the *École Centrale* in the interwar period gained professional experience. They worked in various branches of industry, including electrical and mechanical equipment, administration, education, chemical and automobile manufacture, and transportation. Still, although their formal education entitled them to the same positions as their male colleagues, meaning jobs of greater responsibility than most of their female contemporaries, women's main roles were as substitutes. This relegated them to the lower positions within the engineering hierarchies. Women themselves kept their career objectives in line with this idea. Seeking acceptance and promotion in the world of engineering, most women tried to make their male colleagues forget that they were female: 'To be fair, I think I received much help and support, which entailed more obligations than rights for me. I don't call myself a feminist. Our power is the more effective when it is less visible.'[19]

Women were also taken on by the major traditional *grandes écoles*, such as the *École des Ponts et Chaussées* and the *École des Mines*, as civilian engineering students. These schools which only served as schools of application attended by state engineers until then, also offered programs to the new civilian engineers since the middle of the nineteenth century.[20] Women, however, were not admitted into these schools until the 1960s,

J. David et E. Vallois, phot.-édit. 99, rue de Rennes - Paris

Figure 30. Civil engineering class of 1924 of the École Centrale *among whom two women students, Andrée Sazerac de Forge and Thérèse Danger. Grooming its graduates for engineering jobs in industry rather than state service, the* École Centrale *welcomed women long before the* Ecole Polytechnique *which was closely linked to the French state and the military. Reproduced with Permission and Courtesy of École Centrale, Paris, France.*

half a century after a woman first enrolled at the *École Centrale*. The *École des Ponts et Chaussées* admitted a female student in 1959, and the *École des Mines* followed suit in 1970.

A major step in women's higher education and engineering was realized with the passing of the 'Camille See law' in 1880, which led to the establishment of secondary girls' schools. These schools offered various compulsory courses as well as special training programs for instructors. The law did not, however, allow women to take the *baccalauréat*. The school certificate hence did not open the door to professional occupations, nor did it allow girls to apply to institutions of higher learning or universities. The few women who, in the early twentieth century, did obtain a *baccalauréat* in science had usually taken special, private courses. It was not until 1924 that secondary education for girls was made equal to that of boys.[21] What is more, preparatory classes in science, which opened the way to the *grandes écoles*, remained exclusive to elite men's schools even after the Second World War . These two-year programs were heavily focused on mathematics and physical science. To this very day, they remained a crucial barrier to women who try to gain access to higher engineering studies.

During the second half of the nineteenth century, female students slowly began to find their way into university, taking their places in both lecture halls and laboratories. In subsequent years, their number started to increase rapidly. In 1907, 1317 women students were registered at the Sorbonne, 829 foreigners among them. Schools for engineering, however, especially the major institutions, continued to be very hostile towards women.

In this historical context, women developed two central strategies to force their way into professional engineering that persisted throughout the twentieth century. When they started their training during the war, they claimed their competence as men's substitutes or as 'engineering aides.' At the same time, however, when they gained access to the same formal educational opportunities as their male colleagues, they sought professional employment at equal levels.

The French government began to create incentives to admit women to the engineering schools toward the end of the First World War to replace the men who had gone off to the front. In November 1915, a government instruction was issued to encourage industrialists to recruit women when possible: 'The feminist conclusion of this war is that women are able to succeed in any job, whether intellectual or physical. We'll just have to give them proper opportunities, which we will because we are forced to.'[22] Accordingly, in 1917, faced with a growing need for engineers and dwindling numbers of male students, the Ministers of Commerce and Agriculture granted women admission to several prestigious schools of engineering, including the *École Supérieure d'Electricité*, the *École Centrale*, the *École de Physique et Chimie de Paris*, (where Marie Curie had her laboratory), and the *Institut National d'Agronomie*.

At first, women mainly attended the new schools of engineering in the provinces, for example in French cities of Marseille, Rouen, Rennes, and Strasbourg, which had admitted women from the start.[23] Most of these were private schools that trained engineers for the new industrial sectors.

Among the first female institutions of engineering was the School of Female Technical Education, founded by a woman named Hartzfeld and an electrical engineer Jean Laurent at the end of the First World War. The program provided female students with training in industrial drafting, and was expected to open up job opportunities in different branches of industry. Drafting was a new discipline for women and helped to create new roads of entry for women into industry, especially after the Second World War. Still, although women were allowed to move into drafting departments, they were not entitled to assume management positions. What is more, preparing students for the job market was a new idea in female education, since, up until the First World War, the education of girls was completely disconnected from professional objectives. The new school offered three programs, one for engineering aides and draftsmen, one for chemistry aides and chemists, and one for medical secretaries and X-ray assistants. It was financed with state funds and grants awarded by the Society of Electrical Engineers. In its first year, it had eight women graduates. Three years later, there were one hundred.[24]

The example of social engineering reflects the critical role of newly emerging technical fields in the process of women's professionalization and emancipation even more clearly, and also attests to the importance of those service-oriented professions that were considered to be particularly suited to women. In 1917, the Social School for Women Factory Superintendents was founded. Students were expected to have taken the *baccalauréat* and were offered a six-month training program. In order to upgrade the program, this was extended to one year at the end of the war, and again to three years in 1938. The new specialization was needed in industry when Taylorism began to catch on in France, and a scientific mode of labor organization was introduced into factories. Social engineers were expected to further contacts between workers and managers and to improve the workers' living and employment situations. It was mainly women who worked as social engineers. This new technical occupation opened up avenues for them both in executive positions and technical work in industry.[25] About one hundred female social engineers were at work in 1928. This number had doubled by 1937. About half of them were offered jobs as superintendents in manufacturing departments, while the other half were recruited by both private and public administrative bodies. Unlike most women workers, who were forced to quit their jobs at the end of the war in order to clear the way for returning male engineers, female social engineers were not forced back into the home. This was because the service-oriented nature of their new profession was considered to be in line with contemporary notions of femininity.[26]

Another illustration of how women forced their way into technical fields dominated by men is provided by the famous School for Higher Commercial Studies for Young Girls *École de Haut Enseignement Commercial pour Jeunes Filles).* In fact, this may be the perfect example for showing how women found new opportunities through the expansion of industry. Louli Sanua, an ambitious young Jewish woman who had obtained her *baccalauréat* in 1904, established the School for Higher Commercial Studies, one of the first higher trade schools for women, in 1916.[27] Her father had left his native Egypt to work as a journalist in Paris, while her mother hailed from Alsace, in Eastern France. In 1929, she married Jean Milhaud, founder of the General Commission of Scientific Organization *(Commission Générale d'Organisation Scientifique,* CEGOS), with whom she had two children.[28] In 1925, Sanua attended the Sixth International Congress of Women in Washington DC, USA, where she met several teachers as well as business and industrial executives involved in the technical training of women.[29] An instructor of young girls herself, Sanua was particularly keen on helping women to develop high-level professional skills in order to increase their chances to pursue successful careers. In 1909, she founded the Association of Academic Women Instructors (*Association des Institutrices Diplômées*), and, in 1916, the School for Governesses (*École de Gouvernantes*). Sanua did not follow the older strategy that was based on the Victorian argument that women in trade would serve as substitutes for men. Aware of the need for qualified personnel in sales and marketing, she suggested to the Chamber of Commerce that women be allowed to enter the male School of High Studies in Commerce. This suggestion was turned down, but Sanua decided to found a school for women in the same field.[30] The school was organized along the same lines as its male counterpart, offering the same courses. 'I felt the need,' she said, 'to provide women with an education that would enable them to earn their own living under any circumstances, so as to maintain their personal and financial independence. The situation that developed after the war confirmed my convictions in this respect, and formed the inspiration for my challenging work: educating women for the new era was of the greatest urgency.'[31] The school particularly attracted women who did not want to become instructors or teachers.[32] It was located at the National Institute of Engineering (*Conservatoire National des Arts et Métiers*, CNAM), the unique higher technical institution offering programs in engineering to workers and production supervisors. It was also entitled to grant the title of engineer. Two specific characteristics differentiated the national engineering institute from the *grandes écoles*. First, its training programs were based on practical learning situations and applied knowledge, and, second, its students strongly rejected elitism.[33] The location of women's School for Higher Commercial Studies at the national engineering institute proved an important factor in paving the way for women's training in engineering a few years later.

From the Women's Electromechanical Institute to the École Polytechnique Féminine: An outsider in French Higher Engineering Education

In 1925, the first engineering school for women, the Electro-mechanical Institute for Women (*Institut électromécanique féminin*), was founded by a brilliant young engineer Marie-Louise Paris (1889–1969), according to André Grelon.[34] Her father, a military officer, died when Marie-Louise was only nine years old. Consequently, she was encouraged to continue her education and take up a profession. She obtained her *baccalauréat* in science in Besançon, a small town in eastern France. She then moved to Paris to pursue her studies at university, and graduated in science from the Sorbonne. Marie-Louise Paris and her sister Helene were then admitted to the Sudria school, one of the most renowned private schools in mechanical and electrical engineering, which opened its doors to women during the First World War. Both sisters graduated in 1921 and went on to apply with the Institute of Electrotechnics in Grenoble. They graduated from the Institute within the year, together with two other female and 500 male students. Helene married and stayed in Grenoble, whilst Marie-Louise decided to move back to Paris. During the 1920s, she started working as an engineer in various laboratories and drafting offices, moving among several jobs in the first months of her professional career. Information about her life is rather scant, however, and so it unfortunately does not allow for a more detailed analysis of her early professional experiences.

Marie-Louise Paris' attempt to establish an institute of higher education was not unique in the French context. Similar initiatives were taken by other individuals. In contrast to the university system, the French system of *grandes écoles* were surprisingly free from state regulation. Any exceptional individual with some money, personal relationships, and a strong personality, could establish a school of engineering. Marie-Louise Paris was therefore following an example set by any number of men.[35] As in Russia, the foundation of the women's institute was strongly encouraged by eminent male teachers. Among these were Gabriel Koenigs, professor at the University of Science in Paris and at the CNAM since 1923; Labbé, President of the Technical Education Department at the Ministry of Public Education; Paul Langevin, president of the School of Physics and Chemistry (*École de physique et chimie*); Léon Guillet, president of the *École Centrale*; Léon Eyrolles, president of the School for Public Works (*École des travaux publics*); Paul Appell, dean of the Academy of Science of Paris; Edouard Branly, physicist; Louis Barbillon, Paris' own former professor in Grenoble.[36] She made lengthy attempts to find a location for the new school, and was eventually allowed to use a number of classrooms at the CNAM and at the Sorbonne. A letter she sent out on May 26, 1925, to the president of the CNAM, asking for classrooms, suggests the kind of obstacles she was facing and the strategies she employed to achieve her goal. It is therefore worth quoting at length.[37]

Dear Sir,

I thank you for your kindness and your encouragement. I understand the reasons why you hesitate to welcome me at your Institute. Nevertheless, when, knocking at the door of the Sorbonne, nobody will open it for me, I dare hope that you will do as much as you can to allow me to start up my 'Electromechanical Institute' that will benefit so much to women, next November.

I wrote to the Dean yesterday and asked for a meeting with him. As soon as I learn whether he can provide me with a conference room, I will inform you. My courses would take place either in the mornings or the evenings (preferably the first) for three hours a day and five days a week. I might be able to arrange teaching some courses at the CNAM and some at the Sorbonne. I will accept any combination, whatever the drawbacks... I ask for only one thing, and that is to begin to exist.

This project forms the response both to a dire necessity, the necessity for women to cope with the uncertainties of life, and a vocational need, the teaching vocation, which has always been so attractive to me. In addition it finds its roots in previous struggles, work, and investigations.

I have no doubt, Sir, that you will have understood and felt, like all the men of higher moral and intellectual standards, how important the women's perspective is, and that you will do your best to help me succeed with my project.

Yours faithfully,
M.L. Paris.

In the end, Paris limited the scope of her project to a part-time program. She was given permission by the president of national engineering institute (CNAM) to use some classrooms during the morning, while people with day jobs took evening classes. Like Women's School for Higher Commercial Studies, the Electromechanical Institute for Women highly benefitted from the CNAM's hospitality.[38]

Marie-Louise Paris' school opened its doors on November 4, 1925. The event was very favorably reviewed by journalists, who emphasized the eagerness and dedication of its founder. Although around this time a number of male engineering schools were admitting women, Marie-Louise Paris did not believe that this would open the way to successful careers for women engineers. She thought that women would not be able to enter a world that she considered to be suited only to men. For one thing, higher technical education required time and money, investments only men could afford. For another, according to the prevailing view of engineers as civil engineers working at building sites, professional engineering required physical strength. In Marie-Louise Paris' own words: 'As to professional engineering, are women – a few exceptions notwithstanding – capable of competing with men at the construction sites of railways and bridges, or in large corporations? I don't think so.' Her perception of women's professional roles limited the extent of their progress, for she added: 'Their abilities, on the contrary, naturally allow them to take up positions in drafting departments in which projects are conceived as well as in industrial laboratories.'

Paris also put great effort into reassuring the political, scientific, and industrial establishment by presenting female employment as a necessity since, in a modern society, women too had to cope with the uncertainties of life. The problems brought about by the war were still vivid in everyone's memories, and continued to determine how they thought. Paris stressed that her objective was not to 'encourage women's emancipation and take them away from their homes and the families they were expected to found; on the contrary, we wish to provide them with the required resources to found homes at all.'[39] Chaumat, a professor of electrical engineering at the CNAM, used the same argument to support the new school: 'In view of the fatalities of war, we must help all young women who wish to do so to take up professions in engineering without, by any means, pressuring them into leaving behind their more immediate duties, i.e. raising a family. Experiences at the *École Supérieure d'Electricité*, having allowed women in for a number of years now, demonstrate that the two activities can be successfully combined. Indeed, almost all the young female graduates – with the exception of two – are presently married and have children.'[40]

In the early years of its existence, only a few women graduated from the new Institute. Fewer than ten per year at first, their numbers would increase rapidly to more than twenty annually after the Second World War. The Institute offered a program consisting of theoretical courses, experimental team projects within various electronic branches, and visits to factories. In the early stages, three professors were doing all the teaching: Marie-Louise Paris herself, and two professors enticed from the male *grandes écoles* to teach drawing and mechanics. Paris' aim was to further female progress in the public sphere by integrating women's training into an industrial environment, rather than raising political arguments about equal opportunities. In contrast to the women's school of engineering in Saint Petersburg, which worked for political emancipation, the French institute based its strategy on women's professional development, and their entry into those niches of industry in which they would, thanks to appropriate training in engineering, easily prove their competence. Marie-Louise Paris unquestionably sought to meet the industrial needs of the times. She maintained close contacts with supportive industrialists. Her contribution to the Seventh Congress of Chemistry on 'women's careers in engineering and industry,' presented at the request of industrialists themselves, testifies to the fact that they had a reciprocal interest in the institute for women and Paris' ideas. (Figure 31)

Paris had to fight for classrooms again in 1932 when new training programs were implemented at the CNAM. The school's president considered using the rooms lent earlier to the Institute for Women. But Paris received support from the Minister of Public Education, and was eventually allowed to keep using the borrowed rooms.

In order to heighten the institute's prestige in 1933, Paris changed its name to the Women's Polytechnic School, and expanded its program to three years. The new name reflected new courses that had been introduced to meet demand in newly emerging fields of specialization. Students were then able to study electric lighting, telegraphy, and trade law. Paris showed enthusiasm for these new fields, especially for aerospace technology, by taking up flying lessons. With the support of the President of the Technical Department of Caudron-Renault Airplanes, she succeeded in creating a model airplane that was tested by the Ministry of Air Force.[41] In addition, the curriculum of the institute was expanded to include foreign languages such as English, Spanish, and German – the latter restricted to commercial and technical aspects only. These changes reflected Paris' concern with providing women with the education to allow them for new branches of industry and in international trade. The new name further implicitly refers to the prestigious *École Polytechnique*, as Marie-Louise Paris regarded the female engineers

Figure 31. Marie-Louise Paris (1889–1969) on the left, founder of the École Polytechnique Féminine (1925–1993), the all-women engineering institute established with support of some members of the French engineering elite. Paris is posing here in 1926 with the first women graduates in the court at the Engineering Institute of Arts and Trades, which hosted the women engineering's classroom. Similarities of both the engineering institutes for women and the national engineering cultures in France and Russia are striking. Permission and courtesy of École Polytechnique Féminine, Archives, Paris, France.

she trained as appropriate 'intellectual companions' for their counterparts at the *École Polytechnique.*

In 1923, when the job market was particularly tight, a state commission was established to regulate the engineering profession. It completed its task in 1934.[42] Although the Women's Polytechnic School did not appear on the first list of schools allowed to award the title of engineer, it was included in the 1938 list, since captains of industry, expecting another war, badly needed more engineers. During the Second World War, both women and industrialists appear to have sought women's labor as a substitute for men who had gone off to war. In 1939, the Women's Polytechnic School initiated a new program to train female aircraft technicians, an initiative strongly supported by the Minister of Air Force. About one hundred women enrolled. The program was, however, abolished in its early stages with the onset of war in 1940.[43]

Meanwhile the relationship between Marie-Louise Paris and the president of the CNAM deteriorated during the war. Louis Ragey began to complain about the Paris' independence and ambition, and suspected Paris of using the CNAM's lecture halls without official authorization. Furthermore he objected to her using the prestigious name of the CNAM (for example, by using the same letterhead) as a boost to the prestige of the women's institute.

Paris once again obtained financial support from the Minister of Public Education, and thus for some time managed to postpone a battle. The difficulties re-emerged, however, when the CNAM opened up new programs and faced a concomitant lack of classroom space, as a result of the damage done to various other several schools during the war. The president of the CNAM finally forced the *Women's Polytechnic School* to leave his institute. For almost ten years afterwards, the school led a nomadic existence, moving from the *École Centrale* to the *École Supérieure d'Electricité*, as well as through various Paris high schools. Finally its founder decided to sell her own property in order to buy a building in Sceaux, just outside Paris, where she relocated the institute in 1956. The Women's Polytechnic School has been at the same spot ever since.

Marie-Louise Paris proved to be something of a visionary, capable of anticipating the future needs of industrialists. She established the first classes in computer science, solid-state physics, and aerospace, always surrounding herself with competent, bright engineers and scientists, influential corporate executives, and distinguished public administrators. All of these helped her to secure a position in an otherwise hostile male professional environment. She was also a highly sociable person and ready to listen to others, but firm when it came to defending the Women's Polytechnic School, which she considered her personal territory.[44] She received strong support from various quarters, especially from industrialists whose needs for specialized professionals she recognized.[45] Maurice Berthaume, an eminent professor of aerodynamics, who held a management position at the major French aerospace corporation, *l'Aérospatiale*, supported Marie-Louise Paris and her institute from 1938 onwards, and even left his job to run her school after her death in 1969. Pierre Lhermitte, another distinguished professor of computer science, and the central force behind a radical reform of the (male) *École Polytechnique*, was also a strong advocate of Paris' work. Lhermitte's position as Executive at the *Société Générale*, a major national bank, was useful in enabling Paris to direct her school's programs towards industrial needs.[46] Her financial resources derived from tuition fees, the financial backing she secured from various sectors of industry, and her private capital. Another main source of income was the support she received from the parents of her students. In this way, the French case resembled the situation in Austria, where parents strongly advocated emancipation for their daughters and substantially contributed to it through education and professional development.[45] Paris always gave priority to high scientific standards over and above profit, which meant that she had to pay distinguished professors and maintain well-equipped laboratories. She nonetheless always managed not only to safeguard the school's existence, but also to keep up its academic standards. Once, facing a particularly severe financial crisis in 1966, she even called on President de Gaulle for help.[48]

Paris' success is all the more remarkable in view of the fact that she established a new school of engineering in a period of deep economic crisis, though her ideas had been formed when women were increasingly employable because of the absence of men during the war. By the time her project got off the ground, however, a growing numbers of engineers were facing unemployment. Paris nonetheless succeeded in safeguarding her institute and offering new job opportunities to her students by finding new niches, venturing into new sectors, and anticipating further industrial developments, rather than by trying to force her entry into established (male) bastions of engineering.

Not until the end of the Second World War did the *Women's Polytechnic School* begin to take off in terms of numbers of graduates and professional positions offered to women. Henceforth, and especially during the 1950s and the 1960s, the school's progress became even more closely linked to the process of industrialization and associated with the development of new industries, notably aerospace and electronics. Most of its graduates succeeded in finding jobs in these sectors. Because they were new, such fields opened new opportunities for female engineers as they offered a less discriminatory atmosphere than traditional male-dominated environments, and also because these new industries were swifter to adopt the new Taylor-based model of organization. In contrast to the less rigid system prevailing in traditional industries, the hierarchical structure of segmented work organizations offered clearly defined positions for technical aides and assistant engineers. This fitted nicely with Paris' strategy of preparing women for engineering employment. (Figure 32)

The 1950s and 1960s, however, were less favorable to her strategy of creating a separate institute that would become integrated into the male system of the *grands écoles*. State policies and social change encouraged the development of co-educational institutions. However, Paris again resisted the spirit of the times and preserved her separate institute for women. In the 1970s, the established *grandes écoles* admitted women students and championed integration and harmonization. But industrialists, finding the engineers they needed at the Women's Polytechnic School, once again protected Paris' work.

During the 1970s and 1980s, there was also growing pressure for increasing specialization. This would relegate the women's institute to a lower level in the educational hierarchy, while the more important *grandes écoles* would preserve their 'ownership' of higher-level and conceptual approaches. Paris and her successors gained the support of distinguished professors and former presidents of the *grandes écoles* (particularly from the *École Centrale*). On the basis of their advice and with their help in setting up new science courses, the school's curricula were upgraded. But teaching methods still focused on applied and practical approaches, in order to maintain the key distinguishing feature of the Women's Polytechnic School: that it served the needs of industrialists. The school claimed to provide its students not only

Figure 32. Laboratory class in electromechanical engineering, one of the new engineering branches not firmly coded as male, at the women's engineering institute (École Polytechnique Féminine) in the 1960s. The programs early emphasized experimental engineering work to meet French industrial needs. Permission and courtesy of École Polytechnique Féminine, Archives, Paris, France.

with mathematical insights but also with engineering tools. Paris and her successors did not compete directly with the *grandes écoles*; rather, they believed that there was room for different approaches and that the more practical approach suited young women and helped them to feel comfortable in the profession. Hence, the predominantly theoretical perspective was left to the established *grandes écoles*.

Improving Society and Building a Community: Women Organize

In the course of the twentieth century, female engineers began to organize collectively in several different ways. Some groups followed the example of the associations of male engineers, recruiting their membership from a specific school of engineering. Two of these were the women's association (*Association Amicale des Femmes Ingénieurs*), which was established at the *École Centrale* in 1928, and the women's association established at the Women's Polytechnic School in 1938.[49] These groups were founded a few years after the first women had graduated as engineers. When more women began to receive their degrees, they also started to enter the male Association of Graduate Engineers. Some, such as Claude Guillaume (class of 1936), were members of both associations at the same time. International connections among women engineers were strong. Guillaume, for instance, particularly recalls her participation in an International Congress of Women Engineers organized by the Society of Chemist Engineers in Oxford, England in the 1960s. She was invited as a woman engineer and the Vice-President of Research at the French branch of the British Petroleum

Company, to present her new production process for petroleum proteins. At a later point, she, as well as three other women, were elected to serve as presidents of regional branches of the association of male engineers. There was, however, still a long way to go before women would occupy such positions on a national level. These initiatives reflect the two different strategies used by women in entering the male engineering world. One consisted in setting up a separate female group within a male-dominated institution, while the other was to enroll into a male association through active participation in its organizational structure. Some women, as we have seen, combined the two.

In 1948, a few years after the end of the Second World War, an association of female engineers decided to try to unite all their women colleagues, regardless of the school from which they had graduated. It is interesting to note that, in the nineteenth century, civilian engineers who had graduated from the *École Centrale* had similarly attempted to organize all their colleagues in this manner by founding the Society of Civil Engineers (*Société des Ingénieurs Civils*), so as to protect their interests against state engineers.[50] The Circle of Women Engineers (*Cercle des Femmes Ingénieurs*) was founded as part of the French Society for Academic Women University Graduates. Its first President, Josette Garaix, who had graduated from the *École Centrale* in 1938, strongly supported the project, being joined in her efforts by another graduate from the *École Centrale*, Lucie Danel. As the latter pointed out: 'It is our own duty to improve society and institute a more appropriate division of tasks.'[51] The Circle's aim was not only to realize equal access for women to engineering, educationally as well as professionally, but also 'to help women to express themselves as women and, if they wish to do so, also as mothers.' The Circle encouraged young women to embark on a career in engineering and to pay attention to what women as women had to bring to the profession. Danel expressed her ambition to establish a new role for women in the profession: 'I hope that our daughters won't need an association such as ours, and that our conferences will become redundant in the future; or, at least, that they will have a different purpose. It is up to us, however, to prepare for such a future.'

In 1983, the Circle of Women Engineers became independent and changed its name to the French Association of Women Engineers (*Association Française des Femmes Ingénieurs*), under which name it still exists. After the Second World War, the Association did not acquire a very influential position in the world of engineering, however. From the 1960s onwards, most of the women graduating from the *grandes écoles* did not favor joining an association specifically for women. Their relative invisibility from the feminist community may be a reflection of their full integration into the male-dominated environment. The first woman to be admitted to the *École des Ponts et Chaussées* in 1959 later remarked: 'I have never felt different nor alone; I never stood outside the school's mainstream.' Another woman, who graduated in 1969, similarly commented: 'We were fully integrated into our school environment; besides, we did not need to get together, we did not make up a separate group.' For such women, identification with feminism was unlikely.

Breaking Boundaries: Women Enter the Traditional 'Grandes Écoles'[52]

Unlike the schools of engineering founded after the First World War, which had opened their doors to women from the start, the oldest and most prestigious schools continued to constitute an exclusively male world – although women had never been formally prohibited from entering. It was not until the 1960s that such schools began to welcome women first as civilian engineers and, in the 1970s, as state engineers. Social and political changes, combined with the expansion of co-education in secondary schools encouraged women to pursue an education in engineering. During the years of the Cold War, French industrialists, like their counterparts in Eastern Germany, confronted a growing need for qualified engineers in new technological fields. These needs began to be felt even more strongly in the late 1960s and early 1970s, when President de Gaulle decided to leave the NATO alliance and pursue an independent French military course. Other major changes in the social landscape followed in the wake of the events of May 1968 when different social and political movements claimed liberation and emancipation. Among them women's movements were particularly active in their struggle for equal opportunities for women.

The Institutionalization of Feminist Action Formed an Incentive for Change. Between 1965 and 1975, the Ministry of Education launched various initiatives to encourage women to apply for technology programs. In 1975, the United Nation's Year of Women, a State Secretary for Women was appointed. Françoise Giroud was the first to hold this position. She prepared a list of 101 propositions aimed at encouraging young women to diversify their professional choices, including engineering. In education, new pedagogical methods and new programs were established allowing new social groups to enter among whom were women who took the opportunities to push the gates open of the *grandes écoles*.

Up until the mid-1950s, the number of civilian engineers graduating from the *École des Ponts et Chaussées* had never reached more than thirty-five annually. By 1975, these had grown to 121. However, it was not only the number of students that changed. Aspiring engineers were also increasingly recruited from parts of the population that had not previously supplied technical personnel. In the context of French post-colonialist policies, schools had begun to admit foreign candidates, especially from Algeria and Vietnam. Candidates from different backgrounds and graduates from other lower-level schools and university programs in engineering and science were also accepted and no longer considered exceptions. Although debates on the expanding field of recruitment made much mention of women's recruitment, women were quick to take advantage of the new policies.

New courses were designed in line with developments in new technologies. Moreover, in the face of increasing technological variety, existing disciplines were split into various optional programs, each of which developed in direct relation to a specific field of engineering. Most of the new options were meant to meet new industrial needs. The curriculum also put more emphasis on experimental work, on student internships in industry, and on visits to actual work-sites. An additional change was a growing focus on economics and industrial management, largely as a result of the necessity to catch up with American industrial developments. The French economy was said to be lagging behind because of a lack of efficient labor management. All the reports issued by the Productivity Missions in the United States that had been established by the French Government at the end of the Second World War, stipulated the necessity of a form of 'scientific management' based on Frederick Taylor's theoretical model. Traditional pedagogical methods were increasingly called into question. Large-group lecture courses were reduced in number and replaced by small interactive seminars.[53] Most of the unimaginative scholastic exercises were replaced by work on individual projects.

Such changes met with resistance, especially from those who feared the loss of unity in educational programs, less attention to traditions in engineering, and the disintegration of the engineering community. At the *École des Ponts et Chaussées*, advocates of educational unity were especially concerned with increasing specialization, which they saw as a risk for the school as much as for engineers themselves. They believed that job opportunities would be reduced as a result of over-specialization, since engineers might not be capable of adapting to different technological sectors.[54] They further feared the negative impact such diversification might have on the image of the school. If training programs became too differentiated, the school and its graduates would no longer be able easily identified by captains of industry and recruitment officers on the basis of established profiles, and engineers would no longer be offered appropriate positions. Moreover, it would become extremely difficult to meet the demand for specialized engineers. Thus, variety, usually the result of a multidisciplinary approach, was regarded detrimental in the training of engineers. Differentiation was expected to bring complexity and over-sophistication. The possibility that the quality of knowledge might be improved through the input of people of different backgrounds was simply ignored.[55] In the 1960s an optional program in urban planning was established. Experimental in nature, the program was to be based on multidisciplinary methods that would help students to learn how to work with different kinds of people; neither teachers nor students would necessarily have the same background.[56] At Board meetings heated debates emerged. Opponents to the project were highly concerned with the loss of the engineering community's homogeneity. Also, student internships were regarded as threats to the proper socialization of students and the school's identity. Students would be scattered among a variety of businesses and organizations outside the controlling reach of their teachers. They would not spend much time together and the interaction with their classmates would be less intense, which would negatively affect the social bonds binding the professional community.[57] Women's entry might just be another cause of that loss of homogeneity.

When, in the 1960s, the first women began to apply for admission with some *grandes écoles*, even after having passed the selection procedures and proved they had fulfilled all the requirements, they still needed special authorization from members of the school boards. In this way, finally, the first woman was able to enter the *École des Ponts et Chaussées*. The board as well as the school president argued that recruitment policies did not explicitly exclude women, and decided to admit her. Hence they departed from the course assumed by previous boards and presidents, who had emphasized that the school's recruitment policy did not allow applications from women. At the *École des Mines*, the issue of female students had also been raised at board meetings. Only a few teachers were in favor of admitting women. However, after looking carefully at school regulations, the board had to conclude that women could not be refused.

In the 1960s, the *École des Ponts et Chaussées* did not have more than one female graduate per year. However, from the early 1970s onwards, these numbers started to increase, and since the late 1980s, there have been twenty to thirty female graduates annually (Table 2). A similar development can be seen to have taken place at the *Ecole des Mines* (Table 3). The fact that it appears unthinkable today that not a single woman would apply for admission to a form of higher technical training testifies to the social and institutional changes that have since taken place in the world of engineering. From 1970 to 1990, the percentage of women at all engineering institutions rose from six to twenty. By 1990, there were 3000 female engineers working in different fields and a variety of organizations, whereas thirty years earlier there had only been about 300.[58]

The numbers and percentages of women represented in these tables, indicate the gradual change of the gendered division of labor in engineering in the 1960s and 1970s.[59] This, however, does not mean that the conditions for entering professional engineering became particularly favorable to women.

As the first women entered those schools, new opportunities were offered to them but they had to internalize a single (i.e. masculine)

Table 2. École des Ponts et Chaussées. *Numbers of male and female civil engineering graduates by year, 1962–1998. These* Ingénieurs civils *were trained to work in the private sector mostly in construction, public works, and transportation. The number of women trained as civil engineers are relatively high compared to those trained as state engineers for public administration. Source:* Association des Anciens Élèves de l'École Nationale des Ponts et Chaussées.

	1962	1963	1964	1965	1966	1967	1968	1969	1970	1971	1972	1973
Women	1	0	1	0	2	1	1	3	2	1	3	1
Total	68	62	81	68	74	79	93	92	95	86	94	97

	1974	1975	1976	1977	1978	1979	1980	1981	1982	1983	1984	1985
Women	5	5	10	9	3	8	13	6	5	10	16	14
Total	104	116	111	116	121	118	123	108	112	109	138	98

	1986	1987	1988	1989	1990	1991	1992	1993	1994	1995	1996	1997	1998
Women	11	12	15	16	19	13	9	21	14	17	27	17	24
Total	83	99	117	93	132	119	119	131	130	125	127	138	134

Table 3. École des Mines. *Numbers of male and female civil engineering graduates by year, 1972–1998. These* Ingénieurs civils *were trained for the private industrial sector. The number of women trained as civil engineers are relatively high compared to those trained as state engineers for public administration. Source:* Association Amicale des Anciens Élèves de l'École Nationale Supérieure des Mines de Paris.

	1972	1973	1974	1975	1976	1977	1978	1979	1980	1981	1982	1983	1984	1985
Women	1	4	6	5	6	7	2	6	3	4	11	10	8	9
Total	61	76	73	80	90	80	78	78	78	78	90	85	85	90

	1986	1987	1988	1989	1990	1991	1992	1993	1994	1995	1996	1997	1998
Women	8	14	20	15	13	19	19	24	22	14	19	13	16
Total	83	88	105	99	112	104	100	122	115	119	119	119	119

model of the profession. Diversity could not be recognized, and moving beyond the norms connected with this model remained difficult. As one of the women who graduated at the end of the 1960s comments: 'We were trained to be with men, and to work in a professional male-dominated environment.'[60] This implied having to behave like men. Such accommodation was probably a precondition for women to be accepted into the new system, and the only chance of being offered the same jobs as men.

Figure 33. Marie
France Clugnet, the first
woman civil engineering
student at highly
prestigious École
Nationale des Ponts et
Chaussées, the
engineering institution
which traditionally had
strong ties to the
military, at a 1961
student ball. Although
she figured prominently
throughout the visual
record of the event, she
was erased from the text.
Permission and courtesy
of École Nationale des
Ponts et Chaussées,
Archives, Paris, France.

In addition, the image of women as servants or fantasies, which had
beginning to fade during earlier decades, regained its considerably popularity in the course of the 1960s. When the first female student of the
École des Ponts et Chaussées arrived at the school doors on her very first
day in 1959, she was received with flowers, and an article by two students about students' lives was accompanied by pictures of her.[61]
(Figure 33) These pictures not only revealed the male students' perception of women – the pictures showed her, pretty and elegant, in a ball
gown, at an official reception – they also underlined the fact that female
students were expected to be Woman incarnate. The article in question
did not say anything about the student herself, nor was she interviewed
for the occasion. A woman who graduated a few years later comments:
'We were often invited by the older engineers who had graduated from
the École des Ponts et Chaussées; they were willing to meet and support
us, but we somehow felt that we were objects of curiosity.'[62]

A major change occurred in 1970, when the École Polytechnique's military status was transformed. Its new regulations of 1972 granted women
and men equal access to all programs. A crucial barrier nonetheless
remained in place. Since the school still operated under the supervision of
the Ministry of Defense, male students enrolled as special reserve officers,
whereas women were not allowed to join the military corps and so promote

their careers as officers of the French Army. This formal condition had made entry into the *École Polytechnique* viable for men only for two centuries. To remove this barrier, a special corps of female reserve officers was established in 1972.[63] (Figure 34). Since 1972, more than 500 women have been admitted into the *École Polytechnique* (Table 4); about half of them, like half the men, achieved a sufficiently high-level rank to join a state corps.

The opening of the doors of the *École Polytechnique* to women also implied the opening up of the main avenue into the state corps. Another avenue was opened to them when at the end of the 1970s graduates from the main scientific institution, the *École Normale Supérieure*, were allowed to enter the state engineering corps. In 1975, the first female state engineers to have completed their studies at the *École Polytechnique* entered the *écoles d'application*. Although their numbers have been growing since then within the *écoles d'application* – which offer advanced training to both state and civilian engineers – there is still a line between the sexes. State engineers still remain a largely male-dominated group, while the 'civil engineers' is gradually becoming a mixed community.

Preparing female engineers for the profession was not a main concern of the major engineering school in particular the state schools as

Figure 34. Students at École Polytechnique, major military ceremony in 1995. Since 1994, at their request, women have worn the same hat as men. Recently, they asked for pants instead of skirts. The inscription on the flag reflects the identity of engineers: "For the fatherland, the sciences, and glory." Permission and courtesy of Direction des Etudes Service Audiovisuel École Polytechnique, Paris, France.

Table 4. École Polytechnique. *Number of female engineering students enrolled by year, 1972–1998, with percentages of women. As France's most prestigious engineering institution, the* École Polytechnique *allows only half of its best ranking graduates to enter the state engineering corps. Women were admitted only since 1972, their percentage was low and climbing up to 14 % in recent years. Source:* École Polytechnique

	1972	1973	1974	1975	1976	1977	1978	1979	1980	1981	1982	1983	1984	1985
W	6	11	10	18	21	15	11	18	18	20	25	27	28	26
T	301	278	262	300	302	299	308	309	308	311	311	322	333	336
%	2	4	4	6	7	5	4	6	6	6	8	8	8	8

	1986	1987	1988	1989	1990	1991	1992	1993	1994	1995	1996	1997	1998
W	34	26	26	22	29	34	32	40	47	56	58	57	55
T	337	315	309	339	360	380	400	400	400	400	400	338	401
%	10	8	8	6	8	9	8	10	11	14	14.5	14	14

opposed to the *École Polytechnique Féminine* (Women's Polytechnic School), at which women were trained for specific professions. At a board meeting of the *École des Ponts et Chaussées* in 1975, questions arose about what kind of occupation might be offered to the two female state engineers who were to graduate within two years of time. Nobody had really thought about this question before. At the beginning of the 1970s, one of the board members eventually explained: 'As far as urban design and public planning are concerned, it is not a disadvantage at all to be a woman; in fact, it may even lead to increased motivation.'[64] Women were defined as 'naturally' having certain skills and interests, which defined what jobs were appropriate for them: public relations, urban design, planning, and other service-oriented jobs.[65] Although women were thus offered roles in a professional environment, these roles were limited, and so were career opportunities.

Women's entry into traditional schools of engineering was widely celebrated, but at the same time also still seen as threatening. In 1975, when the first women began to graduate from the *École Polytechnique*, the Society of Mining and Civil Engineering gave a reception at which Françoise Giroud appeared as the guest of honor. In his welcoming address, the President of the engineering society expressed his concerns with reference to women's 'natural' grace thus: 'We are currently worried about having female students around, for if, at this point in time, some "technocrats" are capable of combining both charm and competence, God!, where will we be going next?'[66]

Upon entering the state corps, female engineers were determined to claim equal rights and not to accept any discrimination. A woman who

Table 5. École des Ponts et Chaussées. *Number of women and men state engineers admitted by year, 1975–1998. Highly selective, only a minority of graduates (20%) are state engineers to work at the Ministry of Construction and Transportation. Very few women graduate as state engineers.* Association des Anciens Éleves de l'École Nationale des Ponts et Chaussées.

	75	76	77	78	79	80	81	82	83	84	85	86	87	88	89	90	91	92	93	94	95	96	97	98
W	2	0	1	1	2	0	1	1	4	2	3	2	3	2	2	2	4	1	3	3	1	2	6	7
T	30	33	33	34	29	26	27	28	26	25	27	26	24	21	22	26	30	31	29	29	29	31	30	30

---

[1] These figures merely include state engineers who graduated from the Ecole Polytechnique and from the Ecole Normale Supérieure. These two schools represent the most prestigious and highly selective routes to the state corps of engineers. Another path is available to the *Ingenieurs des Travaux Publics de l'Etat*, who must complete seven years of professional practice in order to be allowed to join ; the requirements are equally highly selective for both men and women. The relative proportions of women and men to enter the corps des ponts et chaussées by means of this possibilty are the same.

Table 6. École des Mines. *Number men and women admitted by year, 1975–1998. Even fewer graduates (10%) are state engineers to work primarily at the Ministry of Industry. Very few women graduate as state engineers at this institution. Source:* École Nationale Supérieure des Mines de Paris.

	75	76	77	78	79	80	81	82	83	84	85	86	87	88	89	90	91	92	93	94	95	96	97	98
W	1	1	0	1	1	2	1	2	0	1	1	2	1	1	1	1	3	1	1	1	0	1	1	0
T	13	13	13	13	12	11	12	12	13	12	13	11	11	9	10	12	11	13	12	12	14	13	13	13

---

[2] These figures include state engineers who graduated from the Ecole Polytechnique and from the Ecole Normale Supérieure only. Just as in the case of the corps des ponts et chaussées, a number of other highly selective routes lead to the corps des mines.

graduated from the *École Polytechnique* in 1976, and who went on to the *Corps des Mines*, recalls that she, along with her female friends, insisted on being able to obtain the same positions as the men with equally high educational levels. She added: 'We did not mind moving to some remote area; however, since we had reached the same academic levels as our male classmates, we demanded the same jobs with the same responsibilities and do not want to be wall flowers.'[67]

Opposing or Complementary Strategies?

Where were the French pioneers to be found: at the Women's Polytechnic School or in the male-dominated *grandes écoles*? Probably at both: the difference between them may be seen to lie in the various strategies they used to gain access to the male-dominated domain of engineering.

Confronted with an anti-feminist backlash at the end of the First World War, Marie-Louise Paris responded by establishing Women's Polytechnic School, and reinforcing the role of female graduates as the assistants to male engineers. Along similar lines, the leaders of the association of female engineers, while claiming equal professional opportunities, did not believe that women would be able to face up to the challenge of male-domination in the profession. Women's chances of advancing their careers on an equal footing with men consequently appeared slight, unless women themselves would help to build a better society – one in which all individuals could fulfill their potential in accordance with their abilities.

Women entering the *grandes écoles* in the 1960s and 1970s employed strategies that were inspired by the more general struggle for equal rights. These students had grown up at a time when women had already won the right to vote (in 1944), and they anticipated the new social

Figure 35. Launching preparation of the rocket, Ariane 5, the twentieth-century national symbol of the successful wedding of French science and the military. If eighteenth-century civil engineering paved the way to the construction of the national state, twentieth-century aeronautical engineering helped launch French national interest into space. Despite the male icon, many women graduates from the École Polytechnique Féminine *found midlevel engineering jobs in aerospace. Permission and courtesy of Archives L'Aérospatiale, Paris, France.*

movements emerging in 1968. Unlike their predecessors, they entered the engineering schools by going through the same selection processes as men (in which mathematics had gained increasing importance) and succeeded in establishing themselves in traditionally male-dominated professional domains in the course of the 1980s and 1990s.

In the wake of the First World War, graduates from the Women's Polytechnic School had tried to gain access to the world of engineering through the foundation of a separate institution for women, as well as by developing female networks. Women of subsequent generations who entered the traditional *grandes écoles*, hoped to achieve equality by integrating themselves into established, male-dominated institutions. In political debates on women's rights in France, these different strategies are now commonly presented as opposed to one another. As a result, tensions between the defenders of each strategy are sometimes seen as an impediment to working collectively for equal rights.

At this point, it seems apt to raise yet another question: how could the Women's Polytechnic School survive in a male-dominated environment throughout the twentieth century? In addition to Marie-Louise Paris' extraordinary character, the involvement of distinguished male professors and the powerful support of industrialists, the Women's Polytechnic School was crucial in the development of industrialization and women's access to engineering education. It was a stronghold capable of ensuring female participation not only in higher education, but also in the world of engineering as a whole. Its graduates thus quietly contributed to the transformation of women's positions in professional engineering, as well as in the public sphere more generally. The presence of these women in the realm of engineering may therefore have functioned as a wedge, allowing subsequent generations of women the means of entry into the traditional *grandes écoles*.[68]

Notes

*   I am particularly grateful to André Grelon for his advice and suggestions and for his invaluable pioneering research on Marie-Louise Paris. I am also most indebted to Evelyne Pisier for her ideas and her helpful insights on women history and policy. I also want to thank Karin Zachmann and Ruth Oldenziel for their great help in reviewing my paper.

1. Terry Schinn, 'The Impact of Research and Education on Industry – A Comparative Analysis of the Relationship of Education and Research Systems to Industrial Progress in Six Countries *Industry and Higher Educations* (October 1998): 270–289, p. 274.

2. Schinn, 'The Impact of Research and Education,' 275–276.

3. See Irina Gouzevitch, this volume.

4. Schinn, 'The Impact of Research and Education,' 277.

5. Ibid., 274.

6. André Brunot and Roger Coquand, *Le Corps des Ponts et Chaussées* (Paris: Editions du CNRS, 1982).

7. Denis Woronoff, Lecture at the *Séminaire Histoire des Techniques*, directed by A. Picon, *Ecole Nationale des Ponts et Chaussées*, November 22, 1995. On the functions of the engineers of the *Corps des Mines*, see: André Thépot, '*Les ingénieurs du Corps des mines – Evolution des fonctions des ingénieurs d'un corps d'Etat au XIXè siècle,*' *Culture Technique* 12 (1984): 55–61.

8. Thierry Vedel, '*Les ingénieurs des télécommunications,*' *Culture Technique* 12 (1984): 63–75.

9. Peter Lundgreen, 'Engineering Education in Europe and in the U.S.A., 1750–1930: The Rise of Dominance of School Culture and the Engineering Professions,' *Annals of Science* 47 (1990): 33–75, p. 38.

10. Lundgreen, 'Engineering Education in Europe and in the USA', 37–38.

11. The *École Polytechnique* has played a major role in educating the French elite ever since. Maurice Bernard, '*Le choix, la formation et le renouvellement des élites en France,*' unpublished paper, August 1996.

12. Thomas Olivier, *Mémoires de géométrie descriptive théorique et appliquée* (Paris: Carilian-Goeury et Victor Dalmont, 1852), iii–xxiii, p. xix.

13. Antoine Picon and Kostas Chatzis, '*La formation des ingénieurs français au siècle dernier: débats, polémiques et conflits,*' *L'orientation scolaire et professionnelle* 21, 3 (1992): 227–243, p. 237.

14. Bruno Jacomy, '*A la recherche de sa mission. La Société des Ingénieurs Civils*', *Culture Technique* 12 (1984): 209–219.

15. Kostas Chatzis, '*La naissance d'une nouvelle figure: l'ingénieur civil et l'École Centrale*', Paris, LATTS-ENPC, 1992, working paper, 10. See also: Picon and Chatzis, '*La formation des ingénieurs français*', 237–238.

16. Glwadys Chantereau, '*Les femmes ingénieurs issues de l'École Centrale pendant l'entre-deux-guerres*', (Paris-Nord University, Master thesis, 1997), 185.

17. Chantereau, '*Les femmes ingénieurs*', 47.

18. Ibid., 185.

19. Ibid., 167.

20. The *École des Ponts et Chaussées* began to offer civil engineering programs in 1851 – whose students were called 'external students' (Decree, October 13, 1851) – while The *École des Mines* only officially awarded the title of 'civil engineer' in 1890 (Decree, July 18, 1890). The *École des Ponts et Chaussées* did not begin to do so until 1922 (Decree, January 7, 1922). For a detailed analysis of the context and the debates preceding this regulation, see: Georges Ribeill, '*Des Ingénieurs civils en quête d'un titre: le cas de l'École des Ponts et Chaussées (1851–1934),*' in *Les ingénieurs de la crise*, edited by André Grelon (Paris: Editions de l'EHESS, 1986), 197–209.

21. Bérard Decree, March 15, 1924.

22. Marielle Delorme-Hoechstetter, '*Louli Sanua et l'École de Haut Enseignement Commercial pour Jeunes Filles (HECJF) – Génèse d'une Grande École Féminine (1916–1941)*', (*École des Hautes Etudes en Sciences Sociales,* Paris, Master thesis, 1995), 59.

23. The other main *grandes écoles* which admitted women at the end of the First World War are the following: *Institut National d'Agronomie, École de Chimie de Paris, École Supérieure d'Aéronautique.* Josette Cachelou, '*De Marie Curie aux ingénieures de l'an 2000,*' *Culture Technique* 12 (1984): 265–271, p. 267.

24. André Grelon, '*Marie-Louise Paris et les débuts de l'École Polytechnique Féminine (1925–1945),*' *Bulletin d'Histoire de l'Electricité, Les Femmes et l'Électricité* (June-December 1992): 133–155.

25. Yves Cohen, '*Le travail social: quand les techniciens sociaux parlent de leurs techniques,*' in *Chantiers de la Paix sociale (1900–1940)*, edited by Yves Cohen and Remi Baudouï (St Cloud: ENS Editions Fontenay, 1996), 117.

26. Cohen, '*Le travail social*', 117.

27. Grelon, '*Marie-Louise Paris et les débuts*', 136.

28. Delorme-Hoechstetter, '*Louli Sanua et l'École de Haut Enseignement Commercial pour Jeunes Filles*', 11, 89, 114.

29. Ibid., 104.

30. 'Industrialists and business men were not favorable to young women who would do tasks considered as specialized.' Marc Meuleau, *Histoire d'une Grande École* (Paris: Dunod, 1981), quoted in Delorme-Hoechstetter, '*Louli Sanua et l'École de Haut Enseignement Commercial pour Jeunes Filles*', 56.

31. Ibid., passim.

32. In the first year, 39 women graduated from Sanua's new school. This number quickly rose to 125 in 1917–1918, and to 137 the following year. After the war, the numbers dropped to level off until the Second World War. The school closed down in 1975 when the HEC, the male school, began admitting female students.

33. Claudine Fontanon and André Grelon, eds., '*Les professeurs du Conservatoire National des Arts et Métiers – Dictionnaire biographique, 1794–1955,*' in *Collection Histoire Biographique de*

l'Enseignement. Institut National de Recherche Pédagogique (Paris: Conservatoire National des Arts et Métiers, 1994), 23.

34. This section is based on: Grelon, '*Marie-Louise Paris et les débuts.*'

35. These included both several male schools, such as the *École Spéciale des Travaux Publics*, a school for civil engineers established by Eyrolles, and the ESIA, which specialized on computer science studies. Another well-known example was the *École Centrale* founded by Léon Guillet. On this issue, I benefitted from a seminar discussion 'Technicians and Society' organized by André Grelon, *École des Hautes Études en Sciences Sociales*, February 5, 1999.

36. Koenigs' daughter Gabrielle was one of his students, who earned her Ph.D degree in natural sciences in 1916 and held a position as Professor at the *École Normale Supérieure*, one of France's most famous school of higher scientific education. Fontanon and Grelon, eds., '*Les professeurs du Conservatoire National des Arts et Métiers.*'

37. 1 DD 15, CNAM Archives, Paris, quoted in Grelon, '*Marie-Louise Paris et les débuts*', 140.

38. The CNAM traditionally provided lecture rooms to new technical schools and institutes, e.g. the National Institute of Agronomy in 1876.

39. Grelon, '*Marie-Louise Paris et les débuts*', 142; Program of admission and studies requirements at the Female Electromechanical Institute, CNAM, 1925.

40. Letter of Chaumat to the Director of CNAM, 29 June 1926, as cited by Grelon, '*Marie-Louise et les débuts*', 143. Chaumat was a Professor at the CNAM and at the *Ecole Supérieure d'Electricité.*

41. Grelon, '*Marie-Louise Paris et les débuts*,' 146.

42. Law, July 10, 1934, regarding the title of Engineer.

43. Grelon, '*Marie-Louise Paris et les débuts*', 148–149.

44. She wrote a poem describing the school as her child, which was published in the brochure commemorating the 50th anniversary of the school in 1975 (Paris: SOREDI, EPF, 1975).

45. Author's interview with Colette Kreder, 22 January 1999. Kreder was a former student of Marie-Louise Paris at the Women's Polytechnic School, and also its Director during the 1980s.

46. Interview with Colette Kreder.

47. See Juliane Mikoletzky, this volume.

48. 1966 letter of Marie-Louise Paris to General de Gaulle, *École Polytechnique Féminine* Papers, EPF, Sceaux.

49. *Association des Anciennes Élèves de l'École Polytechnique Féminine*, 1984, Index.

50. Jacomy, '*La Société des Ingénieurs Civils*', 210.

51. Chantereau, '*Les femmes ingénieurs*', 161.

52. This section is based on the author's survey of the *École des Ponts et Chaussées* Archives, Paris (ENPC Archives, hereafter).

53. Meeting of the '*Conseil de Perfectionnement*', June 11, 1961, ENPC Archives.

54. Picon and Chatzis argue that the mutually exclusive characters of the general and the specialized engineer has been a major source of conflict within the French profession. Picon and Chatzis, '*La formation des ingénieurs français*', 228.

55. '*Conseil de Perfectionnement*', October 8, 1962, ENPC Archives.

56. '*Conseil de Perfectionnement*', December 3, 1975, ENPC Archives.

57. A similar environment has been described by Boel Berner as a 'homosocial' environment. See Boel Berner and Ulf Mellström, 'Looking for Mister Engineer: Understanding Masculinity and Technology at Two *Fin de Siècles*' in *Gendered Practices. Feminist Studies of Technology and Society*, edited by Boel Berner (Linköping: Nova Print, 1997), 54.

58. Catherine Marry, '*Femme et ingénieur: la fin d'une incompatibilité?*' *La Recherche* 241, 23 (1992): 362–363, p. 362.

59. Marry concludes that we have entered a period when being a woman is no longer incompatible with being an engineer. Marry, '*Femme et ingénieur?*'

60. Author's interview April 1996, with a woman engineer who graduated from the *Ecole des Ponts et Chaussées* at the beginning the 1970s.

61. Hubert Roux and Didier Lenoir, '*Les élèves de l'École*', *Regards sur la France*, Special Issue on the *École Nationale des Ponts et Chaussées* (Paris: Department of Propaganda, Publication, 1961), 71–75.

62. Interview with a Female Former student at the *Ecole des Ponts et Chaussées*, April 1996.

63. Decree, January 10, 1972.

64. Meeting of the '*Conseil de Perfectionnement*', 3 December 1975, ENPC.

65. In France, this issue has recently been discussed from a new perspective under the influence of parity and quotas debates in the political arena. See Evelyne Pisier and Eleni Varikas, '*Femmes, République*

et Démocratie. L'Autre dans la Paire?' Femmes en politique, Pouvoirs 82 (1997): 127–143. Women were subject to professional exclusion on familiar and longstanding grounds: because of their supposedly 'incorrigibly feminine nature' any consideration of specific circumstances, situation, or abilities was deemed irrelevant.

66. *'Discours du Président', PCM: Journal of the professional Association of Ingénieurs des Ponts et Chaussées et des Mines* 3 (1975): 43–50, p. 44.

67. Interview with a female state engineer, former student at the École des Mines.

68. On the issue of women's strategy of separatism versus integration, see: Estelle Freedman, 'Separatism as Strategy: Female Institution Building and American Feminism, 1870–1930,' *Feminist Studies* 5, no. 3 (Fall 1979).

Juliane Mikoletzky

6. Precarious Victories: The Entry of Women into Engineering Studies in Austria, 1900–1945

The way educational institutions and occupational systems vary by country tells us how different nations have adapted to modern industrial capitalism. Women's increasing participation in the field appears to be a typical feature of this adaptation. Yet, while technicians of all kinds are key figures in studies of engineering, questions of gender and the entrance of women into the market for technical labor usually receive little attention.[1] Selectivity in higher education has been disadvantageous to women in some places and helpful in others. It has also helped to determine which parts of the labor market women have entered. Therefore, studying the role played by institutions of higher education is crucial to understanding why and when women entered engineering, and how their presence helped to determine the shape of modernizing societies.[2]

In dealing with the problem of women's entry into the engineering profession, we have to take into consideration not only their access to higher education but also how educational systems and labor markets are linked. In this context, the discourses of 'shop culture' and 'school culture' gain new prominence. In a strong 'shop culture' tradition with its emphasis on training in the workplace there may be additional barriers to entering technical studies and thus the better jobs. Such a culture may well encourage women to take entry- or low-level support positions, leading to a 'feminized' set of occupations such as drawing or laboratory work. On the other hand, an entrenched 'school culture' where engineers received their education at academic institutions may ease the path of women into higher technical education, but keep them marginalized in the job market.

In what follows I examine developments in Austria, focusing on two main periods: women's first attempts to enter the institutions of advanced engineering education, and the experience of the first generation of women students at technical colleges up until the end of the second World War. I look at structures and institutions to see how far these go in explaining what happened in Austria, and why it differed from Germany and other countries. There are some particular difficulties in writing about the Austrian case. There were, of course, considerable upheavals due to the two world wars. There were also three major shifts in the political system and the territorial basis of the country – the dissolution of the Habsburg Empire (1918), the incorporation of Austria into the Third Reich (1938 to 1945), and the reconstruction of a second Austrian Republic after the end of World War II. All of these mean that what the label 'Austrian' refers to has not remained stable over the period under investigation. To avoid confusion I use it here to refer to the territory and people of what eventually became the Austrian Republic. However, it must always be kept in mind that the Habsburg monarchy left a powerful legacy for future generations.

Long-term studies of the modern Austrian system of higher education are still rare, and the history of women students at technical colleges and universities has, for the most part, not yet been written.[3] Therefore my argument will be based mainly on the results of my study of the Vienna University of Technology, with some additional material concerning other Austrian technical universities in Vienna (the University of Agro-Forestry), Graz, and Leoben (the University of Mining, Petroleum Engineering, and Metallurgy). Their total enrollments as well as their share of women students were, and still are, much lower than those of the technical university in Vienna. However, using the findings from the Vienna University allows us to develop a more general outline of other institutions. Developments at the technical universities in the eastern crown lands of the former Habsburg monarchy in Prague, Brünn (Brno), and Lemberg (Lviv) also have some relevance for the period preceding women's admission to technical studies in 1919. Meanwhile, higher technical education in the Hungarian area of the Empire, where educational policy was autonomous since 1867, followed its own path. I will discuss it here in comparison with the Austrian example.

Setting the Stage: Austria as a Special Case

In engineering education, Austria followed what I characterized above as the continental-European model.[4] There was an early stress on 'school culture,' and the government strongly influenced credentials and courses. However, this pattern has variations. In many historical studies on the engineering profession and on technical education in the German-speaking world (written by German but also by other non-Austrian authors), Austria is more or less subsumed under the German model and regarded as a sort of German province (which it was, but only between 1938 and 1945.)[5] Certainly the two countries were structurally similar in some respects and had many points of cultural and economic interdependence. Nevertheless, for the most part they developed distinct traditions in politics, education, and the law. Some of these differences had an impact on women's entry into engineering studies.

It is important to realize that the Habsburg monarchy, unlike most other European nations of the mid-nineteenth century, was explicitly constituted as a multi-ethnic state since 1867. Therefore participation in advanced education had not only a social, but also a national dimension.[6] Among the non-German nations, the Czechs and the Poles of Galicia showed an especially strong inclination toward higher education. The Czechs in particular were successful in establishing a complete educational system of their own, and the Czech-speaking branch of the Austrian women's movement played a pioneering role in establishing institutions of higher education for girls. But national groups sometimes appropriated women's struggles for higher education less out of a concern for the principles of sex equality than as part of a project of political and cultural emancipation.

The organization of higher technical education took a different route in Austria than in Germany. The Polytechnic Institute in Vienna, founded in 1815 as a national peer institution, had been designed from the beginning to be strongly academic rather than practical. Its first director, Johann Joseph Prechtl (who framed substantial parts of the statutes), spoke of creating a 'technical university' (*universitas scientiarum technicarum*), whose teaching about the whole domain of technology would be on par with university training. Prechtl therefore wanted high standards for all credentials, and he made sure that the teaching staff was endowed with university status along with the appropriate pay structure. And both students and professors were given freedom of teaching and learning, something that was unusual at the time even for Austrian universities. The 1849 introduction of the institution of private lecturer (*Privatdozent*) – nearly three decades earlier than in Germany – at both the Austrian polytechnic institutes in Vienna, Prague, and Graz and at the Austrian universities in fact helped realize Prechtl's vision. Prechtl's conception differed in many ways from the admired model of the French *École Polytechnique*, but also from the early technical schools in Prussia or in other German states that started with much more modest intentions. Nevertheless, the Viennese Polytechnic Institute had some influence on the organization of the first polytechnic schools in German states, especially in Hannover and at Karlsruhe in Baden.

The early history of Austrian technical education belongs to the period in the early nineteenth century when technology and the sciences were highly valued by aristocratic society. The excellent equipment of the polytechnic institute in Vienna, for example, was due to the personal interest in the project taken by Emperor Franz I. His successor, Emperor Ferdinand II, paid several visits to the Institute to view and wanted to be routinely kept abreast on the latest technical collections. The professors were regularly asked for their counsel in technical and patent matters and also engaged as instructors for the crown princes. Public exams (*Tentamina*) at the Polytechnic Institute in the 1820s were held in the presence of members of the imperial family. In the first years of its existence members of the high nobility even used to enroll for a semester or two. Similarly, the foundation of the *Ioanneum* (the precursor of today's technical university in Graz) in 1811 was the result of Archduke Johann's efforts to promote Styrian industry. Industrial competition with the British and others was, in fact, a key reason that the Habsburgs founded the Viennese institute. The younger German institutions had a similar goal, as did the Royal Institute of Technology (KTH) in Sweden.[7] Like the Swedish Royal Institute, the Polytechnic Institute in Vienna was established with the double purpose of serving public as well as private industrial needs.[8]

It was intended that the spread of technical knowledge be as broad and as quick as possible. Therefore, the institute incorporated a two-year preparatory school (or *Realschule*) and a commercial school department. Until the mid-1860s it (like the polytechnic institute in Prague and the *Ioanneum* in Graz) also held popular lectures on scientific and technological subjects. These lectures were open to everyone; the first presence of learning women within the walls of the Vienna polytechnic institute is directly related to them.[9] During the late 1840s and the 1850s, lectures on chemical and physical subjects and on the history of applied arts were held for mixed groups and 'ladies.' Unfortunately no information about the content of the lectures or the number of women who attended is available. But some professors are known to have admitted women to their lectures. Among these were the chemist Anton Schrötter v. Kristelli, the physicist Ferdinand Hessler, and the art historian and later director of the Vienna Museum of Applied Arts, Rudolf v. Eitelberger-Edelberg. All of these had been involved in popular and vocational education previously, and all belonged to the small group of Austria's liberal bourgeoisie that later supported women's access to higher education.

The institutional reforms of the 1860s and 1870s in Austria upgraded all the polytechnic schools of the Habsburg monarchy into higher technical colleges (*Technische Hochschulen*). This resulted in increasing social exclusion. In the years to come women and all persons of a lower socio-economic background were made less welcome, especially after the end of the short 'liberal era' in the Habsburg monarchy in 1879. In 1871, it was still possible for a Russian women chemist, Maria Manasseina, to work in the laboratory of Professor Julius v. Wiesner at the Polytechnic Institute.[10] Wiesner, when he moved as professor to Vienna's university, became an active promoter of the admission of women to university studies. Some years later, Franz Ševčik, a lecturer of mathematics at the Technical College in Vienna, organized a mathematics lecture course for women (*Mathematisches Damen-Collegium*) which lasted from 1874 to 1880. Some of its participants came from the first school for girl's higher education, founded in Vienna in 1871, where Ševčik taught mathematics. In all, perhaps as many as 700 women attended during the six or seven years of its existence. The *Damen-Collegium* was housed in a lecture room of the Vienna technical college from 1874 to 1877, but after that it had to move. The faculty claimed the women had subjected the male students to unspecified 'inconveniences.' (Figure 36)

The *Damen-Collegium* and women's attendance at the popular lectures at the Polytechnic Institute prove that women at the time were not 'natu-

Figure 36. Translated from German the caption reads: "Building Site Rule: Chatting while working might be deadly." Student caricature drawing conveys the idea that women's entry is threatening to the male engineering world. Permission and courtesy of Dr Alfred Lechner, Private Collection, Vienna, Austria.

163   *Juliane Mikoletzky*   *Precarious Victories*

rally' less good at, or less interested in, mathematical or scientific subjects. Single women also frequented scientific or technological lectures, usually as guests or with some other informal status, at many other technical schools or colleges.[11] As in Austria, their exclusion, more or less rigorously enforced, followed an academic upgrading of the technical schools. In most places this occurred during the 1870s. So in Sweden, the Royal Institute of Technology in 1876 was given more academic statutes that explicitly defined the engineering disciplines as male.[12] In Hungary, in an effort to upgrade the institution the polytechnic school of Budapest was called a 'technical university' as early as 1871.[13] In all the German states, the former polytechnic schools became technical colleges (*Technische Hochschulen*) only from 1879 on.

In Austria, the polytechnical colleges were extensively reorganized in the early 1860s.[14] The Viennese institute was the last to receive its new statutes in 1865, and its name was changed in 1872. The structural reform of 1865 had eliminated the commercial department (the preparatory *Realschule* had been separate since 1851), closed down the popular lectures, and, for the first time, had established five departments. The number of more general disciplines was reduced, and the theoretical and scientific orientation of the curriculum was reinforced. In fact, this was a long-term trend in the history of Austrian advanced technical education: there were several points at which the 'non-technical' subjects were de-emphasized, for example after World War I and in the 1950s. The current efforts to create a corporate identity for the Vienna University of Technology point in the same direction. Austrian technical colleges have never sought to add humanities, social sciences or economic departments or schools for teacher training, as happened in most German states and in Hungary. Even the agricultural disciplines have gradually been transferred to the College (now University) of Agro-Forestry, which was founded in 1872. This trend can be regarded as a heritage of Prechtl's early idea of a 'technological university.' It has had long-term consequences for women students at Austrian technical universities because such subjects have often proved attractive to female students. Up to the present, female students at technical universities enroll for technical or scientific disciplines in Austria. In Germany, where the technical universities also offer 'soft' or non-technical subjects, women were – and still are – more likely to choose them instead.[15]

From the beginning of their modern history engineers had a somewhat higher status in Austria than in Germany. The early interest of the Austrian upper classes in science and technology – though it faded after the mid-nineteenth century – provided a precious dowry. Whatever the reason, the process by which engineers formed a profession started earlier and was somewhat more successful in Austria.[16] Professional organization began during the 1848 revolution with the foundation of the Austrian Engineers' Association (*Österreichischer Ingenieur-Verein* or ÖIV), eight years before German engineers established the Association of German Engineers (*Verein deutscher Ingenieure* or VDI). The Austrian association amalgamated with the Austrian Architects Association (*Österreichische Architekten-Vereinigung* or ÖAV) in 1864 to form the Association of Engineers and Architects (*Österreichischer Ingenieur- und Architektenverein* or ÖIAV). This last organization still exists, and it was, until its suspension in 1938, influential in championing reforms of engineering education. The creation in 1860 of the status of *Privattechniker* or civil engineer, today called *Zivilingenieur*, helped to boost professional self-esteem. It had taken the Austrian engineers more than twenty years to reach it, for the first draft law had been presented in 1834. The code regulated the admission of engineers to free professional practice.[17] The applicant had to have finished his academic studies and to prove he had done five years of practical work in his profession before he was admitted to a special examination. Once he had achieved the title of *Ziviltechniker*, he was authorized to act as a substitute for public building and technical authorities. At first this title was meant only for architects, surveyors, and construction engineers, but by the end of the century it was extended to all technical occupations with advanced education. There was no equivalent status in the German countries.

Other achievements buttressed the sense of professional identity among Austrian engineers. Final exams for technical studies were introduced in 1878. Some years later, in 1901, the Austrian technical colleges were granted the right to bestow doctoral degrees. This was two years after similar moves in Germany and Hungary. Finally in March 1917, Austrian engineers were able to acquire legal protection for the professional title of engineer (*Ing.*), which from then on was restricted to graduates of a technical college. This was more than their German colleagues achieved.

The beginnings of higher education for girls in Austria, as in Germany, were characterized by the efforts of private associations, many of which derived from the bourgeois part of the women's movement. The first high school for girls in the German-speaking parts of the Habsburg monarchy was established in 1871 by the Vienna Society for the Gainful Employment of Women (*Wiener Frauen-Erwerbs-Verein*) in Vienna. This was followed by a girls' *Lyzeum* in Graz in 1873. In Prague a secondary school for Czech-speaking girls was established as early as 1863. None of these schools for girls, founded by members of the local middle and upper-middle classes, sought to offer girls avenue into university education. Rather, they were designed for a more thorough and comprehensive general education. Like elsewhere, the aim was to prepare girls for life as a housewife and mother. Or if the goal of marriage happened to be unattainable, the girls could be trained for the respectable occupation of teaching. The girls' school of the Viennese Society for the Gainful Employment of Women, unique because it was founded and managed by a women-only steering committee, was inspired by a desire to create decent job opportunities for middle class women in trade and industry.[18] In 1870, Marianne Hainisch, one of the founding members of the Austrian women's movement, spoke at the general assembly of the Vienna Society, which had organized vocational courses for girls since the late 1860s. For the first time in Austria, she advocated equal higher education for girls and boys, and proposed to establish a girls' school of the *Realschule* type in Vienna. She also suggested that mathematics and science were good intellectual training for girls.[19] This was presumably a tactical position: Hainisch wanted girls to be taught to 'think like the boys,' which meant learning Latin. But knowledge of Latin was a key qualification for university admission in the Central European educational systems. So to structure Latin into the girls' curriculum would have meant a direct challenge to male educational privileges and social dominance. Hence she probably took the wiser course of proposing mathematics and science instead of Latin, because otherwise her proposal would have been entirely rejected. In the end, it was probably due to the emphasis she placed on these subjects that as much as fifty percent of the classroom time at the Viennese higher girls' school was spent on them.[20] But from the 1880s on, following a general trend toward conservatism in Austrian politics and society (when Count Taaffe was installed as prime minister, 1879–1893), the hours devoted to mathematics and science in women's high school curricula were drastically reduced.

At the same time, a new type of high school for girls, the *lyceum*, was established. It had a curriculum more closely tailored to 'womanly tasks and abilities,' and of course studies did not lead to certification for university entry. The Lyzeum in Graz, mentioned above, was one model for this type of school. The government also looked to examples in Germany and Switzerland, where similar schools had been established earlier.[21] In addition, from 1890 on there were a few secondary schools (*Gymnasien*) for girls who wanted to acquire a certificate that gave them the right to enroll at a university. Again, the Czech-speaking provinces led the way with a private establishment in Prague in 1890. Vienna followed in 1892 with the first private high school for girls in the German-speaking crown lands; it was named *Gymnasiale Mädchenschule*, because the term *Gymnasium* alone was reserved for boys. Nevertheless, it offered the same curriculum to girls as boys had access to elsewhere.

In Germany, special courses for women that led to matriculation standard were established in 1893. And a system of full-scale public high schools for girls developed there too, after the general admission of women to university studies in 1908. In Austria, however, only a very few of the newly established secondary schools for girls had public support before the 1930s. Moreover, the number of these schools in Austria remained very limited until the end of World War I. But it was possible for girls to attend a boys' *Gymnasium* or *Realschule* as visiting students by special permission of the ministry of education. In addition, women graduates from teachers' colleges in Austria were allowed to enroll as special students at universities.

Austria belonged to those countries where women were admitted comparatively early to university studies. Indeed, from 1897 on, women in possession of the appropriate credentials could enroll as regular students, while in Germany, women were admitted only after 1900 (Baden) and as late as 1908 (Prussia). But there is an important point to make: in Austria, admission was first granted only to the philosophical departments. In 1900 and 1901 medicine and pharmaceuticals followed. In Hungary, women gained access to philosophy, medicine, and pharmaceuticals in 1895, though they had to apply individually to the Hungarian ministry of religious affairs and public education.[22] All the other faculties and institutions of advanced learning, including the technical colleges, remained closed to women until the end of World War I. This step-by-step admission was characteristic of the Austrian government's approach; law was opened to women in 1919, the Protestant theological department in 1923, and the Roman Catholic theological department only in 1946. In this respect, the Austrian case differs fundamentally from the German model. There women were first admitted to university studies some years later, but were then allowed to study all subjects at all institutions. In the Habsburg monarchy, the early but partial success of 1897 (by which women were admitted to philosophical and medical areas) gave them the opportunity to study scientific subjects such as chemistry, physics, biology, and so on – since these were taught in philosophical departments. Many women took advantage of this opportunity, as Austrian universities only separated science from philosophy in 1975.[23] Yet, despite the fact that debates on the capability of women and their access to higher education were frequent and passionate, there is no record of them in the documentary evidence of Vienna's technical university. Faculty there did not feel these debates were relevant to them.

The Struggle for Admission

The gradualist policy of the Austrian ministry of education had a lasting effect on how women finally entered technical studies. For the opening of one domain of advanced study to women set off a chain reaction. It was only logical that, sooner or later, all the departments would open. In this sense women's admission to the philosophical faculties in 1897 can also be regarded as the starting point for requests to enter technical colleges. From the late 1890s, applications were made both to the colleges themselves, and to the ministry of religion and instruction. The ministry was restrictive about the number of exceptional cases it would allow, but it began to admit a growing number of girls as guest students for those subjects that were free for them at universities. Sometimes girls who lived in the Monarchy's crownlands without a university were given special permissions to enroll in the local technical college. Moravia for example supported only two technical colleges (the Bohemian and the German Technical Colleges), which offered subjects like mathematics, physics, and chemistry.

The very first women allowed to attend single lectures at a technical college – without, of course, the right to obtain certificates – were Liudmila Sebesta and Elisabeth Kozel, both of whom were teachers at a vocational school for girls. Both received permission to hear lectures at the Bohemian technical college in Brno in 1899. Elise Kozel, who in 1906 also managed to gain admission as a guest at the Vienna technical college for lectures on 'first aid in accidents,' later made a career as director of a vocational school for girls supported by the *Vesna* association in Brno. In 1918, she was an inspector (*Fachinspektor*) for women's vocational schools.[24]

The first woman to attend architectural and economic lectures at the Vienna technical college as a guest, in 1909, was also a teacher. This was Mathilde Hübner, who had struggled for a year and a half to get permission. We do not know what use she wanted to make of these studies. But her interest may have been stimulated by family experience: her brother Carl and her future husband, Ottokar Hanzel, had both enrolled in the technical college some years before she made her application. In her later years, Mathilde Hübner became active in the Austrian women's movement.[25] (Figure 37)

The first step to real progress was an unintended consequence of two ministry of education decrees regarding teacher education. In 1907 and 1911, the ministry declared that to teach in higher commercial schools (*Höhere Handelsschulen*) or to teach descriptive geometry and freehand drawing at a high school, one had to learn a certain number of subjects that were taught only at technical colleges. Women were not yet admitted formally to such colleges. Yet teaching was a respectable occupation for women and these schools were open to female staff. It is not quite clear whether the ministry anticipated the paradoxical effect created by this legislation, but it was certainly quick to remedy the situation.

In 1912 the rector of the technical college in Vienna, Emil Müller, suggested that the earlier decrees be modified. He was a professor of descriptive geometry who taught prospective secondary school teachers, as well as the father of a daughter with a talent for mathematics. (She had just applied for to study at the technical college in Vienna.) The ministry took up his ideas and, from 1913/14, women were allowed to enroll as extraordinary students. This gave them the right to obtain valid certificates at technical colleges for limited number of subjects, as part of their training to be secondary school teachers.

Figure 37. Two women students of architecture pose at Vienna Technical College's building site around 1931 when male students drew their caricature drawings. Note the different gender strategies in dress and habitus of these two aspiring women in negotiating the training ground of the male engineering world. The woman in the white dress might be the future architect Lionore Perin. Permission and courtesy of Vienna University of Technology, University Archive, Vienna, Austria.

The number of women who took advantage of this ruling was small: from 1913/14 to 1918/19, the last year before their full admission, only twenty-seven women enrolled at the Vienna technical college. There were nearly 3,000 male students in the year 1913/14 alone.[26] Meanwhile, the university in Vienna had 796 women in 1913/14, of whom 498 were extraordinary students.[27] These figures demonstrate quite clearly that in Austria, admitting women to technical studies was a question of equality, rather than a result of urgent social pressures as had been the case in the struggle for university admittance: women did not immediately flock to fill the places open to them for structural reasons as happened elsewhere. While most of the first special women students at the Vienna technical college wanted to be secondary teachers for descriptive geometry or freehand drawing, or to teach at a higher commercial school, a small number evidently chose descriptive geometry as a preliminary to later technical studies. Usually they wanted to continue in Germany or another foreign country where the engineering studies were already accessible to women.

From 1913 on, the number of women applying for guest status to visit lectures in the engineering disciplines seems to have grown, as did the number seeking admittance as regular or extraordinary students in technical studies. Most of these sought to enter architecture or mechanical engineering. Some gave as the reason for their studies that they were to take over a family firm. But while guest status seems usually to have been granted, in all the other cases the ministry remained adamantly opposed to the presence of women. This state of affairs lasted until the very point in April 1919 when women were finally given the full right to enroll at technical colleges.

The documentary evidence shows that the first women seeking to enter technical studies acted in an individual capacity. But from 1908 on, groups were organized to push for women's rights of access. Some of these were made up of parents, especially from the Czech-speaking provinces of the monarchy, who petitioned the ministry of education to open the technical colleges to both sexes. Generally they used a legal argument, stating that their daughters deserved an equal opportunity to enter technical colleges, since they were already studying in the *Realschule* system (which prepared graduates for the technical colleges). After the turn of the century, girls who went to a *Realschule* or *Staatsgewerbeschule* (a higher vocational school open to women since 1910) where no Latin was taught, instead of a *Lyceum* or a *Gymnasium*, were at a real disadvantage. Boys were ahead, of course, but so were girls who had learned Latin in school, or privately, for university study. Parents were especially concerned because the number of *Realschulen* was far greater in the eastern crown lands than in the western provinces, and the existing secondary schools for girls were usually expensive, private establishments. The disadvantage was especially felt in Moravia, where there was no university, but instead two technical colleges (German and Bohemian) at Brno. Moravian parents, petitioning the Ministry, argued that they had to send away the daughters, who wanted to study chemistry or mathematics because the nearby technical colleges would not admit them. But to enroll at a university in Vienna or elsewhere required passing additional exams. Increasingly this discriminatory situation seemed illogical and untenable.

Petitions from these parents' groups were numerous, especially in 1912, but the ministry turned them down on 'reasons of principle' or in some cases did not even bother to respond to these urgent requests. However, it was harder to ignore the branches of the women's movement, which addressed the issue starting in 1908. For the first time, these petitions elaborated the argument for opening technical colleges to women. They were sent to the governing bodies of all the technical colleges, so they must have credit for ensuring broad discussion of the subject.

Previously, only the Bohemian technical college in Brno had pleaded (in 1903) for the general right of women to attend as guests. The German technical college there demanded full admission of women in 1904.[28] The reason they were so forward-looking was partly because Moravia had no university. But there was also a strong desire among the Czechs to compete with German-speakers in culture and learning.

At the most important technical college in Vienna, the first open discussion on the subject occurred in 1910. The faculty considered a third petition from the Academic Women's Association (*Akademischer Frauenverein*). The society demanded free admission of women to all kinds of advanced studies, including the technical disciplines, 'under the same conditions as men.' It wanted equal rights for the women who graduated from *Realschulen* and from the *Lyzeum*, arguing that the constitution of 1867 granted freedom of vocational training to all citizens regardless of sex, and allowed citizens, including technology-minded women, to follow their inclinations. The petition also discussed problems of access to the labor market. Firstly, the petitioners pointed out that women graduates from technical colleges, expected to work as free professionals, would have to face the same conditions as men. Because of this, only the best qualified would survive and women would not be able to compete unfairly against men. Secondly, and this seems to have been the main point of the whole petition – the petitioners described the actual labor market for women university graduates as too limited. With only two academic professions open for women at that time, teaching and medicine were already showing signs of overcrowding. Opening technical studies for women was expected to alleviate this situation by creating new respectable occupations.

In discussing this petition, the faculty of the Vienna Technical College split into two factions of nearly equal size. One pleaded for a more or less unrestricted admission of women, while the other insisted on their further exclusion. These two factions remained remarkably consistent before 1918. They were the basis for all later resolutions of the faculty not only in Vienna but at technical colleges throughout the monarchy, with only minor variations. (The Czech colleges, for example, tended to be more liberal in their approach.) All of the arguments for and against the admission of women hinged on the problems of the labor market for male engineers. Around the turn of the century, the profession had begun to fight the 'juridical monopoly' in higher public administration, like their colleagues in Germany. In addition, from 1876 on, there was growing pressure on the graduates of technical universities after the establishment of a new secondary level of technical education, the *Höhere Gewerbeschulen* or *Staatsgewerbeschulen*. Graduates of these new schools also claimed the title of 'engineer.' But there were never as many of these engineers in Austria as there were in Germany, where they soon outnumbered the academically trained engineers.[29] In this situation, women engineers were regarded as potential competitors in the labor market. It was felt that their presence might also damage the prestige of the profession that had been so carefully built up. Therefore, the labor market and career opportunities dominated the discussion throughout the following years. Previously it had been argued that women had insufficient intellectual capacity for scientific study. But that had lost much of its power to convince after 1897. Now other reasons to exclude women had to be found. (Figure 38)

At the Vienna Technical College, opponents of women's admission held the opinion that the special problems of technical studies made their presence 'undesirable.' But they were more concerned with the practical demands of engineering work, which they defined explicitly as 'male.' Women were thought to have the intellectual, but not the personal and social abilities needed to compete with men. Among the heroic virtues required of an engineer were resolution, authority especially in dealing with workers, intellectual and managerial leadership, and even the ability to endure physical hardship in outdoor workplaces. On the other hand, women were thought to be qualified for auxiliary services in the technical field. (This became an important line of argument later). A few moderates, such as the professor of electrical engineering, Johann Sahulka, advised that women should be permitted to enter the colleges, albeit with some restrictions and warnings. He argued that they would be admitted at some point regardless of the faculty's wishes, but that they would probably present little competition to male graduates as their numbers would remain small.

As for the administration, it seems to have begun serious deliberation on the subject in about 1908 or 1909 in the department of vocational training. But it took ten years for anything to happen, and it was the experience of World War I that led to a decisive change. From 1916, a military and political turning point in the course of the war, the Austrian technical colleges themselves began to speak out in favor of admitting women to technical studies. There were several reasons for this new and almost unanimous position. First, the conviction seems to have grown that the admission of women would be inevitable. Second, the minor institutions, like the two technical colleges at Brno faced a dramatic drop in the number of enrollments after 1914; they may have welcomed

Figure 38. "Building Site Rule: A mortar pit is no bath tub." Caricature drawing of women civil engineering and architecture students ridiculing their critical technical judgment and hands-on experience. From: Album Bauhof *and* Technical Research Institute *of Vienna Technical College (circa 1935). Permission and courtesy of Alfred Lechner, Private Collection, Vienna, Austria.*

women as one means to keep their doors open. Third, women's work in the war industries had shown conclusively that women could do 'male' work and in traditionally 'male' workplaces. In particular, when the ministry asked the departments of technical colleges their opinion on the possibility of admitting women students, many cited female university graduates in chemistry who went on to work in wartime laboratories as successful precedents.

And evidently their experiences in the war did inspire some women to think of technical studies. From 1915 there were some cases in which women employed in technical workplaces in the war industries sought admission to technical lectures. At Vienna's technical college, for example, in 1917 Melitta Urbanek applied for permission to attend lectures on architecture as a guest. She had gone to several vocational schools (*Staatsgewerbeschulen*) and was working as a construction engineer (*Bautechniker*) at the time of her application. Another woman was Elisabeth Köhler, who wanted to hear lectures on mechanical technology. She had been working for some time in the steel and ironworks at Ternitz.[30]

Finally, the war experience had evidently shown the academic engineers how to deal with their professional status problems: from the end of 1916, representatives of the technical colleges talked about gender in their debates over women's admission. Hence they began suggest that drearier, 'semi-professional' or less prestigious positions, such as laboratory assistant, draftsman, schoolteacher (but also interior designer and garden architect), were 'female,' or at least compatible with women's essential 'nature.' These were, of course, marked off from the better 'male' jobs. The line seems to have been drawn between technical activities performed mainly 'inside,' which were usually thought suitable for women, and those performed 'outside.' The latter term was, of course, metaphorical as well as literal, indicating public position as well as the field or building site. At the same time once more women entered the occupations of lower status, male technicians would be free to concentrate on their 'proper tasks' as managing engineers. (Figure 39)

This argument was premised on the notion that post-war reconstruction would lead to an increasing demand for engineers. Considering the enormous loss of life among men, proponents feared a shortage of qualified technicians if women remained excluded from engineering studies. Moreover, they anticipated that many young women, especially of middle-class background, would not be able to marry and would have to seek a respectable occupation. Access to new fields of advanced education appeared a sensible way of opening new jobs to women and avoiding overcrowding in those occupations where women already had a significant presence. In 1917, the German technical college in Prague even went so far as to plead that the national good demanded the admission of women: after the war that had destroyed the lives of so many promising German men,

Figure 39. "Building Site Rule: The building site mustn't be too entertaining ('erbaulich')". It portrays women checking their make-up, smoking cigarettes, chatting, and distracting their male colleagues. From: Album Bauhof *and* Technical Research Institute *of Vienna Technical College (circa 1935). Permission and courtesy of Alfred Lechner, Private Collection, Vienna, Austria.*

women engineers of German origin must take leading positions in industry and administration so that Czechs would not do so.[31] The way to a political path, however, seemed possible when a group of Czech members of the Habsburg Reichsrat raised the question of opening technical colleges for women in Parliament in the fall of 1917. In reaction to this political initiative, the ministry of education launched an inquiry concerning women students at the technical colleges of Germany and Switzerland. All reported their experiences of these women students had been favorable.

In June 1918, the conference of the heads of the Austrian technical colleges once again advised that women be admitted under the same conditions as men. After the dissolution of the Habsburg monarchy, the new state office for instruction (*Staatsamt für Unterricht*) of the young Austrian Republic made one more assessment of the views of the technical colleges, provincial governments and regional chambers of engineers. Then it declared (on 7 April 1919) that from the academic year 1919/20 women would be admitted as regular students to technical universities if they met the same standards as men, but on condition that the study of their male colleagues not be disturbed in any way. In explaining this decision and its timing, the decree stated that it was a 'trend of the time' that women, inevitably, would enter the higher technical professions. The decree also made reference to positive experiences with women in the war industry, and the right of women to equal opportunities for personal and occupational development.

The new Czech Republic gave women the right to enroll at technical colleges in the winter of 1918. But in Hungary, the gradualist policy continued: the first initiative to admit women to Budapest's technical university was in 1917, but the dissolution of the Habsburg Monarchy intervened before anything was done.[32] During the short-lived Hungarian Republic in 1918 a few women were admitted to engineering studies, but in the following years access was restricted to the department of economics. From 1925, women were accepted in architecture under certain conditions, with a quota of five per cent. But their number was very small until the end of World War II. They were then deliberately barred from the other technical disciplines. The new restrictions may have led some women to make their way to foreign technical colleges, especially in Germany and Austria. In any case, there was always a small number of Hungarians among the women students at Vienna's Technical College between the wars. Hungarian technical universities only granted full, unrestricted admission to women after 1945.

The First Generations: Women Students at Vienna's Technical University to 1945

After the end of the First World War the history of women engineers falls into two distinct parts each with different political and institutional settings. I therefore shall deal separately with the period of the first Austrian Republic, 1919–1938, and the years of Nazi occupation of what was then *Ostmark* from 1938–1945.

When Austrian women finally won admission to engineering studies in 1919, their victory was, perhaps, untimely. Political, social, and economic conditions had changed so completely that almost nothing fitted the expectations of the pre-war and war years. After the end of the monarchy, most of its German-speaking inhabitants found themselves in a new Austrian Republic. Much of the former Empire's territory was lost in the war, and so was a substantial share of its industry and internal markets. The new state had a huge bill for the war and poor financial resources. In the decades that followed, it suffered from permanent economic, social, and cultural crises.

Despite what had been expected in 1917, there was now no labor shortage. Instead there were too few jobs for engineers. Perhaps because of this, those who had predicted that few women would enter the technical universities proved correct. Until 1938, only in three years out of twenty did the number of women students of all categories at the Vienna technical college ever reach more than 150. During that time the total number of students fluctuated between more than 5,000 in the early 1920s and about 1,800 in the late 1930s. The number of regular full-time women students was much lower; there were twenty just after the war, with a high of seventy-one between 1931 and 1933. A substantial share, especially in the years just after 1919, was of Bohemian or Moravian origin. Having opted to live in the new Austrian Republic they helped to 'import' the stronger inclination towards women's technical studies that had already been a feature in eastern crown lands. The percentage of women rose very gradually from 1.3 per cent in 1919 to a maximum of 7.6 per cent in 1935. Considering the general falling of enrollments, these numbers indicate that attendance patterns developed differently for men and women. German and Prussian technical colleges seem, on the face of it, to have had much higher enrollments of women in technical studies. But in fact these data are skewed because the German institutions had a different department structure, and many women in non-technical disciplines.[33]

It is difficult to know how women students at the technical college in Vienna were treated by their male colleagues once they gained admission. The university's archives hold no evidence of obvious discrimination against female students. When U. Georgeacopol held interviews with a number of women architects from the 'first generation,' she found no indication that students had been inconvenienced by male prejudice.[34] This does not prove that such prejudice did not exist or have some effect. But the fact that it was not remembered as a problem indicates that it may have been experienced as part of personal conflicts rather than obvious sexism, or even that recollections of it might have been suppressed by those

Figure 40. Posed picture of architecture students working at Vienna Technical College's building site near the college machine shop around 1931. Although the students donned working clothes, they did not dress in blue-collar garb. Permission and courtesy of Vienna University of Technology, University Archive, Vienna, Austria.

interviewed. In any case we have to consider the social background of the women in that first, pioneering generation. Engineers in Austria were a tightly-knit community, and many of the women enrolling at Vienna's technical university (no doubt many more than we can trace) were related to male engineering students, engineers, and even professors teaching at the colleges. The way others behaved toward them may have been partly determined by the fact of their place within this engineering network. However, this did not prevent men at the colleges from publishing negative and stereotypical images of women in engineering work: see, for example, the series of caricatures created in the 1930s by male students who were doing the required practical courses for architects and construction engineers. (Figures 40 and 41)

Certainly sexual discrimination came into play in academic work and career opportunities.[35] On the one hand, women were employed surprisingly early as research assistants: the first one, was the chemist Margarete Richter, a graduate from the university in Vienna, in 1919. But she was allowed to take up her position only after the search for an adequate male candidate had proved unsuccessful. She was then forced to leave some months later when a suitable young man had been found. Another chemist, Margarete Garzuly-Janke, employed as a research assistant from 1921, had to leave in 1932 when her husband was appointed professor of

Figure 41. "Building Site Rule: The building site is no fashion show!" The appeal and the ridicule of an attractive woman student walking as if on a cat walk and getting caught on trestle's nail is contrasted with the serious male working clothes in the back. From: Album Bauhof *and* Technical Research Institute *of Vienna Technical College (circa 1935). Permission and courtesy of Alfred Lechner, Private Collection, Vienna, Austria.*

chemistry at Vienna's technical college. She later became the first woman to qualify as a private lecturer (*Privatdozentin*), but she never succeeded in getting regular employment as a scientist. Six women were hired as research assistants at the Vienna technical college before 1938, most of them graduates not of the college itself but of the university. It took some time to educate enough young women in the technical disciplines before they made a sufficient pool from which to draw assistants. This stage of development was only reached in the 1930s.

Usually, about two-thirds of the female students at the college had extraordinary or guest status, which was not unusual in this early period of women's engineering studies. Similar percentages were reported for many German technical colleges.[36] In Vienna, most of those enrolled as special students at the Technical College intended to become secondary school teachers and therefore attended the university as regular students. The regular technical college students clearly preferred architecture and technical chemistry, followed at some distance by actuarial theory and technical physics. In the core disciplines of engineering – civil engineering, mechanical engineering, electrical engineering, and surveying – their number was tiny. Usually there was only one or two, and in some years there were no women at all.

In analyzing these choices, we can discern certain patterns. Women seem to have felt an affinity (perhaps determined by social conditioning) for traditional 'female' areas such as architecture or chemistry. Architecture bore some relation to the traditional task of 'home-making' and the refinement of artistic learning. Chemistry was associated with household tasks such as preparing food, and was sometimes taught in girls' secondary schools for that very reason. After these two fields, women went into new disciplines, such as technical physics and actuarial theory. Moreover, the statistics show that women students seem to have preferred disciplines that took fewer students.

Job opportunities varied by field, of course, but they were poor for both sexes.[37] For architects there was the possibility of working in a public building office or doing freelance work. But women were traditionally not welcomed by the bureaucratic system, so they usually chose the second option. From the late 1920s, the prolonged economic crisis greatly affected the building trade. Nevertheless, women did not stop studying architecture. It is not clear why this is so, except that some seem not to have expected to practice their profession once studies were completed.[38] Technical chemists could get a position in industrial laboratories, where women had been successful during the war years. But in the Austrian Republic, with the chemical industry at an early stage of development and most firms small, opportunities were poor; from the mid-twenties the technical colleges tried to deter students from choosing technical chemistry. (They did this with some success, as shrinking enrollments from 1925 show.) It is difficult to find exact data for the labor market in the inter-war period. However, in about 1930, the number of women chemists working in the fields they had qualified for seems to have been small, and many graduates from universities and technical colleges were employed in other occupations than those for which they had trained.[39] Actuarial theory, however, was in constant demand because the national insurance system in Austria was extended between the wars. Established as a discipline in the universities in 1894, it required only two years of study and one theoretical exam. This made it a less expensive option for study than those disciplines that required longer courses and two theoretical exams. All these features made it ideal for women, and indeed, from 1924, their proportion rose to more than forty percent. This reached more than eighty per cent in 1937, though absolute numbers went down. Women students who entered this field were mostly from the lower middle classes.

Technical physics was a special case. Established in 1923 as a sub-discipline at Vienna's technical college, it took only fourteen new students each year because of limited laboratory space. The students became highly qualified specialists with excellent job prospects which were, it seems, almost unaffected by the economic crisis of the 1930s. Why was this field popular with women? Usually they were three or five per cent of the total number enrolled, though in some years the numbers went up. Physics like chemistry belonged to those scientific disciplines already familiar to women students from the universities, so some may have expected better job opportunities in this new and rather undefined sub-discipline. In around 1930, at least twelve women physicists were employed in industrial workplaces.[40] How many of them were graduates from Vienna's Technical College is not yet known. (Figure 42)

Occupation and War, 1938–1945

The Nazi occupation of Austria in the spring of 1938 had as its first result a sharp drop in the overall number of students. This was due mainly to the rapid expulsion of Jewish students and of political enemies of the new regime by the end of 1938. Exact data on the number and percentage of students excluded for 'racial' and political reasons are not available. But we do know the names of some women students who had to leave. Among them were the would-be architects Judith Gourary, Anna Szabó, and Helene Alexander; Ruth Zerner, a student of electrical engineering; the technical physicists Mary Black and Renate Laserson; and the technical chemists Johanna Frank-Edelmann and Georgine Altmann.[41] We can only record that the small community of Austrian women engineers was decimated.

Student enrollments remained low until the outbreak of World War II in 1939. Germany had imposed a quota of ten percent for women students in December 1933, but this was ended in 1936. It therefore did

Figure 42. Snap shot photo of Vienna University of Technology students of architecture on a field trip with their professors in 1925. Permission and courtesy of Hans Pfann, Collection, Vienna University of Technology, University Archive, Vienna, Austria.

not apply to Austria. When, from 1938, the regime tried to attract women to study fields useful for war industry, the number of enrolled women began to grow again. The first gains were in technical chemistry, while the other disciplines followed somewhat later. The proportion of women in these fields rose too, mainly because there had been a reduction in the number of male students, many of whom enlisted. In the academic year 1944/45, the percentage of women students at Vienna's technical college reached 22.1 per cent, one of the highest levels ever registered. Meanwhile the absolute numbers of students dropped to an historical low of 598, of whom 119 were women. In pure statistical terms, the war led to a substantial 'feminization' of engineering studies in Vienna. The cultural and social consequences of these developments deserve further study. Nevertheless, it would again turn out to be a precarious gain.

Women's preferences for one subject over another hardly changed during the war. There was still a strong concentration in architecture (resulting partly from the heavy influx of foreign students coming after 1942/43 from Bulgaria, Romania, Greece and other countries in south-eastern Europe that were now tied to Nazi Germany), and technical chemistry. Technical physics continued to grow, but women turned away from actuarial theory. Teacher training was now done at separate institutions. The core engineering disciplines remained 'male' domains, though after 1942 construction engineering, electrical engineering, and mechanical engineering all show small increases in women's enrollment.

The composition of the female student body also changed. The number and proportion of foreign citizens rose, as did that of women students coming from Germany (who were not counted as foreign, of course). Both groups tended to stay only for a short time in Vienna, though for different reasons: while students of non-German nationality usually were not allowed to stay longer than four semesters, students from the Reich followed a traditional pattern of student mobility. They would choose to spend one or two semesters at a foreign university and then to return to their home university to finish their studies.

Little is known about the professional careers of the women who graduated from the Technical College in Vienna during this period. Some of the women architects managed to do freelance work or were given administrative posts. Among the generation that took up their studies late in the 1930s or during the war, several took the opportunity to attempt an academic career, replacing male lecturers, demonstrators, and assistants. Twenty-eight of the thirty-four female assistants engaged between 1919 and 1945 were hired after 1938. Women took advantage of this new opportunity thus adjusting to the new political system. Those who had been forced to leave after the collapse of the regime in 1945 did so not just because men had returned to their old jobs but for political reasons. Even though the political measures affected also male assistants and professors at the college, who were suspected of strong ties with the Nazi regime, a substantial number of the men but not the women managed to return to their jobs during the 1950s. This might indicate that policies hostile to women might have been at work. Soo too an unknown number of men and women students were forced to leave the college because of their sympathies after 1945. Not all of whom returned to resume their studies later on.

If these measures were a necessary political purge, they had a serious consequence for the community of women engineers. The result was the loss of part of a generation that might have served as role models for women in the future. And this loss must have been felt all the more because as it came just when women seemed to be consolidating their position in engineering studies.

Conclusion

The period between 1919 and 1945 saw women firmly established within the engineering profession though, except during the Second

World War, they always remained a tiny minority within the technical universities. The two wars had catalytic effect: women's efforts to gain access to technical studies had started well before World War I, but the war brought opportunities to demonstrate that women were able to fill jobs previously reserved for men. And the war also brought about the political climate in which it was possible to grant equal rights to women. The government was always restrictive, and made concessions only step by step. However, the partial admission of women to university studies in 1897 laid an important groundwork. The Austrian women's movement then took up the struggle with some success, aided by parents' groups and a liberal element among the faculty of the technical colleges. The effect of the World War II was similar: for a short time it encouraged growing numbers of women (most of them foreign) to take up a study at a technical college. For some of these women it offered new chances to enter occupations and academic positions that had been almost entirely the province of men. Something similar happened in all the countries involved in the war (including the non-European ones) but there were special conditions in Austria and Germany. The expulsion of Jews from the universities and technical colleges, and also from the labor market, created an additional demand for qualified labor. This opened further opportunities for some women, so long as they were not Jewish or politically 'undesirable.'

The female graduates demonstrated that women were capable of mastering jobs in the technical field that required extensive qualifications. Thus, the next generation of women in technical studies, in the late 1940s and early 1950s, was able to start from a position of relative strength as compared to the pioneers of 1919/20.

The inter-war period seems to have seen the start of long-term pattern in women's preferred courses of study. Certain domains remained more attractive to women for social and cultural reasons until women's participation in higher education expanded during the 1970s and 1980s.

Most of these developments had a parallel in the history of women's engineering studies in other European countries, where the relation of educational and occupational systems was organized in a similar way. What was unique to Austria (here meaning the whole territory of the former Habsburg monarchy before 1918/19) was the gradualist approach to women's admissions, and the greater specialization of the technical colleges, which never diversified to offer non-technical subjects. Related to this was considerable movement of students (and professors) especially in the sciences, between the universities and the technical colleges. This helped lead to the relatively early entry of women into careers as university assistants.

Nevertheless, it seems that Austrian women on the whole were unfortunate. Their first admission to engineering studies happened because everyone expected there to be an enormous demand for engineers after the first World War. Instead, the country suffered an economic and social crisis, with persistently high unemployment especially within the technical professions. This was presumably an important reason why the number of women enrolling at technical colleges after 1919 was so small and grew so slowly. The next time a window of opportunity opened, it was under disastrous circumstances. The emotional and personal consequences of the war were profound, and of course Austria was again on the losing side. Some of the women who would have been the best of their generation were driven abroad or killed. Of those who stayed behind, many made career decisions that were morally repugnant, and found themselves politically compromised after the war. They lost their jobs, though we don't know how long they were ostracized. The political, social, and cultural implications of the Nazi period and the war had a dramatic and lasting effect on women in engineering which deserves further comparative study. This would illuminate the study of engineering in Austria and Germany, of course, but also what happened in other countries: both those that came under Nazi influence, and those that took in foreign technical personnel during and after the upheavals of war.

Notes

1. For a recent example, see: Peter F. Meiksins and Chris Smith, eds, *Engineering Labour: Technical Workers in Comparative Perspective* (London: Verso, 1996), 233–255. With regard to education: Peter Lundgreen, 'Engineering Education in Europe and the U.S.A., 1750–1930: The Rise to Dominance of School Culture and the Engineering Professions,' *Annals of Science* 47 (1990): 33–75.
2. See also Paul Windolf, *Die Expansion der Universitäten. Ein internationaler Vergleich 1870–1985* (Stuttgart: Enke, 1990). For the question of womens' access to academic studies in a comparative perspective see: Ilse Costas, 'Der Kampf um das Frauenstudium im Internationalen Vergleich,' in *Pionierinnen, Feministinnen, Karrierefrauen? Zur Geschichte des Frauenstudiums in Deutschland*, edited by Anne Schlüter (Pfaffenweiler: Centaurus Verlag, 1992), 115–44; Ilse Costas, 'Der Zugang von Frauen zu akademischen Karrieren. Ein internationaler Überblick,' in *Bedrohlich Gescheit. Ein Jahrhundert Frauen und Wissenschaft in Bayern*, edited by Hiltrud Häntzschel and Hadumod Bùssmann (Munich: Beck Verlag, 1997), 15–34.
3. For the history of the Austrian educational system, see Helmut Engelbrecht, *Geschichte des österreichischen Bildungswesens*, esp. vols. 3 and 4 (Vienna 1984/1986); Gary B. Cohen, *Education and the Middle Classes in Imperial Austria, 1848–1918* (West Lafayette: Purdue University Press, 1996). For higher education of women, see: Marina Fischer-Kowalski and Peter Seidl et al, *Von den Tugenden der Weiblichkeit. Mädchen und Frauen im österreichischen Bildungssystem* (Vienna: Verlag für Gesellschaftskritik, 1986). For a concise comparative view on women's higher education in the German speaking countries, see James C. Albisetti, 'Mädchenerziehung im deutschsprachigen Österreich, im Deutschen Reich und in der Schweiz, 1866–1914' in *Frauen in Österreich. Beiträge zu ihrer Situation im 19. und 20. Jahrhundert* edited by David F. Good, Margarethe Grandner, and Mary Jo Maynes (Vienna: Böhlau Verlag, 1993), 15–31. On the history of women students at technical universities, especially at Vienna University of Technology, see: Juliane Mikoletzky, Ute Georgeacopol-Winischhofer, and Margit Pohl, '*Dem Zuge der Zeit entsprechend…' Zur Geschichte des Frauenstudiums in Österreich am Beispiel der Technischen Universität Wien* Schriftenreihe des Universitätsarchivs der Technischen Universität Wien, Vol. 1 (Vienna: WUV-Wiener Universitätsverlag, 1997); and Juliane Mikoletzky, 'An Unintended Consequence: Woman's Entry into Engineering Education in Austria' *History and Technology* 14, no. 1 (January 1997): 31–48.
4. See also Lundgreen, 'Engineering Education in Europe and the USA'.
5. For an example see ibid., 41.

6. See also Cohen, *Education and the Middle Classes in Imperial Austria*, 240–246.

7. See also: Boel Berner, 'Explaining Exclusion: Women and Swedisch Engineering Education from the 189os to the 1920s,' *History and Technology* 14 (1996): 7–29, quote p. 9.

8. Boel Berner in 'Engineering and Economic Transformation in Sweden,' in Meiksins and Smith *Engineering Labour*, 168–195, p. 171 speaks about 'doubel goal.'

9. For a detailed study on the pre-history of women's technical studies in Vienna see Juliane Mikoletzky, 'Frauen, Naturwissenschaft und Technik und die Wiener Technische Hochschule: Eine Spurensuche,' in Mikoletzky *et al.*, *Dem Zuge der Zeit entsprechend…*, 17–40.

10. We know this only from her report on her published studies. See Maria Manassein, 'Beiträge zur Kenntnis der Hefe und zur Lehre von der alkoholischen Gährung' in *Mikroskopische Studien*, edited by Julius v. Wiesner (Stuttgart, 1872), 116–128.

11. For some examples from German technical colleges, see also Mikoletzky *et al.*, *Dem Zuge der Zeit entsprechend…*, 23–24; for the Stockholm Royal Institute of Technology: Berner 'Explaining Exclusion,' 14.

12. Ibid.

13. Lundgreen, 'Engineering Education in Europe and the USA,' 45.

14. Ferenc Szabadváry, ed., *200 Years of the Technical University Budapest, 1782–1982* (Budapest: Technical University Budapest, 1982).

15. See for example Margot Fuchs, *Wie die Väter So die Töchter. Frauenstudium an der Technischen Hochschule München, 1899–1970* (Munich: Technische Universität, 1994).

16. For a brief overview on the professional history of Austrian engineers, see: Juliane Mikoletzky, '"Der österreichische Techniker". Standespolitik und nationale Identität österreichischer Ingenieure 1850–1950,' in *Technik, Politik, Identität. Funktionalisierung von Technik für die Ausbildung regionaler, sozialer und nationaler Selbstbilder in Österreich*, edited by Klaus Plitzner (Stuttgart: Verlag für Geschichte der Naturwissenschaften und der Technik, 1995), 111–123.

17. For the history of 'Ziviltechniker' in Austria see: Ute Georgeacopol-Winischhofer and Manfred Wehdorn, 'Geschichte des Ziviltechnikers in Österreich' in *Ziviltechniker & Wirtschaft. Gestalter der Umwelt*, edited by Erich Schlöss (Vienna: Bau-Verlag Schmutzer, 1983), 37–47.

18. Margarete Friedrich, 'Versorgungsfall Frau? Der Wiener Frauen–Erwerb–Verein' Gründerzeit und erste Jahre des Aufbaus' in *Jahrbuch des Vereins für Geschichte der Stadt Wien* (1991/92), 263–308, esp. 290–293.

19. Marianne Hainisch, *Zur Frage des Frauen-Unterrichts* (Vienna: Erste Wiener Vereins-Buchdruckerei, 1870).

20. Heidemarie Liebhart, 'Das Mädchenlyzeum des Wiener Frauen-Erwerb-Vereins' (University of Vienna, Diploma-Arbeit, 1995), 35.

21. See also Albisetti, 'Mädchenerziehung im deutschsprachigen Österreich,' 20–23; see also Fuchs, *Wie die Väter So die Töchter*, 3–16.

22. See Eva Vámos, 'It has Been 100 Years now that Women can Study in Hungary' in *Universitas Budensis, 1895–1995*, edited by László Szögi and Julia Várga (Budapest: Loránd Eötvös Universität Budapest, 1997), 477–485.

23. See Renate Tuma, 'Die österreichischen Studentinnen an der Universität Wien (ab 1897),' in Heindl and Tichy, *Frauen an der Universität Wien*, 79–128.

24. Mikoletzky *et al.*, *Dem Zuge der Zeit entsprechend…*, 90, n. 93.

25. For information on Mathilde Hübner-Hanzel's later life, I thank Dr. Monika Bernold at the History Department at the University of Vienna.

26. The following figures are based on the students' catalogues of the Vienna Technical College. Juliane Mikoletzky, 'Die Eroberung der Zulassung als außerordentliche Hörerinnen,' in Mikoletzky *et al.*, *Dem Zuge der Zeit entsprechend…*, 29–40.

27. Tuma, 'Die österreichischen Studentinnen'.

28. Juliane Mikoletzky, 'Die Entwicklung des Frauenstudiums an der Technischen Hochschule in Wien 1919–1945,' in Mikoletzky *et al.*, *Dem Zuge der Zeit entsprechend…*, 109–184.

29. Karl-Heinz Ludwig, *Technik und Ingenieure im Dritten Reich* (Kronberg: Athenäum Verlag, 1979), 41.

30. Mikoletzky *et al.*, '*Dem Zuge der Zeit entsprechend…*', 58–61.

31. Ibid., 64.

32. Szabadváry, *200 Years of the Technical University Budapest*, 35; Vamós, 'It has been 100 Years', 480–481.

33. See also Mikoletzky *et al.*, '*Dem Zuge der Zeit entsprechend…*', 178, Table A.4.1.

34. See also Ute Georgeacopol-Winischhofer, 'Sich bewähren am Objektiven'', in Mikoletzky *et al.*, '*Dem Zuge der Zeit entsprechend…*', 185–258.

35. Mikoletzky *et al.*, '*Dem Zuge der Zeit entsprechend…*', 162–166.

36. See also for the Technical University of Munich: Fuchs, *Wie die Väter So die Töchter*, 35–41; for Germany in general, see: Hartmut Tietze, *Datenhandbuch zur Deutschen Bildungsgeschichte*, Vol. I, no. 1 (Göttingen: Vandenhoeck & Ruprecht, 1987), 46, 48.

37. The assessment of labor market opportunities is based especially on the analyses given for every discipline in the Technical University of Vienna's student yearbooks from 1926 to 1942.

38. For the various attitudes of female architecture students at the Vienna Technical University during the interwar period, see: Georgeacopol-Winischhofer, '"Sich bewähren am Objektiven"', 185–215.

39. See also *Handbuch der Frauenarbeit* (Vienna: Kammer für Arbeiter und Angestellte in Wien, 1930), 307–311; Martha Stephanie Braun, *et al.*, *Frauenbildung und Frauenarbeit in Österreich* (Vienna: Bund österreichischer Frauenvereine, 1930), 298.

40. Ibid.

41. It is possible that some found safety in a foreign country, which is the subject of new research.

Annette Vogt[1]

7. Women in Army Research: Ambivalent Careers in Nazi Germany

Within the German Empire, state technical colleges (*Technische Hochschule*) allowed women to enter shortly after the universities had opened their doors to them. The Prussian state was one of the last to do so, in 1909. Prussian legislation concerning women's right to study, unlike that in Austria, applied to technical colleges.[2] Legislation covering the universities included a special section that allowed professors to exclude women from their lectures. However, a similar law governing the technical colleges failed to grant professors there the same right.[3] Yet, despite the loophole in the law, only a few women chose to study at the technical colleges. Once the law was passed, debates in Germany (and in Prussia in particular) shifted their focus from women's education and its dangers to women's employment.[4] Historians Barbara Duden and Margot Fuchs have argued in their studies of technical colleges in Berlin and Munich that most women students preferred the natural sciences to engineering.[5]

Undoubtedly the resistance of educators to the idea of women in engineering accounts in part for women's low participation, as Duden and Fuchs argue. But women's poor employment opportunities in the field of engineering were also a factor.[6] In 1913, just before the outbreak of the First World War, Eugenie von Soden edited a book about professional opportunities for women.[7] Few of the authors recommended that women with academic training consider work in science. Several authors described instead the teaching positions open in various schools, and jobs within health care and social work in Germany's recently-established municipal and state agencies. Because the emerging German welfare state offered new positions, advocates of women's rights considered these particularly appropriate professions for women. Several contributers to the book were pioneer women scientists in their respective fields. Gertrud Woker (1878–1968) from the University of Bern in Switzerland, discussed career opportunities in chemistry, stressing that the profession offered women few opportunities in industry; commercial laboratories and some private research institutions were considerably more favorable in comparison.[8] Emilie Winkelmann described women's employment opportunities in architecture and engineering, pointing out that few openings existed in the industrial sector for qualified women. Winkelmann believed women engineers had an even harder time than independent architects who managed their own firms.[9] One of the contributers found so few opportunities for women scientists in German universities during the Wilhelmian Empire that she went to the United States to study. This was the biologist Rhoda Erdmann (1870–1935), who wrote about the employment situation for women in zoology, botany, math, and physics.[10] Erdmann left Germany for a research position at Yale, an elite university for men that had opened its graduate school to women and African-Americans who wanted to study the sciences at an advanced level. She also managed to get a fellowship at the Rockefeller Institute for Medical Research before returning to Germany in 1919.[11]

The resistance of male colleagues in all the sub-fields of science and engineering was still intense. As those who study modern professionalization have pointed out, in some fields such resistance expressed the desire of an established profession to maintain its relatively high status. In others, male opposition grew out of an emerging profession's search for recognition and legitimacy. In still others, the antagonism was the result of deeply-held beliefs about women's social roles and doubts about their intellectual capacity to be 'good' scientists.[12] In struggling against male resistance, German women engaged in the same kind of strategies historian Margaret W. Rossiter observed among their American counterparts.[13] Similarly, a few recent German studies have explored the kind of opportunities and discrimination women scientists – chemists in particular – faced in Germany between 1900 and 1945.[14] Biographies are also now available on such women as physicist Lise Meitner (1878–1968), her colleague Isolde Hausser (1889–1951), and the pilot and aeronautical engineer Melitta Schiller (1903–1945).[15] Women in academic science found better opportunities and entered the male domain sooner than women in engineering, even if all faced similar difficulties. Women scientists in Germany found employment at private laboratories, industrial research laboratories, and some research centers that had both government and private funding. One such was the Robert Koch Institute in Berlin where bacteriologist Lydia Rabinowitsch-Kempner (1871–1935) worked with her husband. A relatively large number of women found scientific research work at the research institutes of the Kaiser Wilhelm Society (*Kaiser Wilhelm Gesellschaft*).

Why did those research institutions of the Kaiser Wilhelm Society offer better employment and career opportunities to women? Using case studies of three women scientists shows the exceptional role the Kaisser Wilhelm Society's Institutes played in offering women jobs in research work, particularly during the Second World War. I pay special attention here to the Nazi and war periods, when many German women scientists, like their colleagues elsewhere, moved up the promotional ladder more quickly in what were traditionally male disciplines.[16] Thus, the first two women I consider were trained in academic science but worked in engineering for the armed forces. In my third example I look at an engineering graduate who found work as a scientist at one of the Kaiser Kaiser

Wilhelm's Institutes. She was forced from her job when she married. All three women built their careers on a scientific reputation in military rather than in civil research. And all three relied on individual resourcefulness for the kind of problems they faced in working as women scientists rather than on the collective strategies advocated by the women's movement.

Science Research in Germany: The Institutes of the Kaiser Wilhelm Society

The Kaiser Wilhelm Society established its first institutes soon after its foundation in Berlin in 1911. It represented a new type of German scientific research in terms of organization, financial support, personnel policies, and intellectual framework. The initiators of this new network of research institutes sought to build a 'German Oxford' for pure science. Most of the first institutes were established in relatively new disciplines and specializations including physical chemistry, chemistry, and biology. They quickly acquired a national and international reputation. Under the umbrella of the society, each research institute ran its own finances, although funds came both from government and from the commercial sector (for example, from the expanding chemical industry). Some institutes, such the Kaiser Wilhelm Insitute for Coal Research, were sponsored entirely by industry.[17] After 1919, the research institutes were located not just in Berlin but throughout the country, from Heidelberg (the Institute for Medical Research) to Düsseldorf (the Institute for Metal Research). The Kaiser Wilhelm Society established Institutes in chemistry, physical chemistry and biology first. Later Institutes were built for fiber research in Berlin, silicate chemistry in Berlin-Dahlem, leather research in Dresden, metal research in Düsseldorf, and coal research in Muehlheim/Ruhr.

The society's first president, the theologian Adolf von Harnack (1851–1930), sought to create research institutes around the reputation of one leading scientist, who would be given absolute freedom to devise a research program, create new departments, and hire and fire research assistants as he thought fit. The scientists Harnack hired as new directors were, almost without exception, the leading experts in their newly-developing fields of specialization. Under Harnack's policy the directors and the scientific members enjoyed extraordinary financial rewards, working conditions, and intellectual freedom.[18] The other scientists, such as doctoral students, unpaid visiting scholars, and research assistants, worked on a temporary basis. The opportunities for men and women to obtain these temporary research positions were similar. And, in contrast to practice at the universities and other state agencies, men and women scientists had similar pay scales.[19] But men found it much easier to climb the promotional ladder.

Between 1912, when the society was established, and the end of the Nazi regime in 1945, the author found as many as 225 women scientists with doctorates worked at its twenty institutes. Most of these have been erased from histories of science. Some women even became heads of departments (a status comparable to that of associate professors in the American system), yet they have been forgotten.[20] Quite a few women scientists worked at the research Institutes for fiber, silicate chemistry, and leather research. Some of them had received degrees from various German technical colleges, where they had studied the (natural) sciences rather than engineering, but few were trained as engineers. The chemists Deodata Krüger and Marie Heckter, the mathematician Irmgard Flügge-Lotz, and the physicist Maria Heyden were the only four among the more than 225 women scientists appearing in the institutes' records who had received a doctorate from a technical college. This fact underscores just how daunting it must have been for women to finish a doctoral degree at the technical colleges. The author found that twelve women acted as department chairs at different institutes of the society. Among them were three physicists, four chemists, four scientists in bio-medical fields, and finally one mathematician-engineer, Irmgard Flügge-Lotz.[21]

One of the most famous of the Kaiser Wilhelm's Institute was the Göttingen Institute of Fluid Mechanics, founded in 1924. The 1936 *Handbook of the Kaiser Wilhelm Society* lists a staff of twenty-eight assistants, among whom there were only two women: the mathematician Irmgard Lotz (later Flügge-Lotz) a graduate from Hannover's technical college, and the physicist Margot Herbeck.[22] Herbeck (b. 1909) worked at the Göttingen Institute of Fluid Mechanics for about four years. After finishing her thesis in physics at the faculty for mathematics and natural sciences at Göttingen's university, Herbeck worked at the Kaiser Wilhelm Institute from 1934 to 1938.[23] By March 1938 she was already in Berlin. Although she identified herself as a scientist living in Berlin, it is unclear where she worked. She belonged to the Society of Technical Physics, an association of applied-science established in 1918. She had become also a member of an organization of women students affiliated to the Nazi Party.[24] Other than Herbeck and Lotz, whose career is described below, no other women were employed by the Institute for Fluid Mechanics.

The renowned Physical Chemistry and Electro-Chemistry Institute in Berlin-Dahlem was a second Institute of the Kaiser Wilhelm Society. Its director was the German-Jewish and nobel prize winner Fritz Haber (1868–1934). Deodata Krüger (1900–1945), who had finished her thesis in chemistry at the technical college in Berlin-Charlottenburg in 1923, occupied an unpaid position there from 1928 until 1933.[25] When Haber was forced into exile because of the anti-Semitism of the Nazi state, the institute changed entirely. Most of Haber's research assistants, including Deodata Krüger, lost their jobs. It is not clear in what capacity Krüger continued her scientific research, but in 1936 the records show her working in the Kaiser Wilhelm Institute for silicate chemistry. And in 1938 she was a researcher at an industrial lab, the *Sächsische Zellwolle Plauen*. In 1941, she contributed to an important international publication, *Methods of Ferment Research*.[26] At the time she also worked at the technical college in Berlin, where she probably had a part-time, unpaid position at a laboratory while also working in research and development at a factory – an employment pattern that was quite common in Germany at the time. She was shot to death on April 15, 1945 while travelling to Berlin just as the last battles raged between the Nazis and the Allies.[27]

A greater proportion of women were employed at Berlin's Kaiser Wilhelm Institute for Silicate Chemical Research than in the society's other institutes. Its director, the chemist Wilhelm Eitel (1891–1979), worked there from its foundation in 1925 until the end of the Nazi regime in 1945. Eitel, who also lectured as honorary professor at the technical college in Berlin-Charlottenburg, was instrumental in offering women scientists the chance to work at his institute. During his administration, eighteen women scientists worked there, of whom two had graduated from a technical college. Under Eitel's patronage, two women were appointed as heads of a laboratory, which was a position of relatively high status between assistant and department chair. The institute's research laboratories were small; nevertheless it is significant that women were allowed in supervisory positions at all. In both laboratories the women oversaw product-analysis research for different industries. One of the two laboratory supervisors was Maximiliana Bendig (b. 1898), a chemist who had done her doctoral work at the University of Berlin in 1925 before joining the institute, where she worked from 1927 until 1936. The other, Christel Kraft, had received her engineering degree at the technical college before being placed in charge of the lab in 1928. Two years later she was forced to leave the Institute because she married. Her successor, Marie Heckter (1898–1956), had also studied at Hannover's technical college and earned her engineering degree in 1934. Before moving into Kraft's management position as head of the chemical laboratory, she worked as an assistant at her *alma mater*. While leading the lab, she worked on her doctoral thesis in chemistry at her former college in Hannover. She continued to lead the laboratory until 1936, when she was dismissed from the institute because she married. She worked partly with her husband, Chemist Fritz Strassmaan (1902–1980).

Maria Heyden (b. 1912) got a position as a research assistant in the Kaiser Wilhelm Institute for Physics from 1939 to 1943.[28] From its beginnings in 1917 this institute had been run by the German-Jewish scientist and nobel prize winner Albert Einstein (1879–1955), until his forced exile to the United States in 1933. From 1917 until 1936 the institute had neither a building nor assistants but funded a generation of young physicists from different universities through grants. After finishing her thesis in physics at the technical college in Berlin-Charlottenburg in 1937, Heyden had a scholarship for two years at the University of Kiel. There she co-authored an article with her colleague W. Wefelmaier on radioactivity in the journal "*Sciences*" (Die Naturwissenschaften). This radioactivity research was conducted in centers based in Berlin, Vienna, Paris, and Cambridge. Both authors followed similar career paths while they were working at the institute (renamed the Max-Planck-Institute in 1938) and at the physics institute of the technical college in Berlin-Charlottenburg.[29] Nazi law required women civil servants to leave their posts when they married, but as the Nazis began the Second World War and needed women to replace men in the war economy, married women like Heyden were kept on. After her marriage to a physicist in 1939, Heyden was able to continue to work at the Kaiser Wilhelm Institute for Physics in Berlin until 1943, when it moved to Hechingen and Haigerloch/Württemberg in the south. When the institute transferred, Heyden did not follow. Instead she remained in the capital, where she found a position as an assistant at the Institute of Electronics in the technical college until 1953. Heyden's career was highly remarkable because she was able to continue her work despite marrying and having two children during the war.

Women Scientists in the Armed Forces during the Nazi Regime

It is crucial not just to the study of gender and technology, but also to the history of science generally, to ask what forces shaped women's opportunities within military institutions. What can we learn about women's entry into engineering in an autocratic state like Nazi Germany? So far, few studies deal with the history of the Kaiser Wilhelm Society's research institutes during the Nazi period.[30] As late as 1998 did the Max Planck Society establish a research group with the express purpose of examining the society's role in this period. The same year, German scientists at the International Mathematical Congress in Berlin launched the first studies of German mathematicians in exile and the crimes of the Nazi regime.[31]

The Nazis held different notions about women's social position than the supporters of the Weimar Republic. To the Nazis, the ideal German woman was of "Aryan descent" and a full-time mother with many children.[32] This view on women was inherently both anti-Semitic and anti-feminist, and of course it stood in sharp contrast to the values that emerged in the bourgeois and socialist branches of the women's movement.[33] However, Nazi views of women's position were always highly contradictory. In theory, the Nazis advanced the ideal of the "Aryan" full-time mother by seeking to expel career women from a number of professions. In practice, they also offered new public roles but only for "non-Jewish" women who should be moreover politically conformist. This was particularly the case after 1936, when the Nazis expelled those

Germans deemed Jewish and started to prepare for war. The Nazis provided a "legal" framework for the expulsion of scientists through the "Civil Service Restoration Law" of 7 April 1933, which allowed state employees to be dismissed if they were Jewish by Nazi definition, Communist, Social democrats or otherwise politically undesirable. Another law initiated under the Weimar Republik, passed on 30 June 1933, ruled that all women civil servants 'who were not in financial need' had to leave their jobs. Both laws applied to the Kaiser Wilhelm Society's institutes since they received state funding.[34] The Kaiser Wilhelm Society protested only in a few cases and with great caution. Of the twenty-three directors employed by the society, nine were fired, although the legal euphemism 'going into (early) retirement' was used. As a result of the Nazi laws, two of every three women scientific members were expelled. Four of the five women heads of a department had to leave their positions between 1933 and 1938 because they or their directors were Jewish (as defined by Nazi legislation). This changed the position for non-Jewish women. When in 1935 all Jewish students were barred from higher education, more non-Jewish women students could enter. In 1933 a law had set a quota of ten per cent for both Jewish and women students, but was cancelled already in 1934 for women students only.

When the Nazi state prepared for war, some women scientists could move into military research. During the Second World War, as previously, women workers and scientists were offered wider opportunities because they were needed as temporary substitutes for men. In contrast to what happened in the first World War, however, a number of highly qualified women scientists were already trained and available, having weathered the poor employment prospects of the twenties and thirties. They were able to advance their careers through military research. The career opportunities of women scientists, specialists in particular, were much more favorable at military institutes than elsewhere. Since most military work was secret, the state was able to make use of women scientists without disrupting the public valorization of women's domestic roles. There were a number of women scientists who benefited from this window of opportunity and worked in all branches of military research. Because of the secretive nature of military work, it is possible that the number involved was much higher than we will ever know.

Portraits of Women Scientist Entering Military Research

The young Gertrud Kobe (1905–1995), a geophysicist whose career at the University of Berlin spanned more than thirty years, wrote a brilliant thesis in 1934 on the hydrographical structure of the North sea area Skagerrak, north of Germany. Albert Defant and Heinrich von Ficker, both well-known scientists, supported her research and acted as her supervisors. Her thesis helped to interpret the results of an international expedition. Born in Berlin in 1905, Kobe was only able to afford her education because she worked at other jobs while pursuing her studies between 1923 and 1928. By taking evening classes, she finished her *gymnasium* degree outside the regular educational route. She began study in the winter of 1928–29, and took courses in geophysics, meteorology, geography, and mathematics. Among her teachers were such prominent mathematicians as Richard von Mises, Erhard Schmidt, and Lotte Möller (1893–1973), a woman who had been promoted to the Associate Professorship for Hydrography in 1935.[35]

Because of her brilliant thesis Kobe was given a position as an assistant at the Institute of Marine Science right after she defended her doctorate in 1934. This allowed her to continue her work in utilizing the results of the German Meteor Expedition, which had gathered geographical, geophysical, and metereological data from the Atlantic Ocean. From 1932 until 1941, Kobe's advisor Albert Defant (1884) would edit 16 volumes on the explorations of the Meteor expedition. Kobe was an assistant at the Marine institute until 1938 when the director, Heinrich von Ficker (1881–1957), became the head of the Sonnblick Observatory (which was part of the Kaiser Wilhelm Society consortium). Because it was known that she was reluctant to actively support the Nazis, her situation worsened in 1938, when she could no longer depend on the protection that Ficker had provided her during the first years of the Nazi regime. Her mentor therefore advised her to leave the university altogether. Ficker was most likely responsible for finding her a new position in 1938 as an adviser ('referent') at the marine observatory in Wilhelmshaven, a port city near the North Sea. She remained with this institute when it moved its quarters to Greifswald, on the Baltic Sea, in 1940. Eventually she was even promoted to supervise a climate research group there, remaining until the very end of the war. As the Russian allied forces pushed westward into Germany, Kobe fled with the military institute away from eastern Greifswald to western Flensburg in the Spring of 1945. Finally it, and she, returned to Berlin in the summer of 1945, when the Nazis had surrendered to the Allied Forces.

Had there been no Nazi regime, Gertrud Kobe would have continued to work as a scientist at the university's institute for Marine Science. It is rather paradoxical that a scientist who had trouble maintaining her university position because she failed actively to support Nazi policies would not only move into a job at a research institution within the Nazi army, but actually make a modestly successful career from it. Kobe held an important position as head of a research group in geophysics, but because of the secretive nature of her military work nobody read her articles except for her close colleagues.

After the capitulation of the Nazi state, Kobe petitioned for a position with her previous employer, the university's Marine Science Institute. Its director, Hans Ertel (1904–1971), engaged her immediately in October in 1945. After the re-opening, in 1946, of the Marine Science Institute and the University of Berlin under the Soviet Military Administration, Kobe became an assistant at the university. Two years later she became Ertel's deputy but on a temporary contract. Under the political regime of the German Democratic Republic, Kobe's career did not advance so quickly or successfully – as one could expect. It would take her another fifteen years, until 1960, before she secured a permanent position as lecturer eventhough she had been Ertel's deputy at the institute since 1948. After some problems arose in the early sixties, she left the university in 1965.[36] She was then sixty years old, which was the official retirement age for women in the GDR. She died at the age of ninety in 1995 in Berlin. Only the newspaper of the Humboldt University published a modest obituary of her, and its author was clearly troubled by her past, omitting any reference to her work during the Nazi regime or indeed any part of her work between 1938 and 1945.[37] Nor did the author describe the unusually harsh conditions Kobe had faced in the years just before her retirement.

Kobe's career path was not as unusual as it might seem. In October 1949, a young mathematician, Eleonore Schwarz, wrote to the Academy of Science in East Berlin asking for a position which she secured at the Institute of Fiber Research in Teltow, a town in the Brandenburg region. This institute did industrial research for a factory in the village of Schwarza near the Polish border. Although she was overqualified for the tasks of reconstituting the factory library in Schwarza and of researching the efficiency of the production procedure, she accepted the job – the most ordinary entry-level position one could have at a research institution. Her contract stated that she had experience and special expertise in production efficiency processes.[38] Why would Schwartz accept such an inferior contract despite her expertise? Of course, living conditions in 1949 were still dismal, especially in and around the bombed-out city of Berlin. Perhaps she had already petitioned the university for a position, without success. She was certainly an enterprising woman: after taking the contract in Teltow she managed to get a position at the Institute for Mathematics at the Academy of Science in October 1951. There she caught the attention of the mathematicians already working there, who were quick to recognize her abilities.

Eleonore Schwarz was born in Berlin in 1910 to a father who was a high state official. Before the war, Schwarz had studied mathematics and physics at the University of Berlin between 1930 and 1936; there she must have witnessed the major political upheavals taking place. Probably because of the worsening employment situation for women who studied science, she finished her study without completing a thesis. Although she trained to become a teacher, she never worked as one.

In 1948, when applying for a job at the Academy, Schwarz wrote in her *curriculum vitae* that she had belonged to an institute called the

DWM – the German weapons and ammunition manufacturer – from 1936 until it was dismantled in 1945. If this chronology is correct, it would mean that she took a job at this facility, which was located in the northern seaport city of Lübeck, right after finishing her studies. In 1938, she became the head of the institute's mathematical literature department. According to her later letter, Schwarz's department maintained a staff of twenty on average, including three to four who were academically trained. Despite these few details, however, the whole *curriculum vitae* is rather cryptic. It mentions no names, titles, or specific job descriptions. About her mathematical work Schwarz said only that it had been related to questions of ballistics, the calculation of probability, and the field of nomography. Nomography was a specialized field of applied mathematics used for computation in aeronautical and rocket research during the Nazi era. Schwarz also named some of her writings, but again in a vague fashion. For example, she mentioned that she had completed her thesis in 1943, yet did not provide its title or the name of the degree-granting institution. She made reference to the only two lectures she gave in 1942: for the Lilienthal Society and at the air force ministry. Otherwise there was little indication of what she had done before returning to Berlin in 1946 as a free-lance worker, after which she landed a part-time job contract at the Academy's Institute of Mathematics in 1947. Even if one doesn't have special knowledge of the history of military research under the Nazis, it is evident from her cryptical resume that Eleonore Schwarz participated in the most secret field of military research at that time, which concerned air power.[39]

What had she done? The best clues come from post-war assessments by the Americans of German science research. In preparation for the 1947 report *German Research in World War II*, a group of American specialists went through Germany after the capitulation of the Nazis in 1945 to inventory German scientific experts. The resulting book, which was edited by Colonel Leslie E. Simon, desribed the structures of science organizations, research programs and their results, and some of the key figures in German science.[40] Significantly, Schwarz was the only woman among thirty-five scientists whose work Simon deemed worthy of in-depth discussion. He noted that 'An institute for theoretical ballistics at LFA computed a part of the aircraft firing tables for the Luftwaffe and did some work on the probability of hitting.' After reviewing work done at this military institute unfavorably he said that 'much better work was done in these fields at the industrial laboratory of DWM at Lübeck.' For his praise, he singled out Schwarz. 'The high quality of the work at DWM was due in large degree to the excellence of the exterior ballistician who was the section leader, Fraulein Dr. E. Schwarz. Dr. Schwarz showed great enterprise in clarifying concepts of the trajectories of bullets fired from aircraft and in improving computing methods for aircraft ballistic data. She improved on the well-known Siaaci method of calculating trajectories, and was perfecting a system whereby the calculation of air-craft ballistic data would be reduced almost to a matter of reading nomo-graphics charts. Schwarz's exterior ballistic work was closely coordinated with the needs of the armed forces. Interesting work was done on the cal-culation of the optimum range at which to open combat between fighters and bombers.'[41]

Elsewhere Simon assessed positively the academic qualities of the DWM, stating that: 'Some of the industrial laboratories were very large, such as the laboratories of Krupp, [while] some were of high scientific caliber, such as those of DWM in Lübeck.'[42] From this American report, it becomes clear why Schwartz could not have asked for a university job in 1947. Most of her colleagues were taken away to work in France, Great Britain, the United States, or the Soviet Union. Because of her iso-lation, she had to solicit a job for which she was overqualified. Her new director, the mathematician Kurt Schröder (1909–1978) chaired the Department of Applied Mathematics in the Institute for Mathematics at the Academy. They must have known each other because Schröder was involved in the oldest state sponsored German aviation research center (Deutsche Versuchsanstalt für Luftfahrt) in Berlin-Adlershof, another important institute for air force research. He worked on mathematical problems which were important for new developments in the field, and had taught applied mathematics at the University of Berlin since 1939. It is highly likely that they met each other either through the Lilienthal Society or in the air force ministry. Despite his work for the Nazis, Schröder had managed to get a professorship at the Humboldt University in East Berlin when it reopened in 1946, and he became a member of the Academy five years later. He was one of East Berlin's most influential sci-entists and one of the GDR's most important mathematicians. He not only introduced computers to the university and the Academy, but also established their first computer centers.

Schröder recognized Eleonore Schwarz's mathematical abilities and gave her a job collecting computer-science literature.[43] Schröder believed that as a new technology the computer would soon be of importance to the young Communist republic, though by the time it happened he was long gone.[44] In the summer of 1959 Schwarz left her job at the institute and wrote – in brief and characteristically cryptic fashion – that she wished to return to her former colleagues.[45] She went to the German Research Institute for Aviation (*Deutsche Forschungsanstalt für Luftfahrt* or DFL) in Braunschweig-Waggum in West-Germany.[46] Her book an edited collection on nomograms was pub-lished in East Berlin in 1960 eventhough she had left by then the country to emigrate to the West.[47]

Both Kobe and Schwarz, in their own ways, had created outstanding careers under the different political regimes. Irmgard Flügge-Lotz (1903–1974) was exceptional among women scientists for a number of reasons. She graduated from a technical college rather than an university and headed a department in a Kaiser Wilhelm Society institute. She even continued her career after marriage, did research in military institutes, and continued her work abroad in France and the United States after the war.

Irmgard Lotz was born in 1903, in Hameln in western Germany. Her interest in technical subjects derived from both parents, for her father was a mathematician and her mother's family had been involved in the construction industry.[48] She trained first at a primary school and then at the gymnasium in Hannover, where she majored in the sciences. After graduating in 1923, she studied mathematics and physics at Hannover's technical college for the next four years. It was then known as one of the best in the country. The technical college in Hannover maintained good working relations with the Kaiser Wilhelm Institute for Fluid Mechanics, too. She finished her professional training with a degree in mathematics in 1927. Upon graduation she became an assistant to the professor for applied mathematics and projective geometry at her *alma mater*. This was a remarkable appointment, for even fewer women were selected as assistants at technical colleges than at the universities. Two years later, in 1929, she finished her thesis in applied mechanics on heating process. Her supervisor, Professor Prange, was also her employer at the technical college.

In 1929, when Lotz finished her thesis, the job market was increasingly dismal because of the international economic crisis. But because Lotz was known to be brilliant, she was offered a place at the Kaiser Wilhelm Institute for Fluid Mechanics in Göttingen, which maintained good relations with the Hannover's technical college. Lotz must have made a considerable impression because when the young aeronautical engineer Werner Albring (b. 1914) became Professor Prange's assistant ten years later, Prange used to reminisce about his former doctoral student and assistant.[49] By that time Irmgard Lotz had become a well-known scientist in the German aeronautical research institute. She developed what became known as the Lotz method for calculating spanwise (wingtip to wingtip) distribution of a wing's lifting force.[50]

When Irmgard Lotz started working at the Kaiser Wilhelm Institute for Fluid Mechanics under the direction of Ludwig Prandtl, she was the only woman among twenty-five assistants. The institute, which also housed the earlier established Aeronautical Research Laboratory (AVA), had opened its doors in Göttingen in 1924 and became one of the most prominent aeronautical research establishments in Europe. Ludwig Prandtl (1875–1953), its first director, was a well-known mathematician, physicist, and aeronautical engineer. Prandtl had been the head of the aerodynamics laboratory from 1919 and became the director of the Kaiser Wilhelm Institute from 1924 until his death in 1953, except for a brief period between 1945 and 1947. A specialist in the field of fluid mechanics, Prandtl ranked among its most prominent theorists.[51] His successor as director at the aerodynamics laboratory after 1924 was his student and senior assistant Albert Betz (1885–1968), who was on the research staff and the head of the laboratory until 1957. Although the aviation research lab operated independently from the institute after 1937, the close collaboration between the lab and the Institute nevertheless continued.[52] In April 1945 the American army closed the institute and British troops occupied the building. It reopened in 1947.

Lotz later wrote in her *curriculum vitae*, that from 1929 to 1934 she was an assistant at the Aeronautical Research Laboratory.[53] Because this was part of the institute, the annual report of the Kaiser Wilhelm Institute listed her name and publications between 1932 and 1937.[54] According to the 1936 *Handbook of the Kaiser Wilhelm Society*, there were twenty-eight assistants in fluid mechanics, among whom the only two women were Irmgard Lotz and Margot Herbeck (b. 1909).[55]

Apparently around 1934 Lotz began acting as head of the department for aerodynamics theory, though she had no official appointment.[56] In 1987, John R. Spreiter and her husband, the scientist Wilhelm Flügge, confirmed this recollection, saying that Lotz had been 'the head of a group of young men...and a staff of computing women.'[57] The 1936 *Handbook of the Kaiser Wilhelm Society*, however, fails to mention Lotz's supervisory position as a head of department. Possibly the Nazi aversion and policy to women scientists, especially in higher positions, might explain this remarkable omission in the record. As a director of the institute, Ludwig Prandtl was able to give her the position as department head as long as it did not come to official notice.

While at the institute Lotz worked on different problems of aerodynamics. Prandtl organized a colloquium dealing with questions of applied mechanics every Wednesday at the university in Göttingen.[58] Lotz lectured there for five consecutive years in the thirties, which was a considerable achievement. Her colleague (and later husband) the aerodynamicist Wilhelm Flügge (b. 1904) gave eight lectures during the same period, but he also held the prestigious post as 'Privatdozent' at the university in Göttingen.

Over the years Lotz lectured on the complex solutions of potential fluids (1932), measurements in the wind tunnel environment (1933), calculations of special tops (1934), new arithmetical calculating machines (1935), and ray calculations (1936). In her articles she dealt with some theoretical problems of applied mechanics, following in the tradition of the well-known scientist Theodor von Kàrmàn (1881–1963). He was another former student of Prandtl and professor at Aachen's technical college before he was forced into exile after 1933 because of the Nazis' anti-semitic policies. By that point, Lotz had become a respected expert in aeronautical engineering within Nazi Germany's scientific community. In appreciation of her outstanding work, Ludwig Prandtl nominated her for a special kind of professorship – a research professorship ('Forschungsprofessur') – in 1937.[59] This highly prestigious nomination was even more unusual than her earlier appointment as head of the department had been. The research professorship Prandtl had in mind for her had been created by the Ministry of Aviation at the newly established Academy of Aviation. Both the ministry and the academy were closely linked to the German military. However, the Nazi leadership rejected the nomination. Lotz was neither appointed as a research professor nor selected for any other position at a university or technical college during the Nazi era.

Lotz tried to continue pursuing her career after her marriage, in 1939, to Wilhelm Flügge who had also been Prandtl's student and had worked on thin shell construction.[60] Both scientists went to Berlin when Flügge got a position at the German Research Institute of Aviation (DVL) in Berlin-Adlershof. The couple did not have any children.[61] Marriage considerably impeded the career advancement of Lotz, who changed her name to Flügge-Lotz. While Flügge became a head of a department at the German Research Institute of Aviation, his overqualified wife worked merely on a semi-official and consultancy basis. Later in life, she described her position as one of scientific adviser for aerodynamics and dynamics of flight.[62] Even though she was more prominent than her husband, Flügge-Lotz found her professional career was disrupted by the Nazi prohibition against the employment of married women. Nevertheless she remained prominent within the community of aerodynamic specialists. Although Werner Albring never knew her personally he remembered her name and reputation immediately when quizzed about her over fifty years later.[63]

Together with Ludwig Prandtl and Professor Walter Tollmien, Lotz was elected to join the highly prestigious prize committee of the Lilienthal Society, which was associated with Germany's military. Every year the society granted an award for the best work in aerodynamics. Irmgard Flügge-Lotz and her colleagues on the prize-committee were asked to judge a wide range of technical work.[64] Her activities as a consultant also grew more important during the second World War.

The career path taken by Flügge-Lotz, like those of Gertrud Kobe and Eleonore Schwarz, illustrates the contradiction between political practice and official Nazi doctrine concerning the proper role of German women. As these three examples show, the Nazi regime considered the work of women scientists more useful at military institutes than at universities or other government agencies. At the armed forces institutes, women scientists remained invisible because there was no possibility of public recognition of their work and everything they did was classified. Obviously they could only maintain a professional reputation within the select community of male colleagues who were also engaged in secret war work. As a consequence, their employment within the military never publically challenged Nazi ideals of women and their roles.

In short, the Nazi German legacy was paradoxical, to say the least. On the one hand the Nazis actively excluded and fired women scientists from the universities and other research institutions. On the other, they hired them to do military work. Then, when the war escalated in 1939, women scientists were offered jobs at the universities as replacements for their male colleagues. Like their male counterparts, most women scientists worked for the institutes that were closely linked to the military. For example, the German historian Margot Fuchs has examined the life of Melitta Schiller (1903–1945), who became an engineer and worked on the construction of airplanes in the German armed forces. Because she was also a pilot, she was a relatively public figure whose reputation extended beyond the site of her engineering work.[65]

Immediately after the fall of the Nazi regime in May 1945, it was the policy of the Anti-Nazi Alliance to establish a special recruiting program for scientists in every occupied zone. This was part of the effort to dismantle Germany's laboratories and institutes, and to mobilize German scientists for scientific research in the United States, Great Britain, France, and the Soviet Union. As the Allied forces bombed Berlin so fiercely since 1943 the German military moved most laboratories of the German Research Institute of Aviation from the capital to southern Germany, near the Bodensee. Then in May 1945, after the unconditional capitulation of Nazi Germany, an employment program of German scientists was established. All partners in the Anti- Nazi Alliance sought German scientists who had been working in the fields of military research under the Nazis. Perhaps the best-known part of this initiative was 'Action Paper Clip,' the program by which the Americans identified and poached atomic specialists. Other programs were devised to bring aerodynamics specialists, like the group headed by Wernher von Braun, to the United States.[66] The Soviet Union's catch included Werner Albring.[67] Since the laboratories of the German Research Institute of Aviation were located at Saulgau near the Bodensee, which was part of the zone occupied by France, German research teams from there were forced to France to work at the newly established *Centre de Technique de Wasserburg*. As a result, Irmgard Flügge-Lotz and Wilhelm Flügge worked in Paris from 1946 to 1948. There Irmgard Flügge-Lotz was promoted to 'chief of a research group in theoretical aerodynamics' at the French National Office for Aeronautical Research.[68] The couple worked in suitable facilities that enabled them to continue to do their scientific work, but the political climate in France and the general confusion over France's political future made them look for opportunities elsewhere.[69] Returning to Germany was no option since they were acutely aware of research conditions in the newly-divided Germany and the Allied ban on any German research related to the military. In occupied Germany, any research related to the military including atomic physics, rocket technology, aerodynamics, and computer technology was forbidden until 1955. As the Cold War heated up, and the Alliance started to break down, both sides, especially the two superpowers, began to see German science in a different light. In the United States, German scientists who had worked under the Nazi regime were now welcomed with open arms. They had access to superb research facilities, and received considerable financial rewards for their work. This was especially true for those scientists who worked under the direction of Wernher von Braun in the field of rocket research, although under the Nazis he had developed the destructive V2 rockets that were used to bomb many British towns. Under American patronage, von Braun would become famous as head of the Apollo space program.[70]

The position of Flügge-Lotz and her husband changed in 1948.[71] In the early thirties, they had met the Russian emigre and American scientist, aeronautical engineer Stephen P. Timoshenko (1878–1972), who was a friend and a colleague of Ludwig Prandtl. He sat on the editorial board of a journal, which published review articles on research in mechanics, with which Irmgard Flügge-Lotz and Wilhelm Flügge had been affiliated since 1938. It was Stephen Timoshenko, a professor at Stanford University, who arranged the couple's journey to the USA, where they arrived in October of 1948. Their trip from Paris to California was not entirely legal but they were eager to leave the uncertainties of their position in France behind. Although Irmgard Flügge-Lotz was much more important in the field of theoretical aerodynamics than her husband, she only obtained a position as lecturer and research supervisor at Stanford while her husband was awarded a full professorship. Apparently the unfairness of the situation was the subject of international comment. Spreiter and Flügge recalled later on that, 'By the middle 1950s, it seemed evident to almost everyone at Stanford that Flügge-Lotz was carrying on all the duties of a full professor but without official recognition. In fact, it was hard for students to understand why she was a lecturer rather than a professor. The same question arose on the international scene in the summer of 1960, when she was the only woman delegate from the United States at the First Congress of the International Federation of Automatic Control in Moscow. By then the disparity had become apparent to all. Before school opened for the autumn quarter, she was appointed full professor in both engineering mechanics and aeronautics and astronautics. Stanford had been in existence for seventy years, but she was its first woman professor in engineering.'[72]

Conclusion

The women engineers who worked as scientists at some institutes of the Kaiser Wilhelm Society had earlier and better opportunities than those in other agencies or research institutes at the technical colleges and universities. There were several reasons why this was so. First, the Society's institutes were partly public and partly private, so they enjoyed greater freedom in appointments and the employment of educated women than government institutions such as universities. Thus, the Society's institutes could hire women scientists and engineers much earlier, pay them the same money as male scientists in similar positions and, in a very few cases, promote them to be heads of a laboratory or a department. Second, the structure of these institutes and the powerful position held by each diretor allowed them to hire and fire as they liked. When a director had no prejudice against women scientists, he had the complete freedom to employ them and grant them relatively high positions. Ludwig Prandtl and Wilhelm Eitel were two examples of directors who used their power in this way. Third, the institutes had been founded to deal with new scientific problems or even to create entirely new disciplines. As such, they tended to be looser regarding gender roles than organizations or institutes serving long-established fields.[73] When the circumstances were propitious, women scientists gained greater opportunities in such newly-created institutions.

The situation in Nazi Germany was paradoxical indeed. Gertrud Kobe and Eleonore Schwarz studied science at a university, then changed their field to work as engineers in the armed forces rather than as scientists. Flügge-Lotz trained as an engineer but worked as an academic scientist in an institute of the Kaiser Wilhelm Society. She then had to change jobs because of her marriage. The women engineers who worked as scientists met with more approval in the armed forces than in the civil sector, since the Nazis found them more useful in military research centers. Here, however, women researchers became even more invisible than before because their work was covert. They were known only within the community of their male peers who also did secret research. Therefore women scientists could exist both as 'visible scientists' and also as 'invisible scientists'. The women engineers and scientists considered here did not have any relationship to the formal women's movement. They relied on personal solutions in working out their careers as scientists in male-dominated fields. All women engineers and scientists who worked in institutes of the armed forces were offered new opportunities after the war not only in the German Democratic Republic and the Federal Republic of Germany, but also in the United States. Their activities for the Nazi military did not fundamentally affect their later careers in German institutes or international science. Despite their war work, they were respected for their expertise.

Notes

1. The author wishes to thank Tara Nummedal and Patricia Fara from the Max Planck Insitute for History of Science for helping to correct the English in the paper and Ruth Oldenziel and Karin Zachmann for their helpful comments.
2. Juliane Mikoletzky, this volume.

3. Barbara Duden and Hans Ebert, 'Die Anfänge des Frauenstudiums an der Technischen Hochschule in Berlin,' in *Wissenschaft und Gesellschaft. Beiträge Zur Geschichte der Technischen Universität Berlin, 1879–1979*, edited by Reinhard Rürup, Vol. 1 (Berlin: Springer Verlag, 1979), 403–23.

4. See for the development at the Technical Colleges in Germany, Karin Zachmann, this volume.

5. Margot Fuchs, *Wie die Väter so die Töchter. Frauenstudium an der Technischen Hochschule München, 1899–1970* (Munich: Technische Universität, 1994); Duden and Ebert, 'Die Anfänge des Frauenstudiums'.

6. Karin Zachmann, this volume. See also Ruth Schwartz Cowan, 'Research Questions about the History of Women Engineers,' Keynote Address, Frauen (t)raum Workshop, January 15, 1999, Berlin.

7. Eugenie von Soden, ed., *Das Frauenbuch. Frauenberufe und Ausbildungstätten* (Stuttgart: Frank'sche Verlagshandlüng, 1913).

8. Gertrud Woker, 'Die Chemikerin,' in Soden, *Das Frauenbuch*, 100–02; Gerit von Leitner, 'Wollen Wir Unsere Hände in Unschuld Waschen?' Gertrud Woker. *Chemikerin und Internationale Frauenliga* (Berlin: Weidler Buchverlag, 1998).

9. Emilie Winkelmann, 'Die Architektin und die Ingenieurin,' in Soden, *Das Frauenbuch*, 108–09.

10. Rhoda Erdmann, 'Die Zoologin und die Botanikerin,' and her 'Die Mathematikerin und die Physikerin,' both in Soden, *Das Frauenbuch*, resp. 103–06; 106–08.

11. On women scientists at Yale, see: Margaret W. Rossiter, *Women Scientists in America. Struggles and Strategies to 1940* (Baltimore: The Johns Hopkins University Press, 1982).

12. For the comparison with France, Switzerland, United States, Great Britain and Germany, see: Ilse Costas, 'Der Kampf Um das Frauenstudium Im Internationalen Vergleich,' in *Pionierinnen, Feministinne, Karrierefrauen? Zur Geschichte Des Frauenstudiums in Deutschland*, edited by Anne Schlüter (Pfaffenweiler: Centaurus Verlag, 1992), 115–44; and her, 'Der Zugang von Frauen zu Akademischen Karrieren. Ein Internationaler Überblick,' in *Bedrohlich Gescheit. Ein Jahrhundert Frauen und Wissenschaft in Bayern*, edited by Hiltrud Häntzschel and Hadumod Bömann (Munich: Beck Verlag, 1997), 15–34. See furthermore Mikoletzky and Zachmann, both in this volume.

13. Rossiter, *Struggles and Strategies* and her, *Women Scientists in America. Before Affirmative Action, 1940–1972* (Baltimore: The Johns Hopkins University Press, 1995).

14. Renate Tobies, ed., *'Aller Männerkultur Zum Trotz.' Frauen in Mathematik und Naturwissenschaften* (Frankfurt: Campus 1997); Jeffrey A. Johnson, 'Frauen in der Deutschen Chemieindustrie, von Den Anfängen Bis 1945,' in *'Aller Männerkultur Zum Trotz'*, 253–71; and his Jeffrey A. Johnson, 'German Women in Chemistry, 1895–1925 (Part I),' *NTM* 6, no. 1 (1998): 1–21; Brita Engel, 'Clara Immerwahrs Kolleginnen: Die Ersten Chemikerinnen in Berlin,' and Mirjam Wiemeler, '"Zur Zeit sind alle für Damen geeigneten Posten besetzt." Promovierte Chemikerinnen bei der BASF, 1918–1933,' both in *Geschlechterverhältnisse in Medizin, Naturwissenschaft und Technik*, edited by Christoph Meinel and Monika Renneberg (Stuttgart: GfN Verlag, 1996), resp. 403–23 and 237–44.

15. Ruth Levin Sime, *Lise Meitner. A Life in Physics* (Berkeley: University of California Press, 1996). See the essays by Margot Fuchs, 'Isolde Hausser (1889–1951, Physikerin in Industrie und Forschung,' in *Können, Mut und Phantasie. Portraits Schöpferischer Frauen aus Mitteldeutschland*, edited by Annemarie Haase and Harro Kieser (Weimar: Böhlau-Verlag, 1993), 149–64; 'Isolde Hausser (1889–1951). Technische Physikerin und Wissenschaftlerin am Kaiser-Wilhelm/Max-Planck-Institut für Medizinische Forschung, Heidelberg,' *Berichte zur Wissenschaftsgeschichte* 17, no. 3 (1994): 201–15; ''Wir Fliegerinnen sind keine Suffragetten.' Die Versuchsingenieurin und Sturzflugpilotin Melitta Schiller, 1903–1945,' in *Bedrohlich Gescheit. Ein Jahrhundert Frauen und Wissenschaft in Bayern*, edited by Hiltrud Häntzschel and Hadumond Bußmann (Munich: Beck Verlag, 1997), 260–65.

16. Rossiter, *Before Affirmative Action*, 1–26.

17. Rudolf Vierhaus and Bernhard vom Brocke, eds., *Forschung im Spannungsfeld von Politik und Gesellschaft; Geschichte und Struktur der KWG/MPG* (Stuttgart: Deutsche Verlagsanstalt, 1990).

18. Bernhard vom Brocke and Hubert Laitko, eds., *Die Kaiser-Wilhelm/Max-Planck-Gesellschaft und ihre Institute. Studien zu ihrer Geschichte: Das Harnack-Prinzip* (Berlin: Walter de Gruyter, 1996).

19. For an elaboration, see: Annette Vogt, 'Die Kaiser-Wilhelm-Gesellschaft wagte es: Frauen als Abteilungsleiterinnen,' in *'Aller Männerkultur zum Trotz'*, 203–19; 'Vom Hintereingang zum Hauptportal: Wissenschaftlerinnen in der Kaiser-Wilhelm-Gesellschaft,' in *Dahlemer Archivgespräche*, Vol. 2 (1997), 115 – 39, pp. 117–121; 'Die ersten Karriereschritte: Physikerinnen an der Berliner Universität und in Instituten der Kaiser-Wilhelm-Gesellschaft von 1900–1945,' in *Barrieren und*

Karrieren edited by Elisabeth Dieckmann and Schöck-Quinteros (Berlin: Trafo Verlag, 2000), pp. 195–230.

20. Neither Vierhaus and vom Brocke, *Forschung im Spannungsfeld* nor vom Brocke and Laitko, *Die Kaiser-Wilhelm/Max-Planck-Gesellschaft* mention any women scientists.

21. The department chairs were the physicists Lise Meitner, Gerda Laski, and Isolde Hausser; the chemists Maria Kobel, Marthe L. Vogt, Charlotte Frölich, and Louise Holzapfel; the biomedical scientists Cécile Vogt, Getrud Soeken, Else Knake, and Elisabeth Schiemann Vogt, 'Die Kaiser-Wilhelm-Gesellschaft wagte es', 207–208, Table. Since the publication, I found the twelfth woman chair. For Hausser, see: Margot Fuchs, 'Isolde Hausser,' (1993) and 'Isolde Hausser' (1994); for Marthe L. Vogt, see: Susan Greenfield, Marthe Louise Vogt F. R. S. (1903–),' in *Women Physiologists* (London and Chapel Hill: Portland Press 1993), 49–59. for Schiemann, see: Ute Deichmann, 'Frauen in der Genetik, Forschung und Karriere bis 1950,' in Tobies, '*Aller Männerkultur zum Trotz*', 232–236; Anton Lang, 'Elisabeth Schiemann. Leben und Laufbahn einer Wissenschaftlerin in Berlin' in *Geschichte der Botanik in Berlin*, edited by Claus Schnarrenberger and Hildemar Scholz (Berlin: Colloquium Verlag, 1990), 179–189. On Cécile Vogt, who was Marthe Vogt's mother, see work by Helga Satzinger, 'Das Gehirn, die Frau und ein Unterschied in den Neurowissenschaften des 20. Jahrhunderts: Cécile Vogt (1875–1962),' in Meinel and Renneberg, *Geschlechterverhältnisse*, 75–82; 'Weiblichkeit und Wissenschaft – Das Beispiel der Hirnforscherin Cécile Vogt (1875–1962),' in Johanna Bleker, ed., *Der Eintritt der Frauen in die Gelehrtenrepublik. Zur Geschlechterfrage im akademischen Selbstverständnis und in der wissenschaftlichen Praxis am Anfang des 20. Jahrhunderts* (Husum: Matthiesen 1998), 75–93; *Die Geschichte der genetisch orientierten Hirnforschung von Cécile und Oskar Vogt in der Zeit von 1895 bis ca. 1927* (Stuttgart: Deutscher Apotheker Verlag, 1998); Annette Vogt, 'Elena Aleksandrovn Timoféeff-Ressovsky – weit mehr als die 'Frau ihres Mannes' in Karl Friederich Wessel, ed., *Festschrift zum 75 Geburtstag von Ilse Jahn* (Bielefeld: Kleine Verlag, 1999 p. 149–170 On Maria Kobel, see Vogt, 'Vom Hintereingang zum Hauptportal,' 134–139.

22. *Handbuch der Kaiser Wilhelm Gesellschaft*, edited by Max Planck, 2 Vols. (Berlin: Springer Verlag, 1936), Vol. 1, 157.

23. I am grateful to Dr. Renate Tobies for bringing my attention to the existence of Herbeck's thesis. University of Göttingen Archives, Göttingen, Germany.

24. Index card, March 1, 1938, Papers of the Hochschulgemeinschaft Deutscher Frauen in der NS Studentenkampfhilfe des NS-Studentenbundes der NSDAP, Herbeck personal file, Document Center Collection, Bundesarchiv Koblenz, Berlin-Lichterfelde, Germany.

25. Elisabeth Boedeker's, *Mathematik, Naturwissenschaften, Technik und Anhang Medizin* Vol. 4, no. 1159 (Hannover 1935) is part of *25 Jahre Frauenstudium in Deutschland*, 4 Vols (1935–1939). For women students at Technical College in Berlin-Charlottenburg, see: Duden and Ebert, 'Die Anfänge des Frauenstudiums.'

26. Eugen Bamann and Karl Myrbaeck, eds., *Die Methoden der Fermentforschung* (Leipzig: Georg Thieme Verlag, 1941) 3 vols.

27. Poggendorff, *Biographisch-Literarisches Handwörterbuch zur Geschichte der exakten (Natur)wissenschaften*, Vol. VI (Berlin: Verlag Chemie, 1937) and Vol. VIIa (Berlin: Akademie Verlag, 1958), both, s.v., 'Krüger.'

28. The Kaiser Wilhelm Institute's papers list her as 'Frl. Dr. Heiden,' Max Planck Society, Archives, Berlin (*MPA Archives*, hereafter). For the information, I thank Dr. Renate Tobies, who conducted an interview with Maria Heyden Joerges in 1996, for confirming Heiden and Heyden are indeed the same person.

29. *Die Naturwissenschaften* 26 (1938), 612. The article mentioned the date August 31, 1938.

30. Kristie Macrakis's, *Surviving the Swastika: Scientific Research in Nazi Germany* (New York: Oxford University Press, 1993); and her, 'Exodus der Wissenschaftler aus der Kaiser-Wilhelm-Gesellschaft,' in *Exodus von Wissenschaften aus Berlin*, edited by Wolfram Fisher *et al.* (Berlin: De Gruyter, 1994), 267–283, appendix and pp. 627–630. Compare with: Werner Albrecht and Armin Herrman, 'Die KWG im Dritten Reich (1933–1945),' in Vierhaus and vom Brocke, *Forschung im Spannungsfeld*, 356–406.

31. Jochen Brüning, Dirk Ferus, and Rainhard Siegmund-Schultze, eds., *Persecution and Expulsion of Mathematicians from Berlin Between 1933 and 1945*, catalogue (Berlin: Deutsche Mathematiker Vereinigung, 1998).

32. Haide Manns, *Frauen für den Nationalsozialismus. Studentinnen und Akademikerinnen in der Weimarer Republik und Im Dritten Reich* (Opladen: Leske & Budrich, 1997).

33. Marilyn J Boxer and Jean H. Quartaert, 'Overview,' in *Connecting Spheres. Women in the Western World, 1500 to the Present*, edited by Marilyn Boxer and Jean H. Quartaret (New York: Oxford University Press, 1987), 187–222, pp. 211–214.

34. Karin Zachmann, 'Women to Replace the 'Front Officers of Technology'? On the Development of Technical Studies for Women in the Soviet Occupation Zone and the GDR (1946–1971) *History and Technology* 14 (January 1997): 97–112; Macrakis, *Surviving the Sastika*; and 'Exodus der Wissenschaftler aus der KWG.'

35. Gertrud Kobe's Thesis, Phil. Fak. no. 764, p. 31, Humboldt University Berlin Archives, Berlin (*HUB Archives*, hereafter).

36. K 817 (Kobe personal file), *HUB Archives*. Hans Ertel, who supported Kobe, wrote many letters on her behalf since 1953. No documents could be found that throw light on the reasons for the conflict; perhaps Fanselau feared her competition.

37. Hannelore Bernhardt, 'Nachruf auf Gertrud Kobe,' *Humboldt. Die Zeitung der Alma Mater Berolinensis* 39 (July 6, 1995): 2.

38. Eleonore Schwarz, Job description, September 16, 1949, Mag. IIIa, Reg.30, p. 6, Archive of the Berlin-Brandenburgische Akademie der Wissenschaften, Berlin (*BBAW Archives*, hereafter).

39. Eleonore Schwarz's *vita* handwritten, June 13, 1948, Mag. IIIa, Reg.30, pp. 3-4R, esp. p. 3R-4R, *BBAW Archives*.

40. Leslie E. Simon, *German Research in World War II. An Analysis of the Conduct of Research* (New York: John Wiley and Sons, 1947).

41. Simon, *German Research in World War II*, 125.

42. Simon, p. 77.

43. Schröder, 'Jahresbericht' (1952), p. 13, VA no. 17037 (Institute for Mathematics at the Academy), *BBAW Archives*.

44. Schröder '*Jahresbericht*' (1953), p. 32, VA (Institute for Mathematics at the Academy) no. 17037, *BBAW Archives*.

45. Letter Eleonore Schwarz to the Academy's Personnel Department, ('Kaderabteilung der Forschungsgemeinschaft der DAW zu Berlin'), June 29, 1959, Mag. IIIa, Reg.30, p. 23, *BBAW Archives*.

46. Eleonore Schwarz's Testimony (*Zeugnis*), February 10, 1960, typewritten, Mag. IIIa, Reg.30, p. 25, *BBAW Archives*.

47. Eleonore Schwarz, ed., *Nomogramme und andere Rechenhilfsmittel für den Ingenieur* (Berlin (East): Verlag Technik, 1960).

48. *Notable American Women. The Modern Period*, edited by Barbara Sicherman and Carol Hurd Green (Cambridge, MA: Harvard University Press, 1980), s.v., 'Flügge-Lotz.'

49. Author's interview with Werner Albring, November 18, 1998. See also: Werner Albring, *Gorodomilia. Deutsche Raketenforscher in Russland* (Hamburg: Luchterhand Literturverlag, 1991).

50. *Notable American Women*, s.v.,'Flügge-Lotz.'

51. Julius C. Rotta, *Die Aerodynamische Versuchsanstalt in Göttingen, ein Werk Ludwig Prandtls. Ihre Geschichte von Den Anfängen Bis 1925* (Göttingen: Vandenhoeck and Ruprecht, 1990); Johanna Vogel-Prandtl, *Ludwig Prandtl. Ein Lebensbild. Erinnerungen, Dokumente* (Göttingen: Max Planck Institute, 1993); *Encyclopedia Brittanica*, s.v., 'Prandtl'.

52. Eckart Henning and Marion Kazemi, *Chronik der KWG*, Veröffentlichungen aus dem Archiv zur Geschichte der MPG Vol. I (Berlin: MPG Archives, 1988); Glenys Gill and Dagmar Klenke, *Institute Im Bild. Teil I: Bauten der KWG*, Veröffentlichungen aus dem Archiv zur Geschichte der MPG Vol. 5 (Berlin: MPG archives, 1992).

53. *Poggendorff*, s.v., 'Lotz'.

54. *Die Naturwissenschaften* 20 (1932), 454; *Die Naturwissenschaften* 20 through 25 (1932–1937).

55. *Handbuch der KWG*, Vol. 1, 157.

56. *Poggendorff*, s.v., 'Lotz'.

57. John R. Spreitner and Wilhelm Flügge, 'Irmgard Flügge-Lotz, 1903–1974,' in *Women of Mathematics. A Biobibliographic Sourcebook*, edited by Louise S. Grinstein and Paul J. Campbell (Westport, CT: Greenwood Press, 1987), 33–40, p. 34.

58. The program's colloquia has been traced in, postcards (1932–1936) Prandtl Papers, III, rep. 61, nos. 2167 and 2168, Kaiser Wilhelm/ Max Planck Society, Berlin (*MPA Archives*, hereafter).

59. Letter Prandtl to A. Baeumker, February 19, 1937, Reichsluftfahrtministerium, III, Rep. 44, no. 175, p. 28, *MPA Archives*, I. For this reference, I would like to thank Frau Kohl at archives.

60. *Poggendorff*, s.v., 'Flügge'.

61. Helena M. Pycior, Nancy G. Slack, and Pnina Abir-Am, eds., *Creative Couples in the Sciences* (New Brunswick: Rutgers University Press, 1996).

62. *Poggendorff*, s.v., 'Lotz'.

63. Author's interview with Werner Albring.

64. Prize committee correspondence May through December, 1941, III, Rep. 61, nos. 2062–2064, Prandtl Papers, *MPA Archives*.

65. Fuchs, '"Wir Fliegerinnen sind keine Suffragetten"'.

66. Michael J. Neufeld, *The Rocket and the Reich: Peenemuende and the Coming of the Ballistic Missile Era* (New York: Free Press, 1995); Mark Walker, *German National Socialism and the quest for nuclear power 1939–1949* (Cambridge, MA: Cambridge University Press, 1993); and *Nazi Science: Myth, Truth and the German Atomic Bomb* (New York: Plenum Press, 1995); Monika Renneberg and Mark Walker, eds., *Science, Technology and National Socialism* (Cambridge: Cambridge University Press, 1994).

67. For German specialists in the Soviet-Union, see: Ulrich Albrecht, Andreas Heinemann-Grüder, and Arend Wellmann, *Die Spezialisten. Deutsche Naturwissenschaftler und Techniker in der Sowjetunion nach 1945* (Berlin: Dietz-Verlag, 1992); Albring, *Gorodomilia*.

68. Spreitner and Flügge, 'Irmgard Flügge-Lotz,' 35; *Poggendorff*, s.v., 'Flügge'.

69. For an excellent description of the France's political instability, see: Simone de Beauvoir, *Der Lauf der Dinge (La Force des choses)* (Reinbek bei Hamburg: Rowohlt Verlag, 1966 [1963]); see also: Gabrielle Hecht, *The Radiance of France: Nuclear Power and National Identity after World War II.* (Cambridge, MA: MIT Press, 1998).

70. Neufeld, *The Rocket and the Reich*.

71. Correspondence Wilhelm Flügge to Ludwig Prandtl in the year of 1947, III, rep. 61, no.456, Prandtl Papers, *MPA Archives*.

72. Spreiter and Flügge, 'Irmgard Flügge-Lotz', 36. A list of her publications can be found in *Poggendorff*, s.v., 'Flügge'.

73. Rossiter described the same phenomena for American women scientists, she calls 'niche,' p. 259. Rossiter, *Struggles and Strategies*, 259–266.

Karin Zachmann

8. Mobilizing Womanpower: Women, Engineers and the East German State in the Cold War

'Nowadays girls and young women enter life with the same qualifications as their male contemporaries. And they hardly need to prove that involvement with technology does no harm to their charm and their femininity.'[2]

Inge Lange, the head of the women's department within the Socialist Unity Party (SED), said this during an interview with a leading journal for women in the German Democratic Republic (GDR) in 1974. She offered this observation as praise for the state's socialist policy on equality. The underlying message, however, was that technical competence in women need not threaten the stability of the bipolar gender order. This was in fact frequently articulated in the fiction and poetry appearing in the GDR in the early 1970s. Both female and male authors published novels, plays and novellas that dealt with the professional circumstances and actual living conditions of women engineers.[3] None of these stories questioned the technical abilities of the women appearing in them as designers, chemical engineers, architects, civil engineers, or building planners. In almost every one, however, the women's romantic relationships were a failure. Hence the central conflict of these stories was between technical competence and femininity.

These stories emerged at a time when educational facilities in engineering were expanding substantially. As enrollments increased, women formed a higher percentage of those in academic technical training. In this context, authors began to ask whether women could find their own place in engineering when it was so heavily dominated by men. Would women be forced to adjust to the established (male) mores in order to survive professionally? Or would the presence of women in engineering mean that, under state socialism, technical competence (which has always stood for gender difference in the modern era) would cease to be a specifically male trait?

Looking at women in engineering education and professional work in the GDR, this chapter investigates how engineering was restructured under state socialist rule. I do not intend to make an icon of femininity – that is, to set it up as somehow preferable to the meanings assigned to men and male work which have been the dominant story of mainstream historiography. Instead I want to assert that technology is never gender-neutral. Focusing on women brings this out clearly because women have been assigned the status of the 'special case' within Western thought; men have long been seen as the normative, gender-free sex, while women always bear the meanings of gender, and figure as the anomalous group.[4] The history of engineering has, as matter of course, identified the engineer – as the agent of technical change and the creator of the modern world – with masculinity. It has failed to take the dimension of gender into account. Looking at the situation of female engineers enables us to place the notion of the male culture of technology in historical context, rather than seeing it as natural or inevitable.

Clearing the Path: Female Pioneers in Engineering Studies

In the first decade of the twentieth century, all technical colleges (*Technische Hochschulen*) in the German Empire followed the lead of the universities and began to admit women as students. Since educational policies in the various German states have always been under regional control, the women of these states gained the right of access in different years.[5] Before this point there had already been many fierce debates. The first time technical colleges were faced with the question of female students was in the 1870s, when foreign women, especially from Russia – where the universities had closed their doors to women in 1863 – began to seek access to universities abroad. They looked first to Zurich and later also to German universities and technical colleges.[6] Subsequently female teachers tried to take advantage of the educational programs offered by the general studies divisions of the technical colleges.[7] In the days of the Empire, scholars in engineering had made efforts to expand these general studies divisions, which offered programs in both natural sciences and the humanities, in order to gain recognition among the university-trained academic elite.[8] Despite these efforts, the general science departments at technical colleges suffered severely from a lack of students. This is why they were among the first to express support for women's access to technical colleges.[9] The scientific elite in engineering, however, largely disapproved of female students. When, in 1896, the presidents of the technical colleges discussed whether women should be allowed to sit in on classes and attend lectures as guests, the majority voted against the proposal.[10] The image of the engineer as a creative inventor, as a conqueror of the forces of nature, and as a leader of men, was directly opposed to the image of bourgeois femininity that prevailed. Middle-class women were relegated to the worlds of love and family, culture and aesthetics, so they could compensate for lack of any of these elements in the lives of middle-class men.[11]

The founder of kinetics and vice-chancellor of the technical college of Berlin, Franz Reuleaux, addressed this division in gender roles in response to a survey initiated by Arthur Kirchhoff on women's suitability for academic study and careers in 1897. In Reuleaux's view, the crucial difference between the sexes, the one that precluded women from studying mechanical engineering, was of a physical rather than intellectual nature. He claimed that women would have difficulty working at horizontally-placed drawing boards, 'without any physical health risks, but rather at the risk of what we call pure femininity in posture and movement.'[12] He believed that women were physically incapable of completing the required one-year preliminary training in practical engineering before starting their studies in mechanical engineering.

'Their small hands are too weak, their feet of insufficient strength, their rib cages too delicate, their body weight much too low; their entire constitution is not durable enough to allow them to tolerate the enormous exertion of this pre-educational period for even a short time, let alone to maintain it.'

And finally, Reuleaux claimed that the climate in the workshop would cause women emotional harm, since they might be confronted with 'rough' and 'mean' language: 'Only physically robust and emotionally tough young disciples of the field have the stamina to see it all through.' In short, he expressed the prevailing view that to study mechanical engineering required a 'whole man,' while the 'domain of women ... [was] the home, housekeeping, family, child-rearing, and nurturing health.' Significantly, Releaux, who was one of the most vocal advocates of academic standards in technical education, made the physical and social dimensions of practical experience in the workshop an essential preliminary to a successful engineering career.[13] Facing women's demand for access to higher education, he made concessions to the newly developed idea of the engineer as an academically trained but thoroughly practical man. This idealized figure became important at the end of the nineteenth century when the social standing of intellectual pursuits diminished and intellectuality became associated with femininity.[14] German engineers, like their American counterparts, contributed to this by stressing hands-on work experience and making it an indispensable part of the male rite of passage into the profession.[15]

That part of the women's movement that emphasized the differences between men and women, and idealized the effects that femininity would have if only women could enter public life, did not take issue with Reuleaux's statement. These feminists saw emancipation as a route by which all humanity might benefit from 'female' values. For them, formal equality was a means of 'tak[ing] advantage of female difference to promote the renewal of society.'[16] In order to benefit from such 'otherness' in the professional world, however, women had to have access to those fields of study that would enable them to assume positions consistent with their areas of experience.[17] As late as 1920, Mary Bernays questioned whether the women's movement should support women's access to technical colleges, since 'not many actual women's professions have been successfully established on the basis of technical studies.'[18]

It was internal strife within the scientific elite that eventually led to the admittance of women to the technical colleges. When in 1899 the Kaiser gave them the right to grant doctoral degrees, technical colleges acquired the much-coveted status of universities. They therefore felt obliged to follow the lead of the universities in admitting women. Neither the universities nor the technical colleges followed the gradualist approach, which prevailed in Austria, but rather opened all areas of academic study to women at once.[19] German mandarins pursued a science policy that treated academic departments equally, in accordance with the idea of the unity of the sciences that formed the core identity of the academic elite. This identity was at stake, however, due to the segmentation of knowledge, the emergence of specific professional cultures, and the increasing plurality of academic institutions. The technical science elite also felt attached to the unity of the sciences since this idea supported their sense of belonging to the academic elite. However, they simultaneously challenged it as they defined the professional identity of the engineer as an academically trained but thoroughly practical man. As a result the engineering elite both opened their colleges for women and at the same time closed the profession to them.

In the end, male engineers who had considered it perfectly acceptable to admit women to technical training because there was no reason to fear they would enter in large numbers proved correct.[21] While 18,316 women were studying at German universities toward the end of the Weimar Republic, making up almost one-fifth of the entire student population, only 944 women registered with the technical colleges. Women were only five per cent of the student body there. After the National Socialists came to power, various demographic, economic, and political developments led to a dramatic drop in the general number of students, with the per centage of women falling at a higher rate than the average.[22] This trend did not come to an end until the winter of 1937/38. From the beginning of World War II, both the universities and the technical colleges saw a rapid increase in their numbers of female students. In the summer of 1943, shortly after Goebbels proclaimed 'total war' and called for the mobilization of all reserves, 1,537 women were enrolled in technical colleges, making up twenty-three per cent of all students.[23] But even under the exceptional conditions of war, there were relatively few female engineeringstudents.[24]

During the Weimar Republic, women most frequently chose to get their engineering degrees in chemistry and architecture. In construction, mechanical, and electrical engineering, women were always an exotic presence, whereas most male students opted for these traditional, core subjects in technical college.[25] Chemistry and architecture thus paved the way for women into engineering in Germany, just as they did in the US and other parts of Europe.[26] In Great Britain, however, it was electrical engineering that first attracted women.[27] From this we can conclude that it was not the content of the courses and disciplines that determined women's access to different types of engineering. Rather, social forces played a strong role.

What motivated the female pioneers in engineering to enter a traditionally male profession? In drawing a collective portrait of the women who studied at Munich's technical college, Margot Fuchs has shown how successful engineers and entrepreneurs encouraged their daughters to pursue studies in engineering.[28] In such cases, family networks and the prospect of taking over the family business facilitated the entry of women into the professional field of their fathers.

The establishment and increasing specialization of large-scale enterprises in the Weimar Republic and the Third Reich opened up new areas of work for women. Among these were patenting and research divisions and testing laboratories, all of which were jobs considered merely assisting engineers.[29] Hence, the rise of semi-professional technical work for women in new fields of technology fostered the employment of women engineers as well.[30] The emergence of large-scale industry also produced corporate engineers, a development that the scientific elite harshly criticized as making the engineering profession overly bureaucratic.[31] It is noteworthy that this new category of engineers, which the engineering establishment thought lowered the prestige of engineers generally, coincided with the entry of women into the profession. This was true not only in Germany, but also, as Ruth Oldenziel has shown, in the United States.

How did the pioneers themselves view the entry of women into technical fields? The 'poet engineers' Max Maria von Weber and Max Eyth had, in their struggle to improve the status of engineers, maintained that the profession demanded the 'whole person.' In 1930, Hildegard Harnisch-Niessing, who was then training to be a mechanical engineer, defended the presence of women in the profession on the same grounds. As Harnisch-Niessing explained: 'Those who are truly familiar with technology know the creative artistry, the deeply human, emotional and social commitment it encompasses.'[32] By making reference to human, rather than 'male' qualities, she was able to assert that the image of the engineer was partly or potentially feminine as well as masculine. But she categorically rejected the dominant view within the women's movement, which was that women could and would change whatever public arena they entered for the better. Harnisch-Niessing explained that 'the tasks faced by the engineer permit only clear-cut solutions. Technology could never assume a different, new character by way of women's participation.' Thus she denied that feminism had any relevance for those women who sought to enter engineering. She thus took a view that was widespread among women engineers of the first generation in the U.S. as well.[33]

The college-educated mechanical engineer Ilse Knott-ter Meer had a completely different point of view. She attempted to carve out a specific space for women within the larger field of technology.[34] Against the background of the Weimar Republic's overcrowded labor market, she justified women's entry into the engineering field by pointing out that women could not simply ignore the influence of technology on their lives. Rather, technology was entering every aspect of daily life including the domestic sphere. She believed that new professional opportunities were becoming available to women in office work and testing laboratories, but she especially recommended household technology as a new area of engineering appropriate for women.[35] An advocate of the difference-based feminism that prevailed within the German women's movement, she called for the establishment of a female space within the hitherto male-dominated world of technology. In 1930, Knott-ter Meer founded the 'Society of German Women Engineers,' modeled on the British Women's Engineering Society of which she was a member.[36] The new association was short-lived, however, being among the many organizations shut by the National Socialists because it was not under their control. It took more than fifty years for women engineers to establish independent professional organizations in what by then had become the Federal German Republic.[37]

On the whole, however, the effects of National Socialism on the situation of women in engineering were ambiguous. On the one hand, the Nazi state prevented women from pursuing careers in engineering by making it impossible to practice their profession. It did this by using political measures of exclusion and racist discrimination, forcing Jewish and Communist women into exile and making it difficult for women to enter the civil service.[38] On the other hand, in the Third Reich, professional careers for female engineers were available in all fields relating to the weapons industry and other areas where men had left positions vacant.[39] Finally, the rapid increase in the number of female students at technical colleges during World War II resulted in a significant change in the formerly male temples of technology.

Bastions of Hope: Women Engineers in Post-War East Germany and the Identity Crisis in Engineering

'Just by the quality of our work we are able to regain the respect and recognition of all free peoples, especially the great people's democracy in the East, and thereby attain economic and political freedom. This, my dear gentlemen students, is our paramount challenge, and we endeavor to train you to meet it.'[40]

Thus Enno Heidebroek, chair of mechanical engineering and machine elements and its first post-war vice-chancellor, reopened the technical college in Dresden on 18 September 1946. In addressing his speech exclusively to male students, Heidebroek made clear his assumptions about what constituted the pre-war and 'normal' state of affairs to which the post-war college should strive to return. Apart from the considerably smaller Freiberg Mining Academy (Bergakademie Freiberg), Dresden's was the only technical college in the Soviet occupied zone of the country. When, after an interruption of eighteen months, it resumed its teaching activities, 388 male and sixty-five female students were registered there.[41]

The women students, fourteen per cent of the study body, constituted a clearly visible group, even if their number was smaller than it had been during the war when, in 1944, 382 women had registered at Dresden's technical college.[42]

What happened to the many women who had studied at the college during the war? Some no doubt fell victim to the Allied bombing of the city on 13 February 1945, while others left Dresden because of the widespread destruction. When the college reopened its doors, it was under the control of the occupying Soviet forces. Admission lists were subjected to the approval of a Major Pluschnikov who represented the Soviet military administration.[43] It was his task to help the occupying force to recruit a new elite from among the working-class and farming communities. From this point on, preference of admission was granted to applicants from groups that had been at a disadvantage under the Nazi regime.[44] This made it impossible for those women from middle-class backgrounds who had attended the college during the war, or joined the Nazi Party (NSDAP) to complete their studies in Dresden. Moreover, the extensive war damage, and the large number of vacancies in faculty positions, led to the decision that only first and second-year students would be admitted when the college reopened. Consequently, only a few of the female students were able to complete their studies after the war, and many did so only after an interruption of several years.[45]

Female chemists and engineers who had completed their studies and begun their careers during the Third Reich faced a difficult situation in the post-war labor market. The weapons industry had begun to collapse even before the war had ended, and unemployment increased further after the military defeat. Large numbers of men and women lost their jobs as a result of the so-called de-Nazification process by which state government was completely restructured. A national and occupational census, held on 29 October 1946, revealed that in the Soviet zone, six per cent of all male engineers and technicians and ten per cent of all female engineers were out of work.[46] Yet there was a great need for technical expertise in the new economy, the technical colleges, and the administrative structure – and equally, for the research projects set up by the occupying force in the newly established engineering offices. However, the acute lack of men, many of whom had been killed or taken prisoner during the war, meant there was a severe shortage of such expertise.[47] Women who had been trained in the natural sciences and engineering studies were therefore able to take up positions of responsibility. Since some of these women at a later point finished their dissertations, we can pick up their trail by finding their printed or archival résumés. One example is Maria Scheibitz, who studied chemistry in Dresden between 1938 and 1944 and also had two children during that time. In 1944, she began working as a chemist in a research laboratory at a chemical plant. After the war, the plant was transformed into a Soviet joint-stock company. In 1947, Scheibitz became head of the laboratory. A year later, when the firm's research division was moving to another region, she quit her job. She then took up a position as a scientific assistant under Professor Arthur Simon at the Institute for Inorganic and Inorganic-Technical Chemistry of Dresden's technical college. There she completed her doctorate in 1953.[48]

In the post-war period, in which women outnumbered men among the general population, and society was generally unstable, the established gender system was increasingly called into question.[49] In the technological realm, this took the shape of an identity crisis among engineers, which was fully exposed in debates about whether or not the profession had responsibility for the war and subsequent defeat.[50] Two positions emerged. One group of engineers rejected any responsibility for the war by arguing that technology is merely an instrument, so that professional labor in the service of technology is politically neutral. Advocates of this view claimed that technology could be abused, but was not positive or negative in itself.[51] An influential representative of this outlook was Enno Heidebroek, whose position at Dresden's technical college has already been mentioned. He wrote an essay on the training and activities of engineers in 1948 that argued:

'The technical product remains factual and objective at all times; its value is dependent only upon whether its creator is a good or an inferior engineer in terms of expertise. To make proper use of the instrument for the benefit of humanity, and to prevent any abuse by destructive forces: that is an educational task for all constitutive factors of society. [W]e should not succumb to the illusion that we … can imbue technology with more humanity by forcing our students to cope with a great deal of so-called humanities subjects.'[52]

Heidebroek believed that the focus should be instead on qualified expertise. Against his view, however, there were those who acknowledged that engineers shared responsibility for the German catastrophe, and who recognized that lack of critical reflection by the profession had led many to collaborate with the Nazis. This group advocated expanding the engineering curriculum to include cultural science subjects. The chemist Heinrich Franck, the first post-war dean of general studies at Berlin's technical university, and one of the driving forces behind the institution's educational reforms, took the view that societal responsibility must be assumed. In 1949, Franck succeeded Heidebroek as president of the *Kammer der Technik*, the East German association of engineers. He hoped that studying cultural subjects would give 'pure technicians' a greater sense of their human responsibilities and the social dimensions of technology.[53]

The acceptance of political and social responsibility changed the professional identity of engineers, and therefore the relationship between technology and masculinity. This was because a central element of the professional self-understanding of engineers – the role of technology in modern warfare – was redefined in the aftermath of the Second World War. It changed from being a factor in social acceptance to a reproach. Furthermore, as the relationship between technology and warfare was criticized, the connection between technology and masculinity was called into question as well.

Did these changes facilitate women's access to the engineering profession? Images of women engineers that appeared sporadically in the press of the Soviet zone provide some evidence. The German Women's Press Service and the East German Women's Association (Demokratischer Frauenbund Deutschlands or DFD) published frequent reports offering up-to-date (but wildly variable) figures on the number of female engineers in the Soviet Union. They also provided portraits of individual East German women engineers and engineering students.[54] The goal of such writing was twofold. First, these reports used the Soviet women engineer to demonstrate how women could find employment and self-fulfillment under socialism. Second, the women featuring in them were presented as part of a new intellectual elite, one that was entirely opposed to the older establishment by virtue of including women. One example shows both of these ideas at work. A Sorbian student named Ingeburg Nawroth called upon other women to enter the engineering profession. This particular student was a member of the only ethnic minority in Eastern Germany and the daughter of a skilled laborer – her father was a radio mechanic and a Communist. Her case was supposed to impress upon female readers that a new society was being built, with an emerging elite arising from precisely those groups that had been at a severe disadvantage under the old order on account of class, ethnicity and sex.

Ingeburg Nawroth wrote:

'[A] new progressive technical intelligentsia is necessary for the success of our economic plans. A multitude of grand opportunities offer themselves to women, opportunities to work diligently and effectively in decisive positions to help build a new Germany, and thus solve the most urgent problem confronting us: peace. I find again and again that women have not yet sufficiently recognized the positive influence they might have on the economic development of our new democratic Germany, and on establishing lasting peace.'[55]

Figure 43. Dual image of Ingeborg Nawroth, one posing as a woman engineering student and the other wearing a Sorbian ethnic dress illustrating her dual identity. Nawroth wrote an essay "What made me studying engineering?" (1949), in which she explains how she sought to bridge the gap between her traditional gender identity and the modernist engineering profession.
Reproduced from Neues Frauenleben 15 (1949), 9.

Nawroth appealed to women's responsibility to sustain peace in her encouragement of women engineers. This appeal was based on the notion, frequently expressed in the post-war period, of a supposed link between women and peace. At the founding conference of the DFD in March 1947, Emmy Damerius, for instance, enthusiastically proclaimed: 'The female capacity for emotion and true feeling ideally empowers women to become the saviors of peace....Our struggle coexists with the recognition that peace and motherhood are inseparable.'[56] This new connection between women, peace, and technology facilitated the emergence of a professional identity for engineers that now included rather than excluded women.

The numbers of women enrolled in the technical colleges and practicing as engineers rose considerably after 1946 and into the early 1950s. Still, while the number of female students had doubled, they now made up only six per cent of the total student population at the two technical colleges, as against thirteen per cent previously.[57] Census figures of the professions show that the number of women engineers and technicians increased substantially between 1946 and 1950, from 1,695 to 2,984: a rise from 2.8 to 3.2 per cent of the total.[58] To some extent, then, in the years after the war, professional engineering became more accessible to women. However, the sharp drop in the percentage of female students at technical colleges indicates that, even in this phase of expanding accessibility, there were also exclusionary processes at work. German ex-soldiers made up a remarkably high proportion of the newly admitted students at technical colleges, although German post-war society did not celebrate war veterans or openly give them privileges as the Allied Forces did; German soldiers had served a vast project of genocide, and furthermore had lost their war. By their presence, ex-soldiers helped to re-establish a culture of maleness within technical colleges.[59]

Ambitious Yet Excluded: Ambivalent Social Policies and their Effect on Women

In July 1952, at the Second Party Conference of the Socialist Unity Party (SED), party chief Walter Ulbricht announced the 'on-schedule establishment of the basic principles of socialism in the GDR.' Engineers experienced this as a fundamental change to their educational and professional position. What did this entail and how did it affect gender relations in the field of engineering?

After the war, the eastern part of Germany had significantly less capacity for training engineers than the western occupation zones: only two of the thirteen technical education institutions of pre-war Germany were located on the territory of the Soviet zone. Furthermore, the proportions within academic education had shifted to the disadvantage of technical education. During the last years of the Weimar Republic, some one in five students had been enrolled in technical colleges. By 1946/47, the total number of students at the technical college in Dresden and the Freiberg Mining Academy had dropped to a mere four per cent of the overall student population in the Soviet occupied zone, showing a slight rise to six per cent in 1947/48.[60] Progress in rebuilding the economy was already seriously hampered by Russian dismantlings and the transfer of specialist staff to the Soviet Union, as well as by large-scale emigration to the West. Therefore there was a sharp increase in the demand for engineers.

A first step in attempting to overcome the lack of educational facilities was to restore pre-war conditions at the two remaining technical colleges.

The traditional structure of higher education, including departments and chairs, was reinstated, and sections of the old engineering elite returned to their former positions.[61] Academic facilities were then rapidly expanded.[62] The preservation of academic training for engineers working within a state-controlled economy entailed that a model of school culture was adopted that had its roots in the earlier German tradition.[63] The first new facility, the shipbuilding faculty at the University of Rostock, represented a breakthrough in the German history of higher technical education as a whole. Its establishment in 1950 resolved the historically anchored exclusion of technical sciences from German universities for the very first time.[64] In 1952, the College of Transport Engineering was set up in Dresden, while the College of Fine Arts and Architecture in Weimar was reconstituted as the College of Architecture and Building in 1953. That year also saw the institution of six special, monotechnical colleges: the College of Electrical Engineering in Ilmenau, the College of Heavy Engineering in Magdeburg, the College of Mechanical Engineering in Karl-Marx-Stadt, the Technical College for Chemistry in Leuna-Merseburg, and the Colleges of Building in Leipzig and Cottbus.[65]

The establishment of these special colleges meant a break with the German tradition of technical education. The first technical colleges had been based on the model of the nineteenth-century polytechnics, offering a broad range of subjects to help engineering gain recognition within the academic world. The special colleges, however, provided a kind of advanced vocational training and were modeled on Soviet ideas of specialized education. The responsibility for higher technical education in these colleges rested with the relevant industrial ministries, so that specialist engineers could be trained in close accordance with industrial needs.[66] The East German government took a similar course by assigning both the new special colleges and the traditional academic institutions where engineering training was on offer to the responsibility of the corresponding ministries.[67] The emergence of special colleges provoked a mixed response from the established engineering community.[68] Most engineers criticized what they saw as an over-emphasis on specialization, the redistribution of limited resources, and the elimination of assistants. The new colleges responded, rejecting the stigma of 'second-class' status by reinstating bourgeois traditions of academic education.[69]

As has been indicated above, even the older technical colleges underwent structural change. New and more specialized faculties and courses such as engineering economics, production processing, aircraft construction, and nuclear technology were developed.[70] The reconstruction and expansion of East German technical education in the 1950s was thus marked by two opposing developments. The state restored the traditional format and course content of higher technical education, but at the same time also began to adapt both the organization and the structure of the curriculum to a Soviet model. Yet, while in the 1950s up to one third of all engineering students in the Soviet Union were women, the number in higher technical education in the GDR at best amounted to 6.5 per cent between 1953 and 1959.[71] Despite many interventions by the Soviet occupying forces, there was a decline in the share of female engineering students.[72] The GDR state also failed to increase the percentage of women during the years of expansion in higher technical education.[73] In the 1950s, above all the male youth built up the new technical elite. In 1953, twenty-nine per cent of all the male students and six per cent of all the female ones enrolled in East German academic institutions took up technical training. In 1959, the male figure had risen to forty-four per cent, as compared to a mere six per cent of female students.[74] In this way, the male culture of higher technical education was effectively restored.

The situation proved to be similar in the new monotechnical colleges. Due to the very structure of these new colleges, the number of female students enrolled in them was significantly lower than at the technical college in Dresden. (The College of Architecture and Building and the technical college for chemistry in Leuna were the only exceptions.) This reflected the endurance of gender-based preferences for certain subjects that had been visible earlier during the Weimar Republic and the Third Reich. Disciplines such as heavy engineering at Magdeburg college, mechanical engineering at Karl-Marx-Stadt college, electrical engineering at Ilmenau college, and civil engineering at the colleges in Cottbus and Leipzig, formed the core curriculum of the traditional technical colleges and were largely chosen by men. Chemistry and architecture continued to be the subjects most frequently chosen by women. In this respect, not much had changed since the years before the war.

What lay behind the restoration of a male-dominated culture of engineering, in the light of the radical political changes in 1950s? One explanation can be seen in the professional ideology and self-representation of the engineering elite. Since the former technical establishment maintained its powerful position within the scientific elite, the ideas held by its members also held sway at the new technical colleges.[75] This led to the continuing reproduction of a male culture in the education of engineers, a culture that discriminated against women and did not encourage them to take up engineering studies. It is worth examining further the underlying notions of technology that were at work in order to understand women's continued exclusion from the technical world. The views of Enno Heidebroek are again of key importance. As the first vice-chancellor of the technical college in Dresden (from the moment of its reopening to the passing of its first two-year plan), the first president of the *Kammer der Technik* (1946–1949), and an active college teacher until 1954, Heidebroek was a central figure within the traditional engineering elite. He exerted a strong influence on the first phase of the reorganization of the technical education system. Thus he may be considered a spokesman for the old technical establishment.

As early as the Weimar Republic, Heidebroek had been one of the 'philosophers' of the profession. He attempted to define the essence of technology and the meaning of engineering work in response to the negative attitudes towards technology within the wider culture, but also to provide engineers with a professional identity in the face of an industrial world transformed by mass production.[76] Heidebroek's basic theory was that technology is not merely a mode of science, but also an art, operating along creative lines as well as along those of inquiry and research. Art, however, had priority in his concept of technology.[77] The engineer as the active representative of technology was thus an artist, mirroring nature with his inventions. In doing his work, he was following a natural drive, since nature is always striving for greater perfection of form. In the same way, technology, he felt, follows an 'irresistible urge to continually bring forth new forms.'[78] His ideas fit well with the dominant notion of culture within the German academic elite. Thus Heidebroek made concessions to the polarized culture concept of the German academic elite, who set culture over civilization; as he redefined the purpose-oriented technical activity as an aesthetic endeavor for optimum forms; he considered technical creativity as a natural (and therefore sometimes irrational) drive; and focused on invention as an intuitive act of creation which thus fitted the idealistic principle of predominance of mind over matter.

Since the turn of the twentieth century, ideological debates about gender difference had become increasingly intense, especially in light of the fierce yearning for order, safety, and coherent identity after the dislocating experiences of the First World War.[79] Concerns about the roles of men and women in this destabilized society paradoxically resulted in the reinforcement – even reification – of complementary gender roles.

Although Heidebroek did not directly address gender, he expressed his ideas within the prevailing ideological framework of the time. His imagery recalls that of Ernst Jünger in his book *Storms of Steel* (*In Stahlgewittern*, 1920), which depicts the community of men in the wartime trenches as an alternative to German society as it really was.[80] Heidebroek expanded Jünger's utopia to encompass the community of engineers. Where Jünger believed war to be a male form of creation, Heidebroek conceived of it as the motor of technical development, a notion that remains widely held in the engineering world. He celebrated those 'front officers of technology' who had successfully survived the 'struggle for initial existence, the wrestling with the material, the imponderable … the bath of steel of fighting and prevailing.'[81] Heidebroek's views on the essence of technology reinforced the idea that technical competence was a part of male gender identity.

During the post-war period, Heidebroek held to the theory of technology he had formulated in the Weimar Republic.[82] Previously his stress on art and a natural drive had served two purposes. First, it formed part of an argument against those cultural critics who anticipated the end of technology. Second, it was also a warning against eroding the professional status of engineers, as large-scale industry and the development of technical staff seemed to threaten. In the years after the war, by contrast, Heidebroek deployed the same notions against workers' education and the planned economy. In his view, technology, as an art, required that only the most talented individuals be allowed to study. He sought to prevent the introduction of quotas on the basis of social origin.[83] Furthermore, he thought that, as an independent drive governed by natural law, technology could and should not be subjected to planned economic intervention.[84] It is revealing that he stuck to the theory of war as the motor of technical development. In his speech to celebrate the resurrection of the technical college in the ruins of Dresden, he criticized war as 'probably the worst abuse of the abilities given to mankind by technology.' He went on to emphasize that it was, at the same time, 'from this direction that the strongest stimuli for [technology's] further development' derived. Although these statements may look contradictory, they were not, since both rested on the fundamental assumption that technology itself is simply a neutral tool. And both attempted to rescue technology as a component of male identity.

Traditional ideas about technology continued to underpin the study of engineering in the academic world of the GDR. Their persistence accounts for the continuing marginality of women at the technical colleges in the 1950s, even if the new political order was founded on equality for both sexes and actively sought to increase the number of female students.

That these efforts remained quite unsuccessful throughout the 1950s was due not only to the survival of bourgeois traditions, but also to various reforming measures. The party state introduced admission quotas to further the development of a new working-class intellectual elite. A condition for the development of a new 'working-class intelligentsia' was 'class purity' at the point of recruitment; the state trusted that socialization among those of proletarian origin would overcome dread of class alienation, which was associated with intellectuals.[85] The introduction of a class-based quota, however, hampered women, for the daughters of uneducated parents faced higher barriers in entering technical colleges. This, after all, meant a break with both their social background and their accustomed gender role. The class quota therefore made it more difficult to meet the female quota at technical colleges. 'Workers' and 'Farmers' Faculties, which had been set up in line with the Soviet model to provide basic training to students from working-class backgrounds, in fact never filled their female quotas.[86]

Still, these imposed policies, especially those that allowed formerly underprivileged social groups, definitely succeeded in their primary goal of changing the constitution of the student body. Until World War II, it

had been mainly the daughters of entrepreneurs, of mid- and high-level bureaucrats, and of the educated middle classes who had pursued studies in engineering.[87] In the 1950s, by contrast, working-class girls also began to enroll in the GDR's technical colleges. A survey of the class structure of the student population at technical colleges in 1961 shows that two-fifths of the female engineering students came from working-class families.[88]

The entry of working-class women into the engineering profession, even if they were vastly outnumbered by men with the same class background (in 1961, fifty-three per cent of full-time male engineering students were the sons of laborers) made the profession much more heterogeneous. However, the diversity of the profession was not solely the result of the new socio-political order. Rather, it resulted from the process of industrialization as such and the wide variety of opportunities offered in engineering. The restoration of the two-tiered system of engineering education that had existed in pre-war Germany – that is, the split between research-oriented technical colleges and the more practical engineering middle schools – and the entry of some skilled workers and technicians into engineering, opened the profession to graduates from a variety of schools and backgrounds. Engineers found

Figure 44. In 1952, the East German Communist party decreed the foundation of Women's Committees to promote women's interests. Women of one of these committees are here photographed at their conference at the Technical College Dresden in 1956. Permission and courtesy of Media Center, Archive, Technical University Dresden, Germany.

employment in science, government, and many sectors of the economy. They worked as research and development engineers, as builders, project managers, special consultants, technologists, test engineers, production managers, and as the teachers of the next generation. The individual career tracks of engineers were therefore extremely diverse.[89]

The SED policy towards the academic elite also worked to fragmentize engineering socially and professionally. The state granted special privileges to members of the old (middle-class) intellectual elite, in order to prevent them from leaving the country. At the same time, however, it had succeeded in shifting the class boundaries within the professional field by promoting higher education for the working classes, and by enabling skilled workers and technicians to take up engineering careers.[90]

In the long run, the increasing heterogeneity within the profession that resulted from these shifts also helped women to enter engineering. The 1950s thus prepared the ground for the rise in female engineers in the 1960s. By that time, however, there were strong oppositional forces at work: traditional ideas of technology and the enduring norms of the middle-class gender system.

How were women in engineering affected by the changed political system and its effects on the professional situation of engineers? The new power elite took two key decisions. The first was to nationalize and centralize the economy. This strengthened the trend towards bureaucracy that had been evident since the transition to heavy industrialization at the end of the nineteenth century. This, in turn, limited engineers doing freelance or commercial work to those fields that were not under state ownership.[91] In the course of the 1950s, however, chances for a freelance career in engineering were cut back even further because the SED had control over the private sector as well. Engineers and architects were subsequently subjected to accreditation procedures, fee regulations, and taxation.[92] These restrictive policies affected both men and women equally. After all, the generation of female pioneers in engineering had quite frequently been made up of women with access to private capital. Either they had begun their careers by inheriting the family business, or they ran an engineering organization either as sole owners, or in partnerships with husbands or other people.[93] This means of entry into the profession became less common as the economy was nationalized.[94] Now the majority of engineers were employed as civil servants.

The second major change initiated by the state was the establishment of basic and heavy industry within the framework of the First Five-Year Plan.[95] This led to many new jobs for engineers, especially in traditional, 'male' fields such as mining, metallurgy, and mechanical engineering. Since these were areas that helped create the enduring imagery of the engineer as a conqueror of nature, inventor, and leader of men, it is not surprising that they remained male-dominated.

Women Engineers and the 'Technological Euphoria' of the Reform Phase

The 1960s were years of wide-ranging reform in the GDR. The power apparatus, the economy, the educational system, as well as youth and cultural policies were all affected. Party chief Walter Ulbricht put himself at the forefront of the reform movement. With its support, he sought to develop socialism in such a way that East German society would emancipate itself from the Soviet model.[96] The course of reform centered on the hope – significantly strengthened by the 'Sputnik shock' – that using science and technology efficiently would lead to an increasingly efficient socio-political and economic system. Socialism would then triumph over the West. As a consequence, any policies promoting technological modernization that had been implemented in the mid-1950s were expanded.[97] The primary goal was to achieve a sharp rise in productivity with the help of completely new technologies. This would enable the GDR to overtake the technological developments in the West without having to compete directly using current technology. By the end of decade, Ulbricht's ideas had led to a widespread, idealistic obsession with technology.

The promotion of science and technology, and the new policies that it inspired, affected gender relations in the engineering profession. In the course of the 1960s, the field of engineering began to lose its exclusively male character. Between 1960 and 1971, the number of full-time female students in engineering science rose from 753 to 7641, or from four to twenty per cent of the total.[98] Professional statistics also showed that an increasing number of women pursued a career in engineering, though they still did not meet the quotas set for them in engineering studies. Until the early 1960s, the greater numbers of men entering the field accounted for the expansion of the profession. Between 1964 and 1971, however, the female share of college-educated employed engineers increased from 3.1 to 5.6 per cent, while those with mid-level training qualifications rose from 5.6 to 9.3 per cent. In 1964, 757 of the former and 8,442 of the latter were working as engineers. By 1971, these numbers had increased to 3,115 and 21,030 respectively.[99] Thus the pattern was very different from that seen in Greece (as discussed elsewhere in this volume), for mid-level technical training in the GDR led significantly more women to pursue careers in engineering.[100] More than one third opted for chemical engineering or, alternatively, for textile, clothing, and leather technology. Both these fields thus became what Margaret Rossiter has called 'participatory' fields, in which the female share (forty-three and thirty-seven per cent) was substantial.[101] The second largest group of women engineers with mid-level training qualifications was the mechanical engineers. However, they were still only five per cent of the field. Table 17 illustrates the distribution of women and men who took up professional employment after graduating from mid-level engineering schools.

Table 7. Significantly more women gained access to an engineering career through mid-level training. But even in the socialist system engineering developed as a highly gender-segregated profession. Up to 1971 more than one third of all mid-level trained women engineers opted for chemical and textile engineering (inlcuding clothing and leather). Employed engineering middle-school graduates in East Germany in 1971. Source: Wirtschaftlich tätige und nicht wirtschaftlich tätige Wohnbevölkerung Vol 5 of Volks-, Berufs, Wohnraum und Gebäudezählung, 1 Jan. 1971 (Berlin 1972), 115–116.

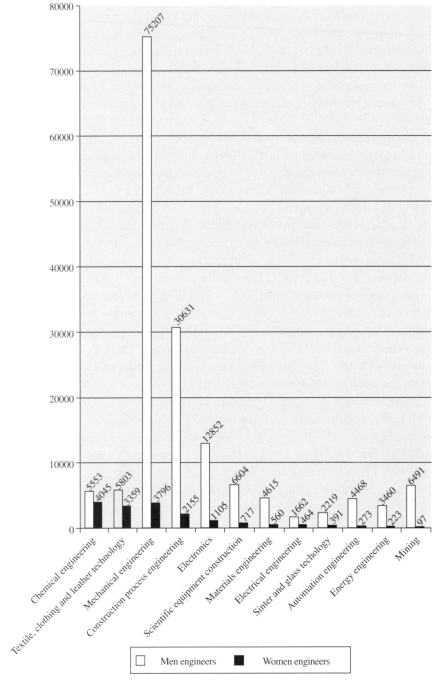

The much smaller group of college-trained female engineers remained a tiny minority amongst professionally employed graduates of the technical colleges. This segment of the profession thus retained its male character until 1971. Women's share in the fields of processing and materials engineering barely exceeded five per cent, while in the other four fields the per centage of women remained even lower. (Table 8)

Still, the end of male dominance could be predicted in these fields too, given the rapidly growing numbers of female full-time students in the engineering sciences. (Table 9).

What led to the erosion of the male character of professional engineering in the course of the 1960s? Since the beginnings of the Cold War, the political elites in both the communist East and the capitalist West placed educated manpower at the center of their strategies for shifting the balance of power in the post-war era. In the US, such an educated population had been referred to as a 'critical national resource' as early as 1945.[102] When the Soviet launch of Sputnik I in October 1957 revealed that science and technology had become the main battlefield in the competition between the two systems, the number of technical experts in each country became a seemingly decisive factor in the contest.[103] In this context, both sides came to see women as an untapped resource, and both tried to mobilize women for engineering. Even the Western media began to present Soviet female engineers as role models, and American publications appealed to the patriotic feelings of bright young women, urging them to place their brains at the disposal of the nation.[104]

Inge Lange, who was in charge of the SED women's policy, took this line. Explaining the new stress on qualifications in the politbüro's communiqué 'Women, Peace, and Socialism,' (December 23, 1961) she stated: 'It is, after all, not merely a question of equality, but also the objectively necessary utilization of all productive forces in society.'[105] In the US, recruitment campaigns to draw women into engineering never moved beyond rhetoric; the highly negative attitudes of both employers and graduate schools towards women remained largely unchanged.[106] The socialist system in the GDR, however, allowed the state to intervene directly to send women into engineering studies.

Beginning in the early 1960s, the East German government had developed policies to enhance the status of higher technical education as well as the number of technical experts. To raise the prestige of academic engineering, for example, the state upgraded the Technische Hochschule in Dresden to a university. The college for heavy engineering in Magdeburg became a full technical college in 1961. In 1963, the special colleges in Karl-Marx-Stadt and Ilmenau received the status of technical colleges. These former special colleges also gained the right to grant doctorates.

The state also reformed existing engineering studies to try to create more and better trained technical experts. In October 1962, the state

Table 8. The much smaller group of college-trained female engineers remained a tiny minority among professionally employed graduates from technical colleges. This segment of the profession thus retained its character at least until 1971. Employed technical college graduates in East Germany in 1971. Source: Wirtschaftlich tätige und nicht wirtschaftlich tätige Wohnbevölkerung Vol 5 of Volks-, Berufs, Wohnraum und Gebäudezählung, 1 Jan. 1971 (Berlin 1972), 117–118.

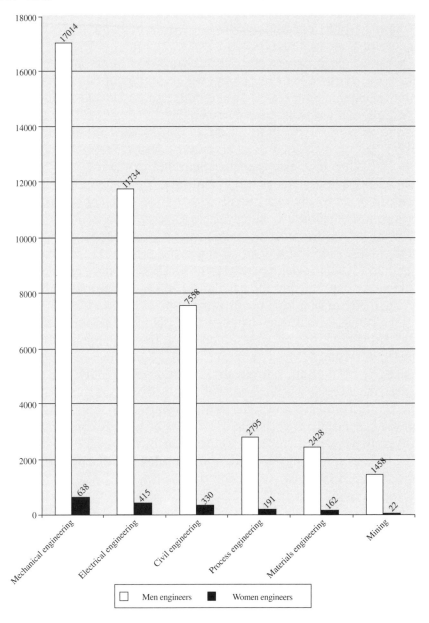

office for college and mid-level education published a set of proposals for redesigning the curriculum of engineering sciences. The imposed reforms called for an end to the traditional form of college study, and the creation of a rotating system, in which periods of study would be alternated with longer-term practical training. While gaining hands-on experience, students would remain in contact with their colleges via a correspondence course. Basic training in mathematics, natural sciences, technology, and economics would be intensified. During their periods of full-time study, students would also focus more on theoretical knowledge than the applied training skills that had dominated the curriculum since the 1950s. State officials for higher education also sought to make the training of engineers fit better with the specific needs of the industries and areas for which they were eventually destined – hence they designed different degrees for engineers in research and development, in design departments, and in manufacturing. At the same time they put great stress on the training of manufacturing engineers, who were to play a key role in implementing increasingly complex mechanization and automation processes. All of these measures, they hoped, would help to modernize the GDR.

These suggestions deviated considerably from the accepted form of engineering education, and representatives of the state office for college and mid-level education feared they would not meet with an enthusiastic reception at the technical colleges. They were afraid, they said, of

'passive resistance ... because individual researchers, overestimating their own competence and assuming that the suggested materials had been prepared exclusively by members of the state apparatus (which is fundamentally correct), believe that the quality of the theses and suggestions would have been significantly better had they had been developed exclusively by scientists from the beginning.'

The state also expected scientists to dislike these proposals as a threat to the 'unity of German higher education.'[107] In actual fact, however, the educational elite was successful in using the suggestions to boost engineering school culture, and in rejecting any attempt to erode the essential principles of academic learning.[108] The expansion of basic theoretical training (the value of which was not disputed) was adapted in such a way that it did not impinge on full-time study. Practical training was limited to set periods of time, and students were to keep their college affiliations while away. Thus the actual effects of the reforms were limited. The idea of specialized training geared to specific fields of future deployment had to be dropped. Finally, the technical colleges retained the right to develop their own curricula. As a result, representatives of the scientific elite who were committed to the tradition of the older German engineering school culture maintained their hold in education.

The state's initial attempt to achieve a rapid increase in the number of technological experts by reforming engineering education was thus quite

Figure 45. This seemingly innocent photography of free-hand drawing practice of students of architecture in a church in 1966 was an act of subversion. Political officials condemned the training of students in churches, punishing professors. Permission and courtesy of Media Center Archive, the Technical University Dresden, Germany.

unsuccessful. It had better luck, however, with the reorganization of mid-level technical education. The principal lever here was the promotion of women. By issuing directives to introduce special women's classes, the state office for higher education partly dissolved traditional forms of engineering education at the middle schools, and at the same time expanded the recruitment potential for professional engineers. The first middle schools began to set up special women's classes in engineering in 1963.[109] The ministry stipulated that only women who were able to prove that they had particular duties toward dependents were entitled to apply for the special classes. In other words, the promotion of women was directly linked to the fulfillment of their traditional duties as mothers and carers. The special classes were really meant for women between the ages of twenty-five to thirty-five, since there was a considerable surplus of women within that age group in the GDR in the early 1960s.[110] Politicians with a special responsibility for education tried to attract women to those classes by offering special promotion contracts, including reduced working hours, and continuous payment plans.[111] Female students were to be commissioned for study by their employers, both as a reward for excellent performance in the workplace, and for appropriate conformity to the system.

Instead of waiting to be delegated by their employers, however, many women applied independently to the special classes to improve their career opportunities.[112] The interest in further study was much greater among women employed in administrative or technical drawing departments than among skilled female laborers. Their employers were also much more willing to commission personnel to study, since skilled female laborers were urgently needed elsewhere to meet production plan targets.[113] In the women's affairs department of the Central Committee, this became a cause for concern. It was another indication that the working-class aversion towards white-collar professions persisted even in the GDR.

Figure 46. Special women's class during the school year 1972/3 in the Braunkohlenwerk Cottbus (brown coal works Cottbus). The women studied engineering economics for mining. Courtesy of Frauenverein Lübbenau e.V. and permission of Dr. Petra Clemens.

Between 1963 and 1975, a total number of 13,000 women took the opportunity to enroll in a special women's class. Most of all graduates from special women's classes earned a degree in engineering economics, whereas only 10 per cent of all graduates took a degree in core engineering subjects.[114]

The special women's classes turned out to be a mixed blessing. They encouraged women to train as engineers, but they also reinforced traditional gender roles. Many women complained that they gained no recognition for their efforts from neither their work team nor the enterprise directors.[115] Although they received privileges such as special leaves of absence and continual wages, the discontent expressed indicates that the promotion of women's position in engineering also created new fields of conflict. The women who benefited from these privileges moreover faced an immense burden, since they had to find a way to balance their job with education and unequal family responsibilities.

At the end of the 1960s, the Ministry of Higher Education sought to extend the special women's classes to the universities.[116] This proved to be less successful, however. Very few special women's classes were established at universities and technical colleges.

Nonetheless, both the restructuring of the system of higher technical education and the efforts to attract women to engineering studies began to gain momentum. In 1962, the state office took explicit control over the admission procedures at the universities. They were now required to report back on female admissions and to justify individual suspensions and rejections.[117] Furthermore, the Ministry of Higher Education ordered the universities to write twice a year on the measures they had taken to promote women's access to study. And finally, in 1965, the scientific department of the

SED's Central Committee decreed that 'in the plan for admissions for the year 1965, the quota to be met for women [be] specified in numbers. The vice-chancellors are to be made fully responsible for fulfillment of this aspect of the plan.'[118]

This meant that the percentage of women enrolled in scientific and technical disciplines was to be dictated from above and their success made a priority. These measures undoubtedly led to a rise in the number of female students in the engineering sciences, although initially the gains were modest.

It was only at the end of the 1960s that a rapid increase in the number of women took place. This was due to a second stage of expansion in higher technical education within a fundamental restructuring of the system of higher education. This has become known as the Third Higher Education Reform.[119] This reform dismantled any residual academic self-administration, so essentially it represented a renunciation of the founding traditions of German academic culture. Faculties and institutes were

Figure 47. Physicist Lieselott Herforth (b. 1916), the first woman appointed to rector of a technical college, heading the all-male university principalship into the matriculation celebrations in 1966 as rector of the Technical University Dresden. Women students made up for a minor but clearly visible group among students in Dresden. Permission and courtesy of Media Center Archive, the Technical University Dresden, Germany.

reorganized into significantly smaller sections and departments with a centralized administration and restrictions on professional power. New study plans were to be organized and directed from the center.[120] A resolution of the State Council finally reduced study time to a maximum of four years.[121]

The centralization of academic power within the institutions of higher education enabled the SED to continue its expansion of higher technical education. On January 14, 1969, its leaders passed a resolution to this effect.[122] This implemented a project referred to as 'overtaking without pursuing,' which Ulbricht had developed. The premise was that the GDR could surpass the production output of the West by developing new technologies and advancing the automation of production.[123] Expanding the number of technical experts was considered an indispensable prerequisite to accomplishing this task. At the ninth plenary meeting of the Central Committee in the previous year, Ulbricht had announced that the GDR needed to attain the highest level of scientific-technical expertise in the world by 1975.[124] In 1969, the state upgraded ten mid-level engineering schools to the status of colleges and founded four new technical sections at the universities of Jena, Berlin, and Leipzig. In addition, it created new curricula and teaching programs for four-year courses, and reformed the system of part-time training so that graduates from the engineering schools could obtain full college degrees under a four-year, part-time program. A total of 14,000 additional students were to receive their degrees by 1976.[125] Without the full participation of women, such a huge expansion in the field of academic engineering was simply unthinkable.[126]

Should the increasing integration of women into engineering be interpreted as above all the result of political intervention? Or were other factors also at play, factors that made it possible to break with the male culture of academic technical studies? Let us first consider how technology was thought about in the wider culture before looking at how these ideas influenced training and employment. In the 1960s, the paradigmatic supremacy of science over engineering culminated in both the East and the West as an euphoria for large-scale science projects prevailed.[127] Thus, the notion of technology as an applied natural science gained great popularity. It was presented as such not only in philosophical discourse, but also in public debates.[128] The president of the GDR research council, Peter Adolf Thiessen, wrote in 1961,

'[the] technology of our days is applied research, it takes in results of research. Technology participates directly in research as well, in fact, in all of its movements.'[129]

And this was, furthermore, a conviction held by sections of the engineering elite, who introduced the discipline of applied mechanics. In 1961, a department of applied mechanics was founded and a study course introduced at the technical university in Dresden, the technical college of Karl-Marx-Stadt, and the technical college in Magdeburg.[130] This, however, profoundly affected perceptions of the careers of engineers, who were now positioned at a drawing board, in an experimental laboratory, or in the control center of an industrial facility devoid of human workers.

The campaign for large-scale research, initiated at the end of the 1960s in connection with 'overtaking without pursuing,' thus produced a significant shift in the view of engineers as scientists. The Minister for Science and Technology announced in 1970 that, 'in the course of the year ... around twenty-eight per cent of university and college personnel will be active in large-scale research centers.'[131] This implied, however, that career paths of many engineers would change radically, encouraging women to enter the world of technology.

Formal education in general also underwent various shifts. The establishment of a unified co-educational school system in the 1950s had already eradicated gender-specific differences in the curricula. The expansion of polytechnic training and the improvement of mathematics and natural science courses in the 1960s led to an increase in that knowledge which was indispensable to students entering engineering studies.[132] While during the first half of the 1960s, one of the principal reasons for the premature drop-out rate of students in the engineering disciplines had been their lack of sufficient preliminary knowledge, such criticism was less often heard by the end of the decade.[133] This helped to dispel the myth that engineering studies were particularly difficult, an idea which had sometimes been used to explain why women allegedly could not master the field. And finally, the abolition of the strict class quota in student enrollment had a positive effect on the number of female students.[134]

All these various factors made it easier for women to enter engineering. The fact that the major breakthrough took place between 1968 and 1971, when the number of female engineering students reached its highest level of growth, underlines the paramount importance of the Third Higher Education Reform. Between 1968 and 1971, one third of all the women who registered at the technical colleges for engineering studies chose such new fields as process engineering and information processing. In 1971, these new realms of study, in addition to smaller fields like architecture and geodetics, showed the highest percentage of female students.

Figures for 1971 show this clearly (Table 9). In four of the ten disciplines appearing in the student statistics for that year, women made at least one third of the total number of students. In seven disciplines they constituted a quarter of the entire student population. At the same time, however, their enrollment in the different engineering disciplines was still apparently determined by sex.

The male-dominated courses preferred by the majority of the male students were now, for the first time, placed alongside several 'participatory' or partially 'feminized' disciplines. Despite the fact that the proportion of women in engineering sciences had already begun to rise in the mid-sixties, it was the structural changes introduced with the Higher Education Reform that allowed the breakthrough in the ratio of men to

Table 9. 'Women's share in all engineering specializations remained low among professionally employed college engineers, but after the Third Higher Education Reform the number of women engineering students grew explosively thus challenging the monolithic male culture of higher education in East Germany. Male and Female Engineering Full-Time Students at Technical Colleges in 1971. Source: Hochschulstatistik *(1971).*

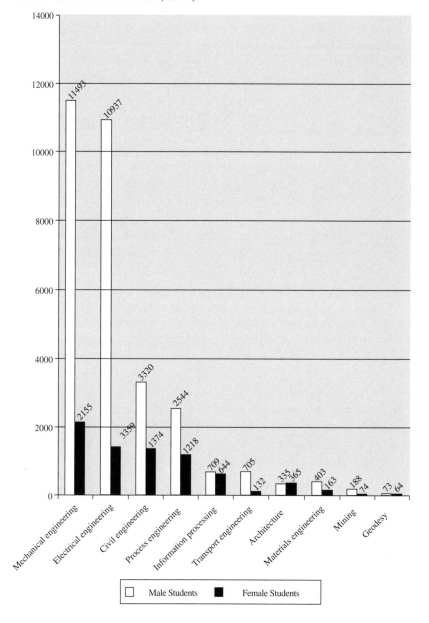

Figure 48. In contrast to France, electrical engineering remained a male dominated field even after the third Higher Education Reform (1967–1968). Photograph of electrical engineering students at Dresden Technical University's workshop in 1969. Permission and courtesy of Media Center Archive, Technical University Dresden, Germany.

women in engineering studies. The dissolution of the teaching faculties and institutes in favor of smaller sections and departments meant individual course disciplines were redefined. The new structures offered scope to absorb the growing numbers of women students, but at the same time permitted gender segregation to remain a fact within the engineering sciences. Thus, it was only the fundamental restructuring of the system of higher education that caused male culture in academic engineering education to disintegrate.

The fact that this break-up simultaneously reinforced the division of labor between men and women in engineering shows that the tendency toward sex segregation in work, a fundamental characteristic of the capitalist gender order, survived even in state socialist society. The significance of the process by which study courses and certain professions were 'feminized,' however, deserves further investigation.

When Erich Honecker succeeded Ulbricht as party leader in 1971, the hope of winning the battle with the capitalist West using new technologies was abandoned. Honecker (who remained in power until 1989) placed his trust instead in the integrating effects brought about by consumption. This led to a turnabout in investment policies, which now came to favor housing, the production of consumer goods, and the energy sector. Large-scale research was abandoned, and the expansion of the educational system was brought to a halt, with new admissions for

Figure 49. Students at the Technical University Dresden in front of the Department of Chemistry's auditorium in 1975. Despite the increasing numbers, male and female students tended to socialize in separate groups. Permission and courtesy of Media Center Archive, Technical University Dresden, Germany.

academic study greatly reduced.[135] At a vice-chancellors' conference in May 1971, only nine days after Honecker and his supporters had forced Ulbricht out of office, the minister for higher education, Hans Joachim Böhme, emphatically proclaimed that 'socialism was essentially built on skilled workers, and that skilled workers would still be its foundation after the year 2000.' He declared that it was time, therefore, 'to end the flights of fancy' that the GDR could have the largest number of scientific-technical experts in the world by 1975.[136]

The reduction in admissions was, however, a source of political conflict, since it meant that some high school graduates would not be able to register at a university for several years. In 1972, for example, 38,877 new students were admitted to all disciplines, while 7,200 applicants were rejected. Among the latter were 5,000 applicants with high grades, so that it was very difficult to explain to them why they had been rejected.[137] The ministry therefore began to look for other ways to reduce the number of admissions. One way was to redirect 3,000 university applicants to mid-level colleges.[138] The most effective measure, however, was a rule that required all male high school graduates to complete their basic military service before pursuing further study.[139] This meant a delay of approximately two years and helped to solve the admissions crisis. As a consequence, the percentage of women among the new students increased dramatically. From forty-nine per cent in 1971, it rose to fifty-nine per cent in 1972, and to sixty-one per cent in 1973.[140] In some disciplines, seventy-five per cent or more of the newly enrolled students were women. Even in the engineering sciences, the number of female freshmen rose to forty-four per cent. In 1974, full-time female engineering students at technical colleges reached their highest number. Fifty-nine per cent of them registered in disciplines in which the women's quota exceeded forty per cent (Table 10).

Table 10. When Honecker replaced Ulbrecht as party leader in 1971, the new administration put an end to the expansion of higher education. In order to reduce the numbers of applicants, the Ministry of Higher Education forced aspiring male students to complete their military service before pursuing college. As a consequence, the percentage of women increased dramatically: women dominated in five engineering specializations and became equals in two others.
Male and Female Engineering Full-Time Students at Technical Colleges in 1974.
Source: Hochschulstatistik *(1974).*

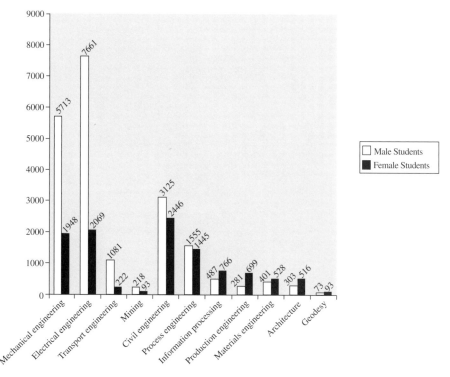

Although politicians involved in education saw this as a unexpected and possibly undesirable development, it was actually brought about by policy makers themselves, since a significant number of female students had been channeled into engineering studies. Having rejected their applications in the first field of their choice, the admissions committees had offered them places reserved for male applicants who were sent on to complete their military service. Many of the women accepted this offer so as to be able to study at all. Deputy Minister Peter Fiedler formulated his reservations about the procedure at the vice-chancellors' conference in November 1973: 'Satisfying as it may be … when women and girls turn more and more to studies in technical disciplines, it

Figure 50. Student couple with child in a dormitory flat at the Technical University Dresden in 1972. GDR leader Honecker replaced the women's policy of the sixties by the pro-natal family policy in the 1970s. In 1972, the state decreed promotion and financial support of women students with children. Permission and courtesy of Media Center Archive, Technical University Dresden, Germany.

must still be emphasized that, with forty-four per cent, we have probably reached a limit that can hardly be exceeded.'[141] The vice-chancellor of the technical university in Dresden, Fritz Liebscher, was even more direct in his comments. He expressed the concern that, even if professors in the engineering sciences had grown accustomed to teaching women, it was very questionable

'whether our partners in industry are prepared for such a high proportion of female graduates in technical disciplines.'[142]

He therefore proposed to re-introduce a women's quota for admission, although this time it would limit rather than promote women's access to the engineering disciplines. Liebscher's confession, that industry was not prepared to place women engineers trained at the universities, illustrates that work and workplaces were still strongly coded as masculine. Liebscher's statements underline the fact that educational planners and scientific engineers had a different rationale for their approach to gender politics than economic and technical administrators in industry. On the basis of their interest in mobilizing all intellectual resources to enhance their power, educational planners and, subsequently, even engineering scientists disregarded gender difference in deciding access to technical knowledge. In industry, however, economic and technical administrators insisted on maintaining a rigid gender hierarchy in the workplace because they thought efficiency depended on it.

This problem was addressed even more directly at a conference of the higher education ministry in November 1972, which focused on the employment of recent graduates. A representative of the building sector criticized the high numbers of women at civil engineering colleges on the following grounds:

'It must be taken into account that it is not favorable to employ female engineering graduates as site managers or technologists on building sites. I can see difficulties, which we will not be able to overcome without problems.... It is my concern to ensure that the women trained at our colleges are offered career opportunities appropriate to their qualifications, since otherwise we may expect difficulties in our political work.'[143]

The minister for higher education attempted to pacify his counterpart at the Ministry of Building by suggesting that:

'The fact that many male students are to be enlisted [in the armed forces] in 1973 will lead to an increased proportion of girls in the engineering disciplines. The male students will then return in 1974, which will improve the situation, both in engineering and in other disciplines, and there will hence be no need to send women out to the building sites.'[144]

With this statement, he conceded that women were being trained in all professions only to enter a restricted employment world. The effects of this state of affairs on the career opportunities of female engineers remain to be investigated.

In more general terms, however, it should be noted that Honecker's accession to power meant that both male and female engineers had fewer chances to use their scientific and technical qualifications. A significant number of graduate engineers had no choice but to accept positions well below what they were capable of, and for smaller salaries than they might have expected.[145] The social prestige of the engineering professions suffered accordingly. It was precisely at this time, however, that women had finally attained a considerable place within the technical intelligentsia. The social recognition of women's technical competence thus only became possible at the very moment when technical competence generally underwent a severe devaluation.

Still, even in the Honecker era the percentage of women engineering students remained relatively high – almost thirty per cent. And in the course of the 1970s, the number of professionally employed female engineers rose to a remarkable level. Thus, the policy that sought to mobilize labor across gender lines in the sixties had gained momentum and remained effective, despite a chance in policy toward technology after Honecker seized power. It was only after the Berlin Wall came down and the GDR became part of the Federal German Republic that the percentage of women among engineering students and professionally employed engineers began to drop sharply. The GDR's process of de-industrialization was one factor affecting all engineers. But others hurt women much more badly than men. Among these were the dissolution of state-organized childcare facilities and the introduction of a new social system (and taxation policy) based on the concept of the single breadwinner. Some Western employers also had old-fashioned views of women and gender roles that determined their hiring practices. Yet just at the time of this precipitous decline, female engineers of the (former) Western part of the Federal German Republic were engaging in a wide range of political activities to further the integration of women into engineering. This shows more clearly than anything else that individual and even collective efforts were insufficient: what really altered the male culture of engineering was the fundamental restructuring of social and economic relations within the GDR. The increase in women engineers in the GDR between the 1960s and the 1980s should not, therefore, be seen merely as part of an international trend toward modernization. It was also the result of very specific policies and conditions. When these no longer existed, the male culture of engineering regained its lost ground.

Notes

1. The author wishes to express her gratitude to all contributors, who participated in the project on the history of women engineers in a cross cultural perspective since 1996. I am most grateful to Annie Canel, Ulrich Wengenroth, and particularly to Ruth Oldenziel for critical readings and

enlightening comments. The research presented in this paper was funded by the Volkswagen Foundation.

2. Inge Lange, *Ausgewählte Reden und Aufsätze* (Berlin: Dietz Verlag, 1987), 99.

3. Using fiction and poetry as a historical source for investigations on the GDR is legitimate and revealing because it permits access to the history of ordinary experience – otherwise accessible only to oral historians – and functioned as an alternative arena distinct from the deformed and state-managed mass media. Although the literary public was also target of centralized control, as a medium fiction sometimes provided a chance for partially autonomous discourses. See also Simone Brack, Martina Langermann, and Jörg Requate, 'Kommunikative Strukturen, Medien und Öffentlichkeiten in der DDR,' *Berliner Debatte INITIAL* 4/5 (1995): 25–38.

4. Gudrun Axeli Knapp, 'Die vergessene Differenz,' *Feministische Studien* 1 (1988): 12–52, p. 18.

5. Women were admitted to the technical colleges (TH) in 1905 (. . München and TH Stuttgart); in 1907 (TH Dresden); in 1908 (TH Karlsruhe); and in 1909 (TH Berlin, TH Hannover, TH Aachen, TH Danzig, TH Darmstadt and TH Braunschweig). See also Margot Fuchs, 'Von der ersten Höherin zur ersten Dozentin. Frauenstudium an der Technischen Hochschule München 1899–1941,' in Ulrich Wengenroth, ed., *Die Technische Hochschule* (Munich 1993), 173–212, p. 174.

6. On Russian women students, see Gouzévitch this volume. Russian women students applied, for example, to the technical college in Dresden in 1873 and 1874. The director Zeuner rejected them, justifying his decision on the basis of his alleged negative experiences with women students in Zurich and Freiberg, where he had previously been a teacher. Ministerium für Volksbildung Papers (*MfV Papers* hereafter) no. 15108, pp. 50–52, Sächsisches Landeshauptarchiv, Dresden (*SLHA Archives* hereafter).

7. On Braunschweig, see Helmut Albrecht, *Technische Bildung Zwischen Wissenschaft und Praxis. Die Technische Hochschule Braunschweig 1862–1914* (Hildesheim: Olms Weidmann, 1987), p. 485. On Dresden, Dorothea Bernholz, 'Probleme Des Frauenstudiums an den Technischen Hochschulen der Deutschen Demokratischen Republik,' (University of Dresden, Diss., 1968), 8.

8. On the emergence of general studies divisions at technical colleges, see: Robert Fricke, 'Die Allgemeine Abteilung,' in *Die Technischen Hochschulen Im Deutschen Reich*, edited by Wilhelm Lexis, Vol. 4 (Berlin: Verlag von A. Asher, 1904), 49–62.

9. In Dresden, the philosopher Schultze and the economist Böhmert were in favor of admitting women. On Schultze, see: Bernholz, *Probleme*, p. 7f. Two years before he left the Zürich's technical college for Dresden's, Böhmert published a booklet in support of studies for women. Victor Böhmert, *Das Studieren der Frauen mit besonderer Berücksichtigung der Medizin* (Leipzig: Wigand, 1872). The title notwithstanding, Böhmert advocated women's admission to all disciplines.

10. *MfV Papers*, no 15122, p. 17ff.

11. For a comprehensive description of the image of the engineer in the early twentieth century, see collection of biographies of the pioneer German historian of technology: Conrad Matschoß, ed., *Männer der Technik* (Munich: VDI-Verlag, 1925). On the prevailing image of femininity, see Ute Frevert, 'Mann und Weib, und Weib und Mann' in *Geschlechterdifferenzen in der Moderne* (Munich: Beck Verlag, 1995), 156f.

12. Franz Reuleaux in Arthur Kirchhoff, ed., *Die Akademische Frau* (Berlin: Hugo Steinitz Verlag, 1897), 287–290, p. 288.

13. Hans Joachim Braun, 'Franz Reuleaux,' in *Berlinische Lebensbilder-Techniker*, edited by Wilhelm Treue and Wolfgang König (Berlin: Colloquium Verlag, 1990), 179–292.

14. For the association between intellectuality and femininity, see: Christina von Braun, 'Der Mythos der "Unversehrtheit" in der Moderne. Zur Geschichte des Begriffs "Die Intellektuellen"', in *Zukunft oder Ende*, edited by Rudolf Maresch (Munich: Boer, 1993), 421–37.

15. I am grateful for this insight to Ruth Oldenziel. On the situation in the US, see Ruth Oldenziel, *Making Technology Masculine: Men, Women, and Machines in America, 1880–1945* (Michigan: Michigan University Press, 1999) as well as her chapter elsewhere in this volume. In 1884, the VDI published a resolution demanding a one-year practical training for all engineering students. See also Kees Gispen, *New Profession, Old Order. Engineers and German Society, 1815–1914* (Cambridge: Cambridge University Press, 1989), p. 152. But the engineering elite faced enormous resistance in industry over the internship. Significantly, the engineering science elite intensified efforts to introduce it prior to war, when the struggle over women's admission to higher education heated up.

16. Barbara Greven-Aschoff, *Die bürgerliche Frauenbewegung in Deutschland 1894–1933* (Göttingen: Vandenhoeck und Ruprecht, 1981), 43. Difference-based or complementary feminism, which dominated debates on the continent until recently, placed the non-hierarchical or companionate couple at the center of its thinking. Equality-based or individuality feminism – strong in the Anglo-American debates – stressed individualism and the liberal notion of equal rights. See also Karen Offen, 'Defining Feminism: A Comparative Historical Approach,' *Signs* (1988): 119–157.

17. In 1904, Marianne Weber, wife of Max Weber and one of the most famous intellectuals within the first German women's movement, suggested that women ought to work in those fields of science in which their female perspective would allow them insights men wouldn't have. Marianne Weber, 'Die Beteiligung der Frau an der Wissenschaft,' in Marianne Weber, *Frauenfragen und Frauengedanken. Gesammelte Aufsätze* (Tübingen: Verlag von J. C. B. Mohr, 1919), 1–9.

18. Marie Bernays, *Die Deutsche Frauenbewegung* (Leipzig: Teubner, 1920), p. 76. Bernays, a student of Alfred Weber, did extensive research on textile workers. She belonged to the first generation of female social scientists. Marie Bernays, *Auslese und Anpassung der Arbeiterschaft der Geschlossen Großindustrie, dargestellt an den Verhältnissen der Gladbacher Spinnerei und Weberei AG zu Mönchen-Gladbach im Rheinland* (Leipzig: Duncker und Humblot, 1910).

19. For the Austrian Salamitaktik, see Mikoletzky in this volume.

20. Fritz Ringer, *The Decline of the German Mandarins. The German Academic Community, 1890–1933* (Cambridge, MA: Harvard University Press, 1969).

21. Albrecht, *Technische Bildung*, 488. The Braunschweig professors, like Reuleaux, assumed very few women would meet the admission requirements for a year of practical training.

22. Hartmut Titze, *Das Hochschulstudium in Preußen und Deutschland 1820–1944* (Göttingen: Vandenhoeck & Ruprecht, 1987), 43, 47. The decrease of women students at technical colleges and at universities was due to a variety of factors. See also Claudia Huerkamp, *Bildungsbürgerinnen. Frauen im Studium und in akademischen Berufen* (Göttingen: Vandenhoeck & Ruprecht, 1996), 80–91.

23. Titze, *Hochschulstudium*, 33.

24. In the summer of 1943, 26% of all female students in technical college took engineering sciences; 88% of them again majored in architecture. Of all female technical college students 38% – of whom chemists made up a majority (74%) – were enrolled in the natural sciences. Of all female technical college students 36% registered in cultural studies. Data derived from *Die Entwicklung des Fachstudium während des Krieges. Beilage zur Zehnjahresstatistik des Hochschulbesuchs und der Abschlußprüfungen*, edited by Charlotte Lorenz (Berlin: Reichsminister für Wissenschaft, Erziehung und Volksbildung, 1944). Since the war statistics were related to the 'Großdeutsche Reich,' they also included data of technical colleges in Austria and Bohemia.

25. By the end of the Weimar Republic, when the number of women students was at its peak, 60% of all female students in technical colleges had registered at the general studies departments. A majority graduated in elementary school education, which was transferred to special teacher training colleges in 1936. In the winter of 1932/33, architecture and chemistry each had 15% of female technical students and economics 7%. Data derived from Barbara Duden and Hans Ebert, 'Die Anfänge des Frauenstudiums an der Technischen Hochschule in Berlin,' in *Wissenschaft und Gesellschaft. Beiträge zur Geschichte der Technischen Universität Berlin, 1879–1979* Vol. 1, edited by Reinhard Rürup (Berlin: Springer Verlag, 1979), 403–23, p. 412.

26. See Berner and Mikoletzky this volume. By contrast, women engineers entered through chemistry in France, where architecture was not even taught at technical colleges. Hence students of architecture did not earn their degrees in engineering. Canel this volume and Barbara Martwich, 'Architektinnen. Frauen in einem untypischen Ingenieurberuf,' in Angelika Wetterer, ed., *Profession und Geschlecht. Über die Marginalität von Frauen in hochqualifizierten Berufen* (Frankfurt: Campus, 1992), 173–186. In Germany, architecture was offered at both technical colleges and art colleges. On the situation in North America, see Ruth Oldenziel, 'Gender and the Meanings of Technology: Engineering in the U.S., 1880–1945,' Ph.D. diss., Yale University (1992), chapter 6 and appendices.

27. For Great Britain, see Carroll Pursell's article this volume.

28. Margot Fuchs, *Wie die Väter so die Töchter. Frauenstudium an der Technischen Hochschule München von 1899–1970* (Munich: Technische Universität, 1984).

29. Mirjam Wiemeler, '"Zur Zeit sind alle für Damen geeigneten Posten besetzt." Promovierte Chemikerinnen bei der BASF, 1918–1933,' in *Geschlechterverhältnisse in Medizin, Naturwissenschaften*

und Technik, edited by Christoph Meinel and Monika Renneberg (Stuttgart: Verlag für Geschichte der Naturwissenschaft und Technik, 1996), 237–44.

30. Leaders of the first German women's movement believed that the job of scientific-technical assistant was an appropriate one for women because of their innate capacity for understanding, servitude, and willingness to work. Marie Kundt, 'Die Frau als Technische Assistentin,' in *Die Kultur der Frau*, edited by Ada Schmidt-Beil (Berlin: Verlag für Kultur und Wissenschaft GmbH, 1931), 228–32.

31. Enno Heidebroek, Technische Pionierleistungen als Träger industriellen Fortschritts' *Zeitschrift des Vereins Deutscher Ingenieure* (ZVDI) 23 (1927): 809–815.

32. Hildegard Harnisch-Niessing, 'Frauen in der Technik' *Technik und Kultur* 9 (1930): 147.

33. Schwartz Cowan and Oldenziel this volume.

34. Ilse Knott-ter Meer, 'Die Frau als Ingenieurin' *VDI-Nachrichten* 24 (11 June 1930): 3–4. Knott-ter Meer, the daughter of an entrepreneur who studied mechanical engineering in Hanover and Munich in the early 1920s, later married the managing director Dr.-Ing. Carl Knott. Fuchs, *Wie die Väter*, 117–122.

35. According to Knott-ter Meer, many women engineers in England and America were already active in this field, a development she expected for Germany also: 'Here [in the field of household technology] there are doubtless many questions which are best solved by women with engineering training, since they clearly fall within the domain of women. They will best be able to judge what technical devices and equipment are truly practical in the household, and what preliminary and developmental activities will be worthwhile. But women engineers will also be able to address problems in the areas of advertising as well as serve as mediators between technology and women. When household technology is widely used, contributing an understanding and feeling for technical things in the broadest circles of women's world will be of significant importance.' Knott-ter Meer. 'Die Frau', 3.

36. 'Weibliche Ingenieure auf der Weltkraftkonferenz in Berlin' *Die schaffende Frau. Zeitschrift für modernes Frauentum*, ed. by Margarete Kaiser 12 (1930): 402; Fuchs, *Wie die Väter*, 120–122.

37. Christine Roloff, 'Frauen in Naturwissenschaft und Technik – Netzwerk und Professionalisierung' in *Frauen untereinander. Dokumentation der 1. Offenen Frauenhochschule in Wuppertal, vom 18. bis 20. Mai 1989* (Wuppertal 1989), 163–181.

38. Research on this issue is only just emerging.

39. For some telling examples, see Margot Fuchs, *Wie die Väter* and her 'Vätertöchter' and ''Wir Fliegerinnen sind keine Suffragetten.' Die Versuchsingenieurin und Sturzflugpilotin Melitta Schiller, 1903–1945,' in *Bedrohlich gescheit. Ein Jahrhundert Frauen und Wissenschaft in Bayern*, edited by Hiltrud Häntzschel and Hadumod Bußmann (Munich: Beck Verlag, 1997), resp. 214–227 and 260–65. See also the article of Annette Vogt in this volume.

40. Enno Heidebroek's speech, 'Die neue Hochschule, Ansprache zur Neueröffnung der Technischen Hochschule am 18.9.1946,' 8f, Rektorat ab 1945 Papers, no. I/11, TU-Archiv, Dresden.

41. Dorothea Bernholz, 'Historische Betrachtungen der Entwicklung des Frauenstudiums an der Technischen Universität Dresden von 1946 bis 1966,' *Wissenschaftliche Zeitschrift der TU Dresden* (1968): 541–52, p. 542.

42. 'Verzeichnis der Studierenden der Sächsischen Technischen Hochschule 1941 bis 1944', Typewritten, MfV Papers no. 15781, SLHA Archives.

43. SLHA Archives, Landesregierung Sachsen (*LRS* hereafter), MfV Papers, no. 1074, p. 200, 221, 229, 225, 254, 266; Alexandr Haritonow, *Sowjetische Hochschulpolitik in Sachsen 1945–1949* (Köln: Böhlau, 1995), 203–204.

44. Herbert Stallmann, *Hochschulzugang in der SBZ/DDR 1945–1959* (St. Augustin: Verlag Hans Richarz, 1980), 102ff.

45. This emerges from student files kept in the archive of the TU Dresden.

46. *Volks- und Berufszählung vom 29. Oktober 1946 in der sowjetischen Besatzungszone* Vol. 4 *Sowjetische Besatzungszone* (Berlin 1949), 62. In these absolute figures, there were 3,534 male and 165 female unemployed engineers. The data did not include Berlin.

47. There was an enormous surplus of women in post-war Germany – even greater in East than in West Germany. In 1946, women in the Soviet Occupied zone made up 57,5% of the population, and 54,8% in the three Western Zones. See also Kurt Lungwitz, 'Die Bevölkerungsbewegung in der DDR und der BRD zwischen 1945 und 1970 – eine komparative Untersuchung' in *Jahrbuch für Wirtschaftsgeschichte* 1 (1974): 63–95.

48. Promotionsakte Papers, TU-Archiv, Dresden.

49. The post-war women's press challenged the established gender system frequently discussing the conflicts women faced when men returned from the war to claim their former positions in both family and the workplace.

50. Many engineers, who had experienced the increased social value of technology during the Third Reich as enhanced professional prestige, took on the task of developing new areas of weapons technology. Karl-Heinz Ludwig, *Technik und Ingenieure im Dritten Reich* (Düsseldorf: Droste Verlag, 1979).

51. Herbert Merthens, '"Missbrauch". Die rhetorische Konstruktion der Technik in Deutschland nach 1945', in Walter Kertz, ed., *Projektberichte zur Geschichte der Carolo-Wilhelmina* Vol. 10 *Technische Hochschulen und Studentenschaft in der Nachkriegszeit* (Braunschweig 1995), 33–50.

52. Enno Heidebroek, 'Problematik der Ingenieurarbeit und -erziehung", *Die Technik* 1 (1948): 1–3.

53. Heinrich Franck, 'Die Neugestaltung der Technischen Universität in ihren allgemeinbildenden Fächern", *Die Technik* 2 (1948): 49–53.

54. 'Einzelne Glieder einer langen Kette' *Deutscher Frauenpressedienst* 12 (1948): 2; 'Eindrücke einer großen Reise' *Deutscher Frauenpressedienst* 22 (1948):8; 'Hilde Hellebrand' *Deutscher Frauenpressedienst* 26 (1948) : 14; Ingeburg Nawroth, 'Was bewog mich, Ingenieurin zu werden?' *Neues Frauenleben. Das Blatt zur Information des DFD Landesverband Sachsen* 15 (1949): 9.

55. Nawroth, 'Was bewog mich,' 9.

56. Emmy Damerius, 'Die Frauen und der Völkerfriede", in *Protokoll des Deutschen Frauenkongresses für den Frieden. Gründungskongreß des Demokratischen Frauenbundes Deutschlands* (Berlin 1947), 71.

57. Data for the technical college Dresden and the Mining Academy Freiberg for the winter semester 1946/7 and the summer semester 1948 from Ministerium für Hoch- und Fachschulbildung (DR3 Papers hereafter) no. 4590, Bundesarchiv Berlin (BA hereafter).

58. *Volks- und Berufszählung vom 19. Oktober 1946*, 46; *Volks- und Berufszählung in der Deutschen Demokratischen Republik am 31. August 1950*, p. 10, Berufszählung 2. The data do not include Berlin. The census group engineers and technicians included graduates of Technische Hoch- and Fachschulen (Engineering Colleges), whose certificates entitling them to work as an engineer. See also, *Die Systematik der Berufe* (Berlin 1950), 73.

59. The Soviet occupying forces prohibited the admission exclusively for politically active members of the former Nazi-Party and war criminals. On preferential treatment of war veterans in the U.S. see Margaret Rossiter, *Before Affirmative Action 1940–1972. Women Scientists in America* (Baltimore: The Johns Hopkins University Press, 1995), 27ff.

60. Hartmut Titze et al, *Datenhandbuch zur deutschen Bildungsgeschichte* Vol. 1: *Hochschulen*, Part 1: *Das Hochschulstudium in Preußen und Deutschland 1820–1944* (Göttingen: Vandenhoeck und Ruprecht, 1987), 30; BA, DR 3 Papers no. 4590, BA.

61. *Geschichte der Technischen Universität Dresden 1828–1988* (Berlin, Deutscher Verlag der Wissenschaften, 1988), 174, 215; *Bergakademie Freiberg 1765–1965. Festschrift zu ihrer Zweihundertjahrfeier am 13. November 1965* Vols. 2 (Leipzig, 1965).

62. 'Ausbildung technischer Kader an den Hochschulen der DDR' *Die Technik* 13 (1958): 393–395.

63. On the tradition of the school culture model in Germany see Peter Lundgreen, 'Engineering Education in Europe and the U.S.A., 1750–1930: The Rise to Dominance of School Culture and the Engineering Professions', Annals of Science (1990), 47:33–75. Although especially during the fifties upward mobility of skilled workers into engineer's positions occurred rather frequently, the expansion of academic engineering education showed, that the new state undoubtedly favored access into engineering by academic training rather than apprenticeship.

64. On the emergence and meaning of this exclusion see Karl-Heinz Manegold, *Universität, Technische Hochschule und Industrie. Ein Beitrag zur Emanzipation der Technik im 19. Jahrhundert unter besonderer Berücksichtigung der Bestrebungen Felix Kleins* (Berlin: Duncker und Humblot, 1970).

65. ZK der SED, Abteilung Wissenschaft (DY 30 IV 2/9.04 Papers hereafter) no. 360, p. 244, Foundation Archive of the Parties and Mass Organisations of the GDR (*SAPMO* hereafter), BA.

66. Ingemarie Neufeldt, *Die wissenschaftlich-technische Intelligenz in der Entwicklung der sowjetischen Gesellschaft* (Berlin: In Kommission bei Otto Harrasowitz Wiesbaden, 1979), 89–90, 93–94; Kendall E. Bailes, *Technology and Society Under Lenin and Stalin* (Princeton: Princeton University Press, 1978), 221–243. The close links between higher technical education and the industrial ministries were abolished after 1933, but the colleges and universities continued to be oriented to industry. Nicolas de Witt, *Education and Professional Employment in the U.S.S.R.* (Washington: National Science Foundation, 1961), 208–209, 216, 226.

67. DY 30 IV 2/9.04 Papers no. 360, p. 315ff., SAMPO, BA.

68. See also Ralph Jessen, *Vom Ordinarius zum sozialistischen Professor*, in *Die Grenzen der Diktatur*, edited by Richard Bessel and Ralph Jessen, (Göttingen: Vandenhoeck & Ruprecht, 1996) 76–107, pp. 89–90.

69. The science department of SED's Central Committee railed against the 'striving of a section of the teaching staff to gain full recognition from the other universities and colleges as quickly as possible by making concessions to bourgeois traditions,' BA-SAPMO; DY 30 IV/2/9.04, Papers no. 360, p. 297.

70. *Geschichte der TU Dresden*, 225–232.

71. For data of the Soviet Union see Norton T. Dodge, *Women in the Soviet Economy* (Baltimore: The Johns Hopkins University Press, 1966), 112. During the war the percentage of Soviet women in engineering was remarkably higher, increasing from 40% (1940) to 60% (1945). For GDR data see *Statistisches Jahrbuch der DDR* (1959), 135.

72. After a steady decrease of the women's quota at preparatory institutions (*Vorbereitungsanstalten*), the predecessors for the worker and farmer faculties founded in 1949, Major Plushnikow of the Soviet military administration threatened the admission committees he would cease to approve their admission lists if the women's quota continued to decrease. Haritonow, *Sowjetische Hochschulpolitik*, 233.

73. During the 1950s, attempts focussed on sending working women from state-owned enterprises to worker and farmer faculties or directly to institutions of higher learning. A May 2, 1952 resolution of the Council of Ministers required all nationally owned enterprises to design plans for the promotion of women and to propose programs and measures enabling women to pursue careers. Key element was firms' obligation to send women to institutions of higher learning. Ministerium der Arbeit, '28.5.1952, Anleitung zur Aufstellung von Frauensonderförderungsplänen in den Betrieben', Freier Deutscher Gwerkschaftsbund Papers (*DY 34* hereafter) no. 39/72/5389, SAPMO in BA.

74. Data from *Statistisches Jahrbuch der DDR* (1959), 134–135.

75. On personnel continuity and restoration at East German universities see Ralph Jessen, *Akademische Elite und kommunistische Diktatur: Die ostdeutsche Hochschullehrerschaft in der Ulbricht-Ära* (Göttingen: Vandenhoeck & Ruprecht, 1999).

76. See also Enno Heidebroek's writing, *Industriebetriebslehre. Die wirtschaftlich-technische Organisation des Industriebetriebes mit besonderer Berücksichtigung der Maschinenindustrie* (Berlin: Verlag von Julius Springer, 1923); 'Technische Pionierleistungen', *ZVDI 71* (1927): 809–815; 'Das Weltbild der Technik", in *Das Weltbild der Naturwissenschaften* (Stuttgart, 1931), 113–135; 'Maschine und Arbeitslosigkeit' *ZVDI 76* (1932): 1041–1048.

77. By additionally adopting a scientific approach to engineering Heidebroek showed, that he was torn between two professional ideals: the pre-modern artist-engineer and the classical modern techno-scientist. He thus found himself in middle of a paradigm change that Wengenroth describes as a secular trend and a preliminary stage for the emergence of the reflexive modern age in engineering. 'Weltbild', 121; Ulrich Wengenroth, 'Der aufhaltsame Aufstieg von der klassischen zur reflexiven Moderne in der Technik,' in Thomas Hänseroth, ed., *Technik und Wissenschaft als produktive Kräfte in der Geschichte* (Dresden, 1998), pp. 129–140.

78. Heidebroek, 'Weltbild', 123.

79. For further details, see Geneviève Fraisse, 'Von der sozialen Bestimmung zum individuellen Schicksal. Philosophiegeschichte zur Geschlechterdifferenz", in *Geschichte der Frauen* Vol. 4 *19. Jahrhundert*, edited by Geneviève Fraisse and Michelle Perrot (Frankfurt: Campus, 1994), 63–95, pp. 94–95, and Francoise Thébaud, 'Der Erste Weltkrieg. Triumph der Geschlechtertrennung' in *Geschichte der Frauen* Vol. 5 *20. Jahrhundert*, edited by Francoise Thébaud (Frankfurt: Campus, 1995), 33–92, pp. 88–89.

80. Ernst Jünger, *In Stahlgewittern* (Stuttgart: Verlag Klett-Cotta, 1997 [1920]. See also Thomas Kühne, '"… aus diesem Krieg werden nicht nur harte Männer heimkehren." Kriegskameradschaft und Männlichkeit im 20. Jahrhundert', in Thomas Kühne, ed., *Männergeschichte, Geschlechtergeschichte. Männlichkeit im Wandel der Moderne* (Frankfurt: Campus, 1996), 174–192. The similarity between Jünger's and Heidebroek's thinking was based on a spiritual affinity articulated by reactionary modernist idealogues. See Jeffrey Herf, *Reactionary modernism* (Cambridge: Cambridge University Press, 1984).

81. Heidebroek, 'Technische Pionierleistungen', 813–814.

82. See also: Enno Heidebroek, 'Die neue Hochschule, Ansprache zur Neueröffnung der Technischen Hochschule am 18.9.1946', in: TU-Archiv, Rektorat ab 1945, Nr.I/11; 'Problematik der Ingenieurarbeit und -erziehung' and 'Die Verantwortlichkeit des Ingenieurs' Lecture held at the 125th anniversary celebrations of the technical college Dresden on 4.6.1953, Berlin 1954 *Berichte über die Verhandlungen der Sächsischen Akademie der Wissenschaften zu Leipzig, Math.-naturwiss. Klasse* Vol. 101, no. 5.

83. Heidebroek foresaw 'the danger that we are steering towards a kind of education inflation, alongside which the value of the actual productive work of the workers, farmers and craftsmen will be pushed into the background. ... Every capable skilled worker aware of his value is ten times more welcome than a poorly academically educated also-engineer.' Heidebroek, 'Die neue Hochschule', 7.

84. 'Technical progress is also to a great extent self-sufficient and is not dependent on planned proposals. The idea must be there first of all. One cannot plan its implementation beforehand.' Heidebroek, 'Die Verantwortlichkeit', 5.

85. On ambivalence of the working classes towards the intelligentsia, see: Dietz Bering, *Die Intellektuellen. Geschichte eines Schimpfwortes* (Stuttgart: Klett-Cotta Verlag, 1978).

86. Each year in his annual reports to welcome freshmen through the years from 1950 to 1954, the vice-chancellor of the technical college Dresden Kurt Koloc complained about the low proportions of women at the college's worker and farmer faculty reaching only between 9 and 18 per cent, which he believed prevented a real advancement of women in technical studies generally. Rektorreden, Rektorat nach 1945, 14–17, Archiv TU Dresden.

87. Fuchs, *Wie die Väter*, 230.

88. Data from Staatssekretariat für das Hoch- und Fachschulwesen, Sektor Planung und Statistik, *Ergebnisse der Jahreshauptstatistik*, Part I *Universitäten und Hochschulen* (Berlin 1961) (further cited as Hochschulstatistik), 87.

89. On heterogeneity as a characteristic of engineering, see: Konrad Jarausch, *The Unfree Professions: German Laywers, Teachers, and Engineers, 1900–1950* (Oxford: Oxford University Press, 1990).

90. On the recruiting of technical personnel see Thomas A. Baylis, *The Technical Intelligentsia and the East German Elite* (Berkeley: University of California Press, 1974), 23–27.

91. This work was limited to sectors where small and middle-size business prevailed such as the textile and clothing industry and mechanical and construction engineering. On the private sector in the East German economy, see Wolfgang Mühlfriedel and Klaus Wiesner, *Die Geschichte der Industrie der DDR bis 1965* (Berlin: Akademie Verlag, 1989), 129–143.

92. Joerg Roesler, 'Wirtschafts- und Industriepolitik,' in *Die SED. Geschichte, Organisation, Politik. Ein Handbuch*, edited by Andreas Herbst, Gerd-Rüdiger Stephan, and Jürgen Winkler (Berlin: Dietz Verlag, 1997), 277–293, p. 284.

93. In 1925, the percentage of self-employed female engineers (23 %) was significantly higher than that of males (14%). *Statistik des Deutschen Reiches* Vol. 402 (Berlin 1929), 415. For examples see Fuchs, *Wie die Väter*.

94. On the sitatuation in the U.S., see Oldenziel in this volume.

95. See also Lothar Baar et al., 'Die Gestaltung der Industriezweigstruktur der DDR durch die Wirtschaftspolitik der Partei der Arbeiterklasse und der staatlichen Organe' *Jahrbuch für Wirtschaftsgeschichte* (1988), Special Issue, 7–46.

96. Monika Kaiser, *Machtwechsel von Ulbricht zu Honecker. Funktionsmechanismus der SED-Diktatur in Konfliktsituationen 1962 bis 1972* (Berlin: Akademie Verlag, 1997), 457.

97. Beginning in the mid-1950s, the SED put the technological modernization of the economy at the center of its concept for restructuring society. See also Sigrid Meuschel, *Legitimation und Parteiherrschaft in der DDR* (Frankfurt: Suhrkamp, 1992), 183–210. It resulted in the promotion of chemical manufacturing procedures, the high-pressured development of new technological fields like semi-conductor technology and steering and control engineering, the expansion of the production of industrial consumer goods, and the industrialization of the construction industry. As a consequence, both the areas of where engineers worked as the social perception of the engineers' mission began to change in the 1960s.

98. Data from *Hochschulstatistik*.

99. *Volks-, Berufs-, Wohnraum- und Gebäudezählung am 1. Jan. 1971* Vol. 5 *Wirtschaftlich tätige und nicht wirtschaftlich tätige Wohnbevölkerung* (Berlin 1972), 114–119.

100. On Greek case see Kostas Chatzis and Efthymios Nicolaïdis in this volume.

101. Data from *Wirtschaftlich tätige und nicht wirtschaftlich tätige Wohnbevölkerung*, 115; Margaret Rossiter, 'Which Science? Which Women?', *Osiris* (1997), 169–105.

102. On the U.S., see Rossiter, *Before Affirmative Action*, 50f.

103. Publications, either triumphantly or anxiously comparing the respective numbers of Soviet and American engineers, appeared in both the East and in the West. For example, in a lecture for the national engineering association ('Kammer der Technik') vice-president of the GDR council of ministers Fritz Selbmann proclaimed in November 1957: 'When we bear in mind that, in 1957, 28,000 engineers graduated from American colleges whilst 70,000 engineers did so from Soviet colleges, then it comes clear that the priority moved to the Soviet Union, i.e. the socialist system. That is the guarantee of a steady superiority of socialism over capitalism in the fields of science and technology.' Fritz Selbmann, 'Die Entwicklung der Technik im zweiten Fünfjahrplan und die weiteren Perspektiven in der Deutschen Demokratischen Republik,' *Die Technik* 2 (1958): 131–134, p. 134.

104. Rossiter, *Before Affirmative Action*, 64–65 and Käthe Ahlmann, 'Ingenieurmangel und Ingenieurstudium für Frauen,' in *Information für die Frau* 7–8 (Bad Godesberg 1958), 17–19. Interestingly, a West German publication, in a curious demonstration of loyalty to the Western alliance, maintained that 40% of engineering students in Russia were women as opposed to a mere 20% per cent of American engineering students.

105. Inge Lange, 'Die Aufgaben der Frauenausschüsse,' *Neues Deutschland* (6 February 1962): 3.

106. Rossiter, *Before Affirmative Action*, 68.

107. 'Material für die Kollegiums- und Senatssitzung an der TU Dresden am 2. und 3.11.1962', DR 3 Papers, no. 6039, BA.

108. Beschlußentwurf vom 15.1.1964, *Verfügungen und Mitteilungen des Staatssekretariats für Hoch- und Fachschulwesen* 17–10 (5 October 1963), DR 3 Papers, no. 5815, BA. B. Knauer, 'Zu Problemen der Forschung und der Veränderung der Ingenieurausbildung", *Die Technik* 18 (1963): 65–68; H. Jäckel, 'Neues Ausbildungsprogramm für Diplomingenieure in der Bewährung', *Die Technik* 21 (1966): 5–8; *Geschichte der Technischen Universität*, 282–285.

109. Dieter Adelmeyer, Ernst Kurt, and Helmut Kunze, *Erfahrungen und Ergebnisse des Frauensonderstudiums an den Ingenieur- und Fachschulen der DDR* (Karl-Marx-Stadt: Institut für Fachschulwesen der DDR, 1977), 4–5.

110. *Der Einfluß wichtiger demographischer Prozesse auf die Entwicklung der Hoch- und Fachschulausbildung in der DDR* (Berlin: Insitut für Hochschulbildung und Ökonomie, 1969), 167.

111. Adelmeyer, *Erfahrungen*, 5.

112. 'Abteilung Frauen im ZK der SED (*DY 30/IV 2/17* hereafter) no. 10, 'Protokoll über die Sitzung der Frauenkommission beim Politbüro vom 16.12.1169', SAPMO, BA.

113. ibid.

114. Adelmeyer, *Erfahrungen*, 6.

115. "Beratung über Frauensonderklassen 23.1.1168", DY 34 Papers, no. 7758, SAPMO, BA.

116. Wulfram Speigner, 'Bildung für die Frauen und Mädchen", in *Zur gesellschaftlichen Stellung der Frau in der DDR*, edited by Hertha Kuhrig and Wulfram Speigner (Leipzig: Verlag für die Frau, 1178), 103–228, p. 212.

117. 'Maßnahmeplan des Staatssekretariats für Hoch- und Fachschubildung zur Frauenförderung, April 1962,' Rektorat nach 1945 Papers, no. 71, p. 248, TU-Archive, Dresden.

118. 'Kempke an Lange 11.7.1965', DY 30 IV A 2/17, no. 16, SAPMO, BA.

119. The Third Higher Education Reform was passed in 1967. For details and the somewhat apologetic discussion, see: Roland Köhler et al, *Geschichte des Hochschulwesens der DDR (1961–1980), Überblick* Part I (Berlin: Institut für Hochschulbildung, 1987), 57–66.

120. 'Maßnahmeplan zur Entwicklung der Universitäten und Hochschulen bis 1975. Neugestaltung der Ausbildung', p. 10, DY 30 IV, 2/9.04/ no. 28, SAPMO, BA.

121. DY 30 IV 2/9.04 no. 44, SAPMO, BA.

122. ibid.

123. Ulbricht 'Überholen ohne einzuholen – ein wichtiger Grundsatz unserer Wissenschaftspolitik', Speech at the Reception of the *Kammer der Technik*, 23 February 1970, *Technischen Gemeinschaft* (1970), Appendix to Vol. 3, p. 10.

124. Köhler, *Hochschulwesen*, 79.

125. 'Maßnahmeplan zur Entwicklung der Universitäten und Hochschulen bis 1975', DY 30 IV A 2/9.04 no. 44 and 28, SAPMO, BA.

126. A study on the demographs influencing higher education during the years 1950–1966 concluded that the required increase in the numbers of math, natural sciences, and engineering

students could only be accomplished by the better recruitment of women. If not, it expected to overtax the regular male age groups with the opposite result of a decreasing success rate. *Der Einfluß*, 109–110, 142.

127. Wengenroth, *Der aufhaltsame Weg*, 132. On big science projects, see M. Szöllösi-Janze and Helmuth Trischler, ed., *Großforschung in Deutschland* (Frankfurt: Campus, 1990), and Agnes Charlotte Tandler, 'Geplante Zukunft. Wissenschaftler und Wissenschaftspolitik in der DDR, 1955–1971' (Europäisches Hochschulinstitut Florenz, diss. 1997), 320–331.

128. One influential philosopher, who adhered to the idea of technology as applied natural science was Herrmann Ley. Herrmann Ley, 'Zum Verhältnis von Philosophie und Technik,' *Die Technik* 9 (1958): 587–593. Bernd M. Lindenberg, *Das Technikverständnis in der Philosophie der DDR* (Frankfurt: Peter-Lang-Verlag, 1979), 100. This notion was increasingly popular after 1963 when the concept of science as a productive force was anchored in the program of the SED. On these ideas, see Gerhard Kosel, *Unternehmen Wissenschaft. Erinnerungen* (Berlin: Henschel Verlag, 1989), 335. Kosel, who was an architect and deputy minister of building and construction industry, had published a book entitled *Productive Force Science* in 1958. Therefore he claimed this concept as his own invention.

129. Peter Adolf Thiessen, 'Die Perspektiven der Technik in der DDR in den nächsten Jahren,' *Die Technik* 1 (1961): 1–6, p. 1.

130. 'Anweisung NR. 7, 1961', DR3, Papers no. 2355, BA; *Die Technische Hochschule Karl-Marx-Stadt in der Zeit von 1963 bis 1975* (Leipzig 1978), 58.

131. Günther Prey, 'Schlüsselproblem für Pionier- und Spitzenleistungen Sozialistische Wissenschaftsorganisation,' *Technische Gemeinschaft* 4 (1970): 26.

132. See also Dietmar Waterkamp, *Bildung und Erziehung. Das Einheitsprinzip im Bildungswesen der DDR* (Köln/Wien: Böhlau, 1985), 139–144. Mathematics and science education improved because of the increasing investments in teacher training courses. From then on, every teacher was qualified in two subjects. Teacher training for the fifth to the tenth grade levels was extended to encompass six semesters.

133. 'Maßnahmen zur Verbesserung des Studiums und zur Einhaltung der Studienzeit.' Rektorat nach 1945 Papers, no. 673, TU Dresden Archives. 'Einschätzung der Studienleistungen der Studierenden der TU im Studienjahr 1964/5", DY 30 IV, A 2/9.04, Papers no. 273, SAPMO, BA.

134. On the abolition of social class quota, see: Christoph Kleßmann, *Zwei Staaten, eine Nation. Deutsche Geschichte 1955–1970* (Bonn: Budeszentrale für Politische Bildung, 1988), 354.

135. Dietrich Staritz, *Geschichte der DDR 1949–1985* (Frankfurt: Suhrkamp Verlag, 1985), 198–203; Meuschel, *Legitimität*, 221–229; Baar, Müller, and Zschaler, *Strukturveränderungen*, 54.

136. 'Rektorenkonferenz 14.5.1971, Schlußwort Böhme.' DR3 Papers, no. B1359/2, BA.

137. 'Rektorenkonferenz 7.5.1971,' DR3 Papers, no. B1359/2.

138. 'Protokoll über Beratung der Koordinationsgruppe am 25 November 1971', DR3 Papers, no. B1360/1, BA.

139. 'vice-chancellors' conference 3.10.1973', DR3 Papers, no. B 1359/3, BA.

140. 'vice-chancellors' conference 2.11.1973', DR3 Papers, no. B 1359/3, BA.

141. Ibid.

142. Ibid.

143. 'Koordinierungskonferenz 9.11.1972 Leipzig', DR3 Papers, no. B 1360/2, BA.

144. ibid.

145. Werner Wolter, 'Wissenschaftlich-technische Bildung und personelles Forschungspotential in der DDR", in Hansgünter Meyer, ed., *Intelligenz, Wissenschaft und Forschung in der DDR* (Berlin: Campus Verlag, 1990), 85–96, p. 89.

Konstantinos Chatzis and Efthymios Nicolaïdis[1]

9. A Pyrrhic Victory: Greek Women's Conquest of a Profession in Crisis, 1923–1996

In the 1964 Greek movie *Miss Director*, the female protagonist temporarily replaces her former mentor and engineering professor as manager of a large construction company. Falling in love with her handsome male assistant, she believes that, as a civil engineer, she cannot court him until she has recovered the femininity she sacrificed to her technical studies. He is in love with her too, but his respect for her and the engineering profession paralyzes him to act on his tender feelings. The comedy's scenario explores the dilemmas faced by a young Greek woman engineer questioning the loss of her femininity.[2] If in the 1920s, educated women's burden had been to prove their ability to be competent and qualified in education and on the job, in the 1960s the onus shifted to their ability to be attractive as well. Being both female and a manager of a private construction company was considered to be a contradiction in terms. In the period between 1890 and 1964 when the movie was made only 4.1 per cent of all engineering and architectural graduates were women many of whom sought refuge in the public sector, where work was more steady and secure, rather than in the private sector.[3] The Greek film marked a moment when a woman engineer on the construction site was eccentric enough as a comic stage set to explore some enduring gender stereotypes. Since the making of *Miss Director* the position of women engineers has changed dramatically. Today women engineers represent twenty per cent of all engineers in Greece, and this percentage is increasing. Women are still more likely to work in the public sector than their male counterparts, but today a great number of women also are employed in the kind of private-sector work portrayed in the 1964 film.

The essay explores how Greek women gained access to engineering education and professional status in the period between 1923 when the first woman graduated from an engineering program until the present day. It also centers on the intriguing question why women entered engineering in such unprecedented numbers even though, on the face of it, Greece was in several respects a traditional, patriarchal society during that period. It describes how today – in the decades following the making of *Miss Director* which had exploited gender stereotypes for comic release – women in Greece are at the forefront of engineering education leading their sisters elsewhere.

*Figure 51. Played by the Greek
actress Karezi, the woman
engineer arrives at the building
site and meets her male assistant, the actor Alexandrakis.*
Miss Director, *a popular 1964
comedy film. Permission and
courtesy of Mrs. Finos and Mr.
Ioannidis,* Miss Director
*(1964), directed by K.
Dimopoulos and produced by
Finos Film.*

*Figure 52. Scene: He puts up
the drawing table for her.
Permission and courtesy of
Mrs. Finos and Mr. Ioannidis*
Miss Director *(1964), film
produced by Finos Film.*

Figure 53. Scene: he is about to discard his cigarette: a gentleman does not smoke in front of an engineer! Permission and courtesy of Mrs. Finos and Mr. Ioannidis, Miss Director *(1964), film produced by Finos Film.*

Figure 54. Scene: how to work as an engineer and still be a woman? Falling in love with her assistant, she realizes she wants be a desirable woman. Permission and courtesy of Mrs. Finos and Mr. Ioannidis, Miss Director *(1964), film produced by Finos Film.*

Figure 55. Scene: asking an expert's opinion on being a woman. Permission and courtesy of Mrs. Finos and Mr. Ioannidis, Miss Director (1964), film produced by Finos Film.

Figure 56. Scene: Rehearsing a woman's role. It's so difficult! Permission and courtesy of Mrs. Finos and Mr. Ioannidis, Miss Director (1964), film produced by Finos Film.

From Girls' Private Schools to Public Schools and Equal Educational Rights, 1830–1930

The Ottoman Empire, to which Greece belonged until the beginning of the nineteenth century, lacked universities or other secular institutions. The new Greek state, established in 1830, followed from a revolutionary movement that appreciated the Englightenment ideal of education for all.[4] A royal decree in 1834 declared that primary education would be free and obligatory for both sexes.[5] Even so, girls found their way to primary education at a much slower rate than boys. In 1837, only nine per cent of all primary school students were girls; forty years later the percentage had reached just twenty per cent. Several education ministers were concerned about the issue, but a 1852 decree prohibiting co-education created new barriers to primary schooling for girls. Most cities could not afford to create separate educational institutions for them. Even when such a possibility was offered, many poor families (who formed the greatest part of the population) did not wish to send their daughters to primary schools because they needed them to work on the farm or at home.[6] Those few primary schools for girls that were established provided an education that was increasingly less than that offered to boys since it was supposed to be 'suitable for the fairer sex.' In 1894, geometry and experimental physics were eliminated from the girls' curriculum, and a 1882 decree put needle-work and crochet in place of physical education.

Although Greek politicians paid lip service to the idea of creating state secondary schools (elliniko scholio) for girls, secondary education for girls depended on private initiative until 1910.[7] The Society for the Friends of Education (Philekpedeftiki Eteria), founded in 1836, created special schools for girls that were known as 'virgins' schools' (Parthenagogion). Run according to the dictates of the ideology of separate spheres, 'virgins' schools' sought to mold 'quiet home-mistresses, good wives and excellent mothers', while boys were educated to be 'virtuous citizens.' These schools developed a curriculum based on the notion that girls were incapable of serious intellectual effort. Consisting mainly of French, music, and classic philology, this training was far removed from that which helped men enter the professional world. The schools offered fewer courses for girls, especially in Greek and mathematics. Their curriculum took only four to six years, whereas secondary education for boys took seven years. Chemistry and physics were not taught to girls until 1870, and the degrees awarded by the 'virgins' schools' did not allow graduates to enter university. As a result, girls were excluded from higher education. (Private secondary education was, in the nineteenth century, a privilege only daughters of upper-class families could afford.) The Society for the Friends of Education also established a limited number of Didaskalia, which were intended to train middle-class women as teachers.[8] These schools also awarded some scholarships, which were partly sponsored by the Society for the Friends of Education, the municipalities and the national government.

This support for women's education, however, soon resulted in a glut of female teachers, despite the fact that women gained the right to teach in boys' primary schools in 1886. By the end of the nineteenth century, unemployed women instructors had become a 'social problem' worthy of public debate. Would women have greater success in employment if they were permitted to enter university? Not coincidentally, this was also the same period when women started to organize and established new women's journals.[9] In 1890, the University of Athens admitted its first woman student, Ioanna Stefanopoli, a journalist's daughter, to the philosophy department. In the 1919–20 academic year, the number of women students reached seventy-seven out of 1759. These women encountered some resistance to their presence, as is summed up in this anti-feminist caricature of an intellectual woman in 1908: 'a little short-sighted…most often absent-minded, a little hunch-backed and a little neurotic.'[10] Despite lacking the active support of the state, Greek women gained the right to university education in the 1920s.[11] Finally, after an intense struggle and long after most other European countries, a progressive government issued a decree in 1929 ending the long debate over women's entry into university.[12] This created a secondary education system for girls that was equivalent to the one for boys. Up to that point, young women who sought to enter university had to obtain the necessary certificates by enrolling illicitly in men's colleges. (From 1907 the Ministry of Education had quietly accepted such irregularities because they were established fact.)

By the beginning of the 1920s, career choices of middle- and upper-class women had changed significantly. Greek women entered medicine, secondary teaching, and art.[13] In 1923, the last male bastion in higher education fell when the first woman engineer earned her degree in architecture from the Technical University in Athens. This was possible because of the new acceptibility of women's participation in higher education, but it also came after a century-long transformation in engineering education.

The First Women Engineers Scaling the Male Bastion, 1923–1950

The Greek state began to institute a national education system to train technicians as soon as it gained independence. It took hunderd years, however, before the system was well-established. In 1830, just after the revolution and the creation of Greece, the young nation lacked even the most basic structures that characterize a modern state. Greece's economy was pre-industrial and technical skills were local, fostered by the guild tradition.[14] Modernizers believed that local resources would not be sufficient in creating a modern infrastructure, so the Greek state began employing European engineers recruited from Bavaria and France, as well as Greek engineers who had graduated from European universities.[15] While Greece called on engineers from abroad, the state also began to establish a national educational system. The Military School, founded in 1828 by Governor I. Kapodistrias and organized by the French military J. Pauzié, remained the only training institution for engineers for the next few decades. By awarding scholarships, the Greek government sought to encourage the school's graduates to continue study at European engineering schools, especially French institutions such as the Ecole Polytechnique, Ecole des Ponts et Chaussées, and Ecole des Mines.[16]

During the nineteenth century the Military School was responsible for training most of the engineers in Greece. Over time, another institution took on that role. In 1837, a young German Bavarian officer in the engineering corps, F. R. von Zentner, was given the task of founding a school (Polytechnico Scholio) for the training of the kind of skilled construction workers and foremen needed to transform Athens into a capital city worthy of the new state. Opened only on Sundays at first, the school developed into a genuine engineering institution, which later became the National Technical University of Athens.[17] From 1840 the new institution held technical and fine arts courses on a daily basis. As the school's director between 1844 and 1862, the architect L. Kautantzoglou emphasized its artistic vocation. After his retirement, the Greek government upgraded and reorganized the school to meet the country's new technological and industrial needs. From the 1860s on, the state required technical personnel to oversee the first urban plan of Athens and the construction of public buildings and infrastructure in other Greek towns. As Greek industry began to develop, mechanics were in much demand for the repair and maintenance of machines and tools, especially the new steam machines.[18] The Polytechnico Scholio was upgraded to face this challenge. It became an upper-level secondary institution and was reorganized into two different daily schools: the school of fine arts and the school of industrial arts.[19] D. Scalistiris, a military and a graduate from the Ecole Polytechnique in France, succeeded Kautantzoglou as director of the Polytechnico Scholio from 1864 to 1873, and upgraded the standards of the technical branch. Under Scalistiris, this branch was divided into three departments: architecture, which trained technicians in civil engineering, surveying, and mechanical engineering. He also extended the curriculum. After the departure of Scalistrisis, the school adopted its present name, *Metsovio Polytechnio* to acknowledge the emigrants from the small northern town of Metsovo who had sent large donations from their new home in Egypt.

In the 1880s, when the Greek state pushed forward an unprecedented program of public works, the *Metsovio Polytechnio* was once again reorganized by a French-trained military. A. Theophilas, its director from 1878 to 1901.[20] In 1887 the Polytechnio was upgraded; a new four-year curriculum, which included differential and integral calculus, now led to a degree in civil or mechanical engineering.

At the end of the nineteenth century, the *Metsovio Polytechnio* underwent a major transformation that gave it a central role in shaping Greek engineering. First, its military personnel were replaced by civilian administrators. Second, professors who until that time had been trained in the major French engineering schools, began to study in Germany. Third, because the fine arts branch became a separate institution, the *Metsovio Polytechnio* became an exclusively technical institution. Fourth, at the suggestion of director A. Ginis, a German-trained civil engineer, a 1914 decree granted university status to the institution. Finally, a 1917 decree gave the school its present form. From that point on, the technical university was fashioned on German models and offered students a five-year course in mechanical and electrical engineering, chemical engineering, civil engineering, architecture, and surveying.[21] Through the 1917 decree, the state upgraded engineering education and, in 1923, Greece's Engineering Association was established to replace the existing engineering societies.[22] All engineers had to join this professional organization to be allowed to work. Greece's Engineering Association collaborated closely with the *Metsovio Polytechnio* in shaping the educational system for engineering. It prevented military engineers from working as architects or civil engineers, and through a 1932 decree, it reserved the title of engineer exclusively for the *Metsovio Polytechnio* graduates. An examination was established for Greek engineers who studied at foreign universities. This was organized under the aegis of the *Metsovio Polytechnio* until 1982, after which the Engineering Association took over the task of certification.

While the technical school was emerging as a key institution, Greece was undergoing serious upheaval. In the decade after 1912, both the territory and the population of Greece doubled. As a result of the wars against Turkey, over 1.5 million refugees, many of whom belonged to upper- and middle-class families of the Diaspora, resettled in Greece. The country entered a phase of industrial development with intense social conflict.[23] The Socialist Party, later changing its name to Communist Party, was founded in 1918, while in 1936 the army installed the dictator Metaxas in order to 'restore order' after a rapid succession of governments.[24]

The interwar years – an era dominated by the modernizer E. Venizelos, three times prime minister in the 1910s and 1920s – were also a key period for the Greek feminist movement. Women formed numerous associations and founded several feminist journals at the end of the nineteenth century, but they had to cease these activities temporarily because of the Balkan War (in 1912–13) and the First World War (which lasted until 1922 in Greece). In the twenties, the women's movement resumed its activities and took off dramatically. At the same time, scores of Greek women entered the labor market; whereas fifty thousand women had worked in the industrial and service sectors in 1907, there were about 210,000 in 1928.[25] As the number of women in employment increased, so did the range of jobs in which they worked. As women took their places in the public arena, they began to clamor for equal rights in education and the professions. Women were also very active in politics, fighting for civil and family rights. By the 1930s, however, the Greek women's movement lost its autonomy and momentum. It became linked more closely – but also subordinated – to other social, political and ideological groupings over the next forty years.[26]

At the turn of the century, a female consciousness arose. In the same historical period, the *Metsovio Polytechnio* was increasingly loosing its male military culture. Also in this period, the school gained a major role in both the educational and professional sides of engineering, since it was the only university-level engineering institution from 1914 until the 1950s. At this historical juncture, two main converging factors have had a direct impact on the choices of women in favor of the *Metsovio Polytechnio:* first, the upgrading of the school to a university level status which attracted women, more willing to enter universities than lower technical institutes, and second, the creation of the Departments of Architecture and Chemical Engineering, in 1917 and 1918. However, women students were very few in the first decades, and moreover, almost invisible. It appears that these pioneers were admitted without significant debate. The board of professors and the board of administrators, preoccupied as they were with reorganizing the institution after the 1917 decree, barely noted their entrance into this exclusively male milieu. Some comments concerning women appear in the record of proceedings for the 26 May 1923 administration board meeting but the focus was not on women's admittance. Without much debate, the minutes simply mention that the first woman student in architectural engineering, Helen Canellopoulou, and her classmates, were asked to explain to the board why they had left the classroom during their metallic constructions exam. Canellopoulou is identified as the student who stated that the material used in the test had not been dealt with properly by the professor in class. There was barely a ripple in the press about the women students or their problems even among women's papers such *The Greek Woman* (*Ellinis,* 1921–1940) and *The Hellenic Review* (*Elliniki Epitheorisi,* 1907–1942).

Only *The Hellenic Review* published a biographical sketch of Helen Canellopoulou.[27] In 1923, Canellopoulou graduated from the *Metsovio Polytechnio* in architecture as the only woman of the seventy students.[28] With her degree in hand she found a job for a year with A. Zachariou and Company. Later she worked for the ministry of transportation.[29] Two other women followed in her footsteps and graduated a year after she did. Virginia Thomaidou, who trained as an architect, neither joined Greece's central certifying agency (Greece's Engineering Association) nor practiced her profession. Sophia Zarafidou, who earned her degree in surveying, worked as an administrator at Salonica's municipal planning office. Three other women graduated from the *Metsovio Polytechnio* in the 1920s. After getting her degree from the architecture department in

Figure 57. Portrait of architect Helen Canellopoulou in the year she graduated as the first Greek woman from the National Technical University of Athens in 1923. Reproduced from Elliniki Epitheorisi *193, (November 1923): 13.*

1926, Margarita Leontidou worked at the city planning office in Athens. Eriketi Ioannidou, an architect, who graduated in 1927, did join the Engineering Assocation but failed to register any professional address. Another graduate during these years was Chariklia Xanthopoulou, a chemical engineer from the class of 1928, whose name was not entered in rolls of the association.

In contrast to the late entry of French women to the Ecole Polytechnique in 1972, Greek women were admitted to the National Technical Univerisity of Athens (*Metsovio Polytechnio*), Greece's sole institution awarding the title of engineer, as early as the First World War. It was no coincidence, perhaps, that while the French Ecole Polytechnique was still under military influence as late as 1972, the Greek *Metsovio Polytechnio* and engineering profession had distanced themselves from the military during World War I. Comparison with the Swedish and Russian cases shows that women's access to engineering was often directly related to whether or not military culture was important within engineering institutions.

As elsewhere the number of women graduates leveled off during the 1930s. Between 1930 and 1939, only five women in 134 students

Figure 58. First year students of the department of Architecture of the National Technical University of Athens, in front of the central building of the university in 1931. Among the 17 freshmen, one woman: Ethel Prantouna. Reproduced from Technika Chronika *181, (July 1939): 206.*

graduated in architecture, and three from the 607 in civil engineering. But by the forties, their number increased again: there were seventeen women among the 124 architecture students, three women among the 609 civil engineering students, and seven women among the 167 chemical engineering students.[30] Dora Kokkinou, who graduated in mechanical-electrical engineering as the only woman in 322 students, found employment with the public telecommunications utility upon graduation in 1949.

Even if these numbers are too small to draw conclusions of precise statistical significance, the numbers show trends which were later confirmed. As in other European countries, young Greek women favored architecture first and chemistry second. Of the forty-two women who graduated from the *Metsovio Polytechnio* between 1923 and 1949, twenty-six (sixty-two per cent) were architects, eight (nineteen per cent) were chemical engineers, and six were civil engineers. Only one woman graduated in mechanical-electrical engineering, and there was only one in surveying.

We speculate that women were guided in their choices both by the subject matter of engineering education and the actual working practices of the engineering profession. In Greece, women had been accepted in the world of art since the end of the nineteenth century and therefore might have been more easily drawn into the field of architecture, whose advocates defined the architect as part-engineer and part-artist. The relative popularity of chemical engineering among women in the technical fields can be attributed to the fact that chemical engineers worked in laboratories, which were believed to be more suitable places for women than the factory floor. Moreover, chemistry was a newer field of specialization, so its gender patterns were not yet set. By contrast, mechanical and civil engineers had to enter male-dominated areas, such as the workshop and building site, from which women were normally excluded.[31] No matter what Greek women engineers chose to study, however, they strongly preferred to work in the public utilities: some eighty per cent of them, according to the 1939 and 1951 indexes of the Engineering Association. It seems that public administration offered career opportunities to women engineers.[32] Architect Thaïs Roussou (class of 1948), the first woman to hold the highest position in the Greek administation as general director at the Ministry of Construction and Public Works exemplifies women's careers in engineering for the state.[33]

A Country under Transformation: The Triumph of Women Architects and the Slow Rise of Women Civil Engineers, 1950–1980

Major political changes, industrialization and urbanization dramatically transformed the country in the 1950–1980 period, contributing to an increasing entry of women in engineering, massively in architecture in the 1950s and slowly but steadily in civil engineering from the 1960s on.

The Liberation in 1944 was followed by civil war, which ended with the defeat of the Communist Party in 1949.[34] In spite of the parliamentary system then established, the authoritarian Right sought to isolate the extreme Left. By the early 1950s, the Greek armed forces, which had a strong affinity to the anti-parliamentary groups, emerged as the crucial actors in the post-war state. In 1967 the country entered a period of military dictatorship.[35] Not until 1974, when constitutional rules were reinstated, did Greece become a full democracy and did the Communist Party, which had been outlawed since 1947, become legal. During this difficult political transition, Greece went through major economic changes, marked by a rapid shift from agriculture to industry.[36] Political events in the 1940s and industrialization in the 1960s also transformed the demographic structure of the country. During the Civil War of 1947–48, 700,000 people (ten per cent of the total population) fled from their villages to the cities. The rural exodus continued with the economic changes during the sixties and seventies. As a result, the urban population exploded, rising from 37.7 per cent in 1951 to 58.1per cent in 1981. In spite of the industrialization process, unemployment rose dramatically at the beginning of the 1960s, resulting in the emigration of 830,000 Greeks (who went mainly to Germany) during the 1961–1970 period. This was a quarter of the working population and a tenth of the total population of the country. However, these workers still contributed to the economic development of the country. Their absence helped to keep unemployment very low in the period between 1965 and 1980, which contributed to social equilibrium, while their remittances contributed significantly to economic growth.

The large-scale emigration of the sixties played a major role in exporting poverty, and unprecedented economic growth benefited the majority of the population. Yet the social success of the 'Greek model of development' cannot be explained without reference to the unique structure of land ownership in Greece. The extension of Greek territory after the wars against Turkey resulted in two important agrarian reforms (1870–1871 and 1917–1924), each of which distributed property among rural families. These parcels of land became a new source of income for the middle and lower classes during the economic boom of the sixties and seventies. Increasing urbanization, which followed the rural exodus, as well as tourist development, fueled a wide change in land use from farming to speculative building. Many small landowners profited greatly from the resulting increase in land values.

Economic development and urbanization made for an increase in school attendance. While in 1961, only 2.9 per cent of the working population obtained a university degree, the figure was 10.8 per cent twenty years later.[37] In the same period, Greece, even more than most other democratic societies, threw open its higher education system to students of all classes.[38] As a result, university education became a major factor in social mobility during the 1960s and the 1970s.[39] Because of the dramatic increase in the number of candidates, university entrance examinations became more selective.[40] In the mid-seventies, the Greek state sought to develop secondary and higher education in technical fields and the professions: both to meet economic and industrial needs, and to satisfy the increasing demand for higher education.[41] However, parents and students disdained technical and vocational education, which then became the province of weaker students or those from poorer backgrounds. Numerous candidates who failed the Greek entrance examinations enrolled at foreign universities.[42]

Women, who won the right to vote only in 1952 (a century later than men and thirty years later than most other European women) benefited hugely from this new cultural obsession with education.[43] Families focused on higher education as the main vehicle for upward mobility. From the beginning of the 1970s, young girls made up half of the students in secondary schools. Ironically, because they performed better than their male colleagues in academic settings, they still remain less well-represented in technical and vocational programs, because these have low status and are not socially desirable.[44]

Women's participation in higher education also dramatically increased.[45] As in the secondary schools, young girls tended, even more than boys, to avoid the higher technical institutes in favor of the high-status university programs.[46] Constituting twenty-three per cent of the student body in the academic year 1960–1961 and thirty-one per cent ten years later, women represented 42.4 per cent by 1980–1981. Their number has risen continually since, and proportionally they have outstripped men, making up 53.7 per cent of the university population in 1991–1992.[47] This progress was characterized by an under-representation of women in the scientific and technological branches, which, while it has decreased, still exists today. Women made up 52.8 per cent of students in the humanities in 1960–1961 and 74.9 per cent in 1980–1981. Against this, women's participation in the sciences was 19.6 per cent in 1960–1961 and 33.5 per cent in 1980–1981 (rising to 40.8 per cent in 1991–1992). In engineering, women represented 11.4 per cent of the student population in 1960–1961 and 20.6 per cent in 1982–1983 (rising to about twenty-eight per cent in 1991–92). If architecture is omitted, women represented 5.9 per cent of the engineering students in 1969–1970 and 16.2 per cent in 1982–1983.[48] Overall, the huge influx of highly-educated women, starting in the 1960s and resulting in high participation by women in the Greek economy (especially in the service sector), caused the sociologist Lambiri-Dimaki, to identify this new group of women as a new 'female middle-class elite.'[49]

Data from the only Greek institute to grant engineering degrees during the 1950s, the Technical University of Athens, show the same patterns among women as were evident in the 1923–1950 period.[50] In the fifties, women still preferred architecture first and chemical engineering second. During this decade, sixty-seven women, or 25.7 per cent of the total number, graduated in architecture, while twenty-three (or 6.1 per cent) graduated in chemical engineering. There were still few women civil engineers – only five women or 0.4 per cent of the total. The same holds for electro-mechanical engineers, as three women (also 0.4 per cent) graduated in this field. Surveying also had few women only one (or one per cent). The civil engineer Efrosini Karidi who defended her doctoral thesis in structural engineering in 1956, was the first woman to be awarded the title of doctor at the Technical University of Athens.[51]

Engineering was one of the most prestigious professions and the most difficult to enter in Greek society until the early 1980s. It was considered to have a high status comparable to medicine and law. Because it was so exclusive, few university students studied engineering.[52] Mostly as free-lancers, engineers along with lawyers and physicians, held the best positions within the professional world.[53] Most Greek engineers earned high incomes when self-employed (as free-lancers or the heads of design offices), while some enjoyed large salaries as senior executives in private industry. The minority who worked in the public sector still earned higher salaries than the other university-trained professionals working for the state.[54]

The high proportion of self-employed engineers makes engineering community in Greece different from that in other European countries, where engineers mainly work as managers in the private sector. This resulted directly from the course of Greek economic development in the sixties and seventies. The development of heavy industry in the 1960s created demand for mechanical, electrical, chemical, naval, and mining engineering. But because the economy was mainly based on the construction sector, it was civil engineering that shaped the Greek technical world most profoundly (Table 11).[55] House building was especially important, and it developed in a unique fashion in Greece.[56] The state's participation was relatively weak in contrast to the pattern in other European countries. In Greece, housing was built by a multitude of small design offices, usually managed by one person (a civil engineer or an architect) that were scattered throughout the country. Again, the peculiarities of land ownership and agrarian reform, as outlined above, explain this.[57] In the postwar period this resulted in the antiparochi system, by which the owner of the construction plot (or of an old building) cedes the site to the local engineer or architect. He or she then erects a block of flats to be sold and pays for the site by exchanging it for apartments in the new building. Such a system of speculative building offered great opportunities to self-employed engineers.

The economic development of the sixties and seventies, when the status of engineering as a profession went up, brought an increase in the number of graduates who entered the labor force. The rates continued to grow until the early 1980s (Table 11). Women profited from the greater demand for engineers, and their participation in the technical world increased more rapidly than that of their male colleagues. Engineering studies, which were highly selective and held attractive employment prospects, challenged women students to be as good as the best male students in mathematics. While women represented 9.7 per cent of the total of engineering graduates in the 1962–1966 period, this proportion almost doubled (to 17.6 per cent) in the 1977–1981 years (Table 12). The increase was even greater if we omit architecture; over the same time period, percentages of women non-architect engineers grew from 2.8 per cent to 8.6 per cent of graduates (Table 12).

In the 1960s and 1970s, women preferred to major in architecture and chemical engineering. Electro-mechanical engineering remained a strongly male-dominated field (Table 12). However, the proportions of women as well as their numbers increased; they were 0.9 per cent of the electro-mechanical engineering graduates in 1962–1966 and 3.3 per cent of in 1972–1976 (Table 12).

Although the main trends of the 1950s continued, there were also new developments in the gendering of engineering between 1960 and 1980. Previously women's choice about what field to study was less clearly linked to the labor market. The development of construction and public works in the sixties and seventies offered new opportunities to women. From the early 1960s, women enrolled in civil engineering in greater numbers, making it their favorite engineering specialization after architecture. 12.5 per cent of graduate women in 1962–1966 were

Table 11. Number of new graduates registered with Greece's Engineering Association with the number of women in brackets, 1962–1996. The Association licensed engineers. During Greece's construction and public works expansion in the 1960s and 1970s, women engineering graduates entered the labor market in record numbers and at a greater pace than their male colleagues. Source: Aleka Vichert and Sophie Migadi, Ifistameni epaggelmatiki katastasi ton ginekon michanikon *(Athens, 1997).*

Specialty	1962-1966	1967-1971	1972-1976	1977-1981	1982-1986	1987-1991	1992-1996	1962-1996
Civil Engineering	1092	2078	2568	3298	3068	3058	3032	18194
	(42)	(138)	(220)	(389)	(645)	(903)	(945)	(3282)
Mechanical and Electrical Engineering	1279	1461	1895	2770	3946	4288	3838	19477
	(11)	(22)	(63)	(90)	(265)	(398)	(538)	(1387)
Architecture	578	1236	2272	4109	2624	1533	1174	13526
	(255)	(515)	(782)	(1462)	(1013)	(710)	(661)	(5398)
Chemical Engineering	169	326	408	777	1324	1325	1033	5362
	(15)	(36)	(57)	(103)	(325)	(377)	(365)	(1278)
Surveying	191	587	846	644	635	686	765	4354
	(9)	(53)	(88)	(100)	(153)	(283)	(288)	(974)
Electronic Engineering	16	43	111	210	198	489	295	1362
	(3)	(3)	(12)	(14)	(28)	(109)	(73)	(242)
Mining Engineering	102	195	159	223	365	299	257	1600
	(1)	(4)	(4)	(13)	(65)	(79)	(96)	(262)
Naval Engineering	35	41	93	307	284	208	131	1099
	(0)	(1)	(0)	(0)	(4)	(17)	(19)	(41)
Total	3462	5967	8352	12338	12444	11886	10525	64974
	(336)	(772)	(1226)	(2171)	(2498)	(2876)	(2985)	(12864)

civil engineers; this percentage rose to eighteen per cent in 1977–81 (Table 13). Another main trend in the 1960s and 1970s was women's access to surveying. Women surveyors ranked third among women engineering graduates in 1967–1976 (Table 13). Surveyors were needed for public works projects, as close collaborators and, often, substitutes for civil engineers. Because it was a four-year course until 1974, surveying was a lower-level discipline, less selective than other engineering fields.

Although statistical data is still inadequate for the period between 1960 and 1980, initial estimates seem to show that a great number of women graduates turned to the public sector. Working as a civil engi-

Table 12. Percentage of women engineers among new graduates registered with Greece's Engineering Association, 1962–1996. Almost absent at the beginning of the 1960s, the number of women currently represent 20 percent of the total and is still increasing. Source: Aleka Vichert and Sophie Migadi, Ifistameni epaggelmatiki katastatsi ton ginekon michanikon *(Athens, 1997).*

Specialty	1962-1966	1967-1971	1972-1976	1977-1981	1982-1986	1987-1991	1992-1996	1962-1996
Civil Engineering	3,85%	6,64%	8,57%	11,80%	21,02%	29,53%	31,17%	18,04%
Mechanical and Electrical Engineering	0,86%	1,51%	3,32%	3,25%	6,72%	9,28%	14,02%	7,12%
Architecture	44,12%	41,67%	34,42%	35,58%	38,61%	46,31%	56,30%	39,91%
Chemical Engineering	8,88%	11,04%	13,97%	13,26%	24,55%	28,45%	35,33%	23,83%
Surveying	4,71%	9,03%	10,40%	15,53%	24,09%	41,25%	37,65%	22,37%
Electronic Engineering	18,75%	6,98%	10,81%	6,67%	14,14%	22,29%	24,75%	17,77%
Mining Engineering	0,98%	2,05%	2,52%	5,83%	17,81%	26,42%	37,35%	16,38%
Naval Engineering	0,00%	2,44%	0,00%	0,00%	1,41%	8,17%	14,5%	3,73%
Total Engineers	9,71%	12,94%	14,68%	17,60%	20,07%	24,20%	28,36%	19,80%
(without architects)	(2,81%)	(5,43%)	(7,30%)	(8,62%)	(15,12%)	(20,92%)	(24,85%)	(14,51%)

neer in the private sector at this time mainly meant working as a free-lance. This demands long working hours and was very difficult to combine with women's family burdens as wives and mothers. For, in spite of the profound social changes in this period, Greek women still saw themselves in these traditional domestic roles.[58] As a result, women preferred to go into public administration, which offered regular hours and good career opportunities to engineers during these years. This choice by women revealed women's subordinate position in the social division of labor. Although women engineers earned relatively high salaries in comparison to other working women, they still made less than their male colleagues in the private sector.

Table 13. Women engineers by specialization as registered with Greece's Engineering Association, 1962–1996. Women have entered all engineering subfields. Source: Aleka Vichert and Sophie Migadi, Ifistameni epaggelmatiki katastasi ton ginekon michanikon *(Athens, 1997).*

Specialty	1962-1966	1967-1971	1972-1976	1977-1981	1982-1986	1987-1991	1992-1996	1962-1996
Civil Engineering	12,50%	17,88%	17,94%	17,92%	25,82%	31,40%	31,66%	26,20%
Mechanical and Electrical Engineering	3,27%	2,85%	5,14%	4,15%	10,61%	13,84%	18,02%	11,07%
Architecture	75,89%	66,71%	63,78%	67,34%	40,55%	24,69%	22,14%	43,09%
Chemical Engineering	4,46%	4,66%	4,65%	4,74%	13,01%	13,11%	12,23%	10,20%
Surveying	2,68%	6,87%	7,18%	4,61%	6,12%	9,84%	9,65%	7,77%
Electronic Engineering	0,89%	0,39%	0,98%	0,64%	1,12%	3,79%	2,45%	1,93%
Mining Engineering	0,30%	0,52%	0,33%	0,60%	2,60%	2,75%	3,22%	2,09%
Naval Engineering	0,00%	0,13%	0,00%	0,00%	0,16%	0,59%	0,64%	0,33%
Total	100%	100%	100%	100%	100%	100%	100%	100%

The Conquest of a Profession in Crisis, 1980–1996: A Pyrrhic Victory?

Statistical data indicate that the eighties were a turning point in the history of women engineers. There were twice as many women graduates between 1982 and 1986 as there were in the late 1970s. This held for almost all specialties, with the exception of architecture (Table 12). During the following two decades, the proportion of women graduates constantly increased, whereas the total number of graduates stabilized or even slightly diminished (Table 11). Co-education might have been a significant factor as girls were allowed to study in the same high schools as boys from 1979.[59] Another major factor was the way social and economic change altered the engineering profession in the 1980s.

Significant changes occurred in the profile of women students in engineering. First, the number of women architects decreased (relatively speaking) between 1987 and 1996 period: they were about twenty-three per cent of the new women graduates, whereas the largest block of women engineering graduates for the period 1987–1996 was in civil engineering (31.5 per cent) (Table 13). Second, between 1992 and 1996, women made significant progress in the male-dominated field of electro-mechanical engineering, a specialty in which an increasing number of students, both men and women, graduated after 1980 (Tables 11 and 13). A major factor in the popularity of this field may have been the development of new technologies, such as computer science, which brought it into the class of white-collar occupations. Electro-mechanical engineering accounted for eighteen per cent of women graduates between 1992 and 1996. This made it the third choice for women after civil engineering and architecture (Table 13). Finally, women also increased their presence in chemical engineering, mining engineering, surveying and electronic engineering, where they represented between one third and one fourth of the new graduates in 1987–1996.

These important changes have not yet affected some of the well-established patterns of women's education and professional status. Between 1962 and 1996, architecture was still their preferred field of study. Forty-three per cent of women engineers graduated in this, and they represented forty per cent of the wider pool of architectural engineers.[60] Also women engineers were employed more often in the public sector than men. Just over thirty per cent of the women engineers registered by the Engineering Association in the mid-1990s were employed in the public sector, while men were just under twenty-four per cent.[61]

Can we call this a conquest of the profession by women? Percentages of women among graduates in 1992–1996, while still low in electro-mechanical engineering, indicate that engineering profession is losing its exclusively male character. However, this conquest has occurred at a time when the prestige that the profession enjoys has been steadily decreasing. This can be explained both in relation to the economic situation in Greece and by the way the engineering profession is organized.

From 1980 to the mid-1990s, Greece was economically stagnant.[62] However, the number of engineers looking for employment has increased. From 1962 to 1976, engineering enjoyed what might be called a 'golden era.' About 18,000 new graduates entered the labor market, while there were another 47,000 between 1976 and 1996. Since 1980, engineering seems to have been a victim of its own earlier successes. The higher the prestige of engineers, the more students it attracted. They did not hesitate to study abroad when they did not pass university entrance examinations in Greece. From 1973 to 1987, forty per cent of engineers who registered with the Engineering Association had graduated from foreign universities.[63] The number dropped to ten per cent in the 1990s, as the crisis in the profession became more visible.[64] Meanwhile, although the number of foreign graduates has decreased since the mid-

1980s, the number of graduates from Greek universities has increased dramatically since 1981, when the Socialist Party (PASOK) came to power. Under increased pressure to open their doors for social reasons, universities have been forced to admit a greater number of candidates. University admissions increased from 14,500 in 1981 to 23,000 in the mid-1980s. Since then they have leveled off. Meanwhile the government also decided to increase admissions in higher technical education. As a consequence, the number of students rose from 9,300 in 1981 to 22,300 a few years later. Engineering departments in universities were also affected by this general trend. While in the academic year 1981–1982, about 1,500 young Greek students entered engineering, this number doubled in 1984–1985.

Inevitably, given the economic crisis on the one hand and spectacular rise in the number of engineers on the other, the profession entered a serious crisis.[65] Several studies have shown that prosperity among engineers is declining, which will affect the young people who are just starting in the profession. Because the crisis affects employment prospects, the public sector, with its greater job security, is becoming more attractive to men who formerly might have shunned it. As a result, more women have been forced to work on a free-lance basis. Today, 42.2 per cent of women engineers work this way as against 38.4 per cent of their male counterparts.[66]

Has this change in patterns of work for men and women resulted in better or worse working conditions for women? Free-lance work today does not have the same meaning as it did during the years of increasing professionalization in the sixties and seventies. Whereas job opportunities have been decreasing in both the commercial and public sectors since the mid-1980s, many young engineers of both sexes may not have the professional and social networks required to avoid free-lance work in this period of crisis.[67] Since they are just starting to make their presence felt in the profession, women engineers seem to be feeling the devaluation of the profession more keenly than their male colleagues. They have higher unemployment levels than male engineers (in 1996, 6.2 per cent and 2.4 per cent respectively). Overall they hold fewer degrees (18.7 per cent of women engineers hold a post-graduate degree versus 23.9 per cent of men) and they are less frequently employed in senior positions (five per cent versus 21.9 per cent). They also are much less likely to be in charge of drafting departments (7.2 per cent of the total), earn lower incomes, and work fewer hours than their male colleagues.[68] This may be a result of having multiple social roles as workers, wives, and mothers in a country where public child-care still needs considerable improvement.[69] More generally speaking however, women belong to the rank and file of the profession. Overall, then, it could be that, in this very tight job market, there is less work for younger women engineers because they may be comparatively lacking in experience or qualifications.

Conclusion

The film with which we began centers on a unique person: dissatisfied with the idea of becoming a schoolteacher, the female protagonist dares to study at an engineering school at a time when the engineering profession was almost exclusively male. She studies not just engineering, but gets the opportunity to become a chief executive. Rejecting the matrimonial expectations of the time she wishes to wed a man of a lower rank. The protagonist thus upsets the classic gender power relations both on the shop floor and in the bedroom. In the movie, she is seen as a source of danger by some, a figure of fun by others. For comic purposes the movie explores the contrast between the choices of the character and the life experiences of Greek women in 1964.

However fictional the film creation might have been, *Miss Director* still has something to say about women in engineering. Running a major construction firm is still unusual if not inconceivable even though working in the private sector is common for women engineers today. Like the character in the movie, whose opportunity arises when her teacher goes on vacation, only those women who have access to useful social connections may reach a high managerial position. So too marrying someone with a lower salary remains very unusual for women: 65.3 per cent of women engineers today earn less than their spouses, whereas this percentage is only 21.1 per cent for men engineers.[70]

Women engineers may no longer confront the same social pressures that made the main character in the film worry about her femininity. Today women engineers inspire neither the same fear nor the excessive respect the movie's female protagonist experienced. In Greece, to be a woman engineer is simply not as noteworthy just as male engineers, who witnessed their profession gradually devalue, are not as special as they once were. Yet if the film were made again today, the protagonist would still cast an extraordinary figure.

Although Greek women entered engineering during the twenties, it was not until the 1950s that they managed to make their presence visible within the world of architectural engineering. After having gained access to the Architecture department, women forced their way into civil engineering, benefiting from urbanization and economic development during the 1960s and 1970s. As in other Western countries, architecture and chemical engineering offered the paths by which Greek women entered engineering. But in Greece they also took to civil engineering before advancing into other fields. This was largely due to rapid urbanization and construction developments during the sixties and seventies. The Greek case also confirms what the French and Austrian studies highlight: the important role parents played in encouraging young women to pursue higher education as a means of climbing the social ladder. Finally, the 1920s proved in Greece as elsewhere to be a key period in which women pried their way into the profession.

During the sixties, women's participation in engineering increased again. But the most significant change in the numbers and proportion of women occurred in the 1980s. Since the eighties, the number of women students has steadily expanded in all other engineering areas. As a result women engineers now represent twenty per cent of the engineering community and twenty-eight per cent of the engineers who registred with the Engineering Associaton between 1992 and 1996. This increase, however, has occurred at a time when the engineering profession no longer enjoys the same prestige once did. It poses the difficult question whether women's conquest of Greek engineering might not to be considered a pyrrhic victory in the end.

Notes

1. The authors would like to thank Efi Avdela for her insights on Greek women's history; Lia Tsialta at the Greek Association of Women Engineers for kind assistance; the women engineers who provided the basis for their thinking on the subject through a number of interviews; the staff at the National Technical University of Athens Archives; and Annie Canel and Ruth Oldenziel for their great help.

2. For the place of women in Greek society as represented in the cinema, see: Constantina Safilios-Rothschild, '"Good" and "Bad" Girls in Modern Greek Movies', *Journal of Marriage and the Family* (August 1968): 527–531.

3. Estimation by the authors based on the registers of graduate students (Archives of the National Technical University of Athens). Except otherwise mentioned, statistical data come from the National Hellenic Bureau of Statistics.

4. For a quick overview of modern Greek history, see: Richard Clogg, *A Short History of Modern Greece* (Cambridge: Cambridge University Press, 1979); Nicolas Svoronos, *Histoire de la Grèce moderne* (Paris: PUF, 1980 [4th ed.]); Thanos M. Veremis and Mark Dragoumis, *Historical Dictionary of Greece* (Metuchen, N.J.: Scarecrow Press, 1995).

5. On Enlightment ideas on girls' education in Greece, see: Paschalis M. Kitromilides, 'The Enlightenment and Womanhood: Cultural Change and the Politics of Exclusion,' *Journal of Modern Greek Studies* 1, 1 (1983): 39–61.

6. Greece has been predominantly agricultural: in 1856, the urban population represented 13 per cent of the total increasing to 27 per cent in 1920. Efi Avdela, *Dimosii ipallili genous thilikou. Katamerismos tis ergasias kata filo sto dimosio tomea, 1908–1955* (Athens: Idrima Erevnas ke Pedias tis Emporikis Trapezas tis Ellados, 1990), 17, 21.

7. Most data comes from Sidiroula Ziogou-Karastergiou, who has written extensively on the issue, see: *I mesi ekpedefsi ton koritsion stin Ellada, 1830–1893* (Athens: Istoriko Archio Ellinikis Neoleas, 1986); 'I exelixi tou problimatismou gia tin ginekia ekpedefsi stin Ellada'; 'I mikti ekpedefsi sta defterovathmia scholia stin Ellada: prooptikes ke dimosies sizitisis apo tis arches tou eona mas mechri simera)' in *Ekpedefsi ke filo. Istoriki diastasi ke sichronos problimatismos* edited by Basiliki Deligianni and Sidiroula Ziogou (Salonica: Vanias, 1994): 71–125, 194–207 respectively.

8. As elsewhere in the nineteenth century, Greek women could only work as teachers, servants, workers, midwives, and nurses. Avdela, *Dimosii ipallili genous thilikou*, 17–24.

9. The most important journal was *I efimeris ton Kirion* (*The Journal of Women*, 1887–1918) edited by Kalliroï Parren. For a discussion on the journal and the Greek feminist consciousness, see: Eleni Varicas, *I exegersi ton Kirion: i genesi mias feministikis sinidisis stin Ellada, 1833–1907* (Athens: Katarti, 1996 [2nd ed.]).

10. Quoted by Sidiroula Ziogou-Karastergiou, 'Pro ton propileon: i exelixi tis anotatis ekpedefsis ton ginekon stin Ellada, 1890–1920' in Deligianni and Ziogou, eds., *Ekpedefsi ke filo*, 333–415, p. 368.

11. Prior to the first registration at the University of Athens, some Greek women had been studying at the University of Paris. On education during that period, see: *Istoria tou Ellinikou Ethnous* (Athens: Ekdotiki Athinon, 1977–78), Vol. 14 (1882–1913), 409–413; Vol. 15 (1913–1941), 489–494. See also, Alexandra Lampraki-Paganou, 'I ginekia ekpedefsi ke i nomothetikes rithmisis stin Ellada, 1878–1985' in *Ekpedefsi ke Isotita Efkerion*, edited by Nontas Papageorgiou (Athens: Geniki Grammatia Isotitas, 1995), 84–98.

12. On modernization under Prime Minister E. Venizelos (1864–1936): Georgios Mavrogordatos and Christos Hadziïossif, eds., *Venizelismos ke astikos eksichronismos* (Iraklion: Panepistimiakes Ekdosis Kritis, 1988).

13. Out of 392 women students registered at universities by 1920, the Faculty of Medicine had the highest number of them with 134 registered women and 115 women registered at the Faculty of Philosophy. At the Faculties of Sciences registered only a total of 26 women. Proportionally women's presence was particularly visible at the School of Fine Arts: 23 women in 145 students in total in 1904. Ziogou-Karastergiou, 'Pro ton propileon', 365.

14. Georgios Papageorgiou, *I mathitia sta eppagelmata (16os–20os e.)* (Athens: Istoriko Archio Ellinikis Neoleas, 1986).

15. Many other Bavarians followed Greece's first King Otto (1833–1862), son of the King of Bavaria in important positions until the 1843 revolt which was to transform Greece into a constitutional monarchy. For an overview, see, Aliki Vaxevanoglou, *I kinoniki ipodochi tis kenotomias* (Athens: Kentro Neoellinikon Erevnon, 1996).

16. On the Greek students of the Ecole Polytechnique (Paris), see, Efthymios Nicolaïdis, 'Les élèves grecs de l'Ecole polytechnique (1800–1921),' Paper presented at Seminar on Greek Diaspora in France, French School of Athens, 1997.

17. On the history of the National Technical University of Athens, see, Kostas Biris, *Istoria tou Ethnikou Metsoviou Polytechniou* (Athens, 1957); Helen Kalafati, 'To Ethniko Metsovio Polytechnio sto girisma tou eona: Epaggelmatikes diexodi ton apofiton ke thesmiko kathestos tou idrimatos,' *Pirforos* 7 (1993): 18–27; Aggeliki Fenerli, 'Spoudes ke spoudastes sto Polytechnio, 1860–1870,' *Istorika* 7 (1987): 3–18; Helen Kalafati et al, *I palees silloges tis bibliothikis tou Ethnikou Metsoviou Polytechniou* (Athens: Ethniko Metsovio Polytechnio, 1995); Ethniko Metsovio Polytechnio, *Ekaton exinta chronia, 1837–1997* (Athens 1997).

18. Christina Agriantoni, *I aparches tis ekviomichanisis stin Ellada ton 19o eona* (Athens: Idrima Erevnas ke Pedias tis Emporikis Trapezas tis Ellados, 1986); Georges B. Dertilis, ed., *Banquiers, usuriers et paysans. Réseaux de crédit et stratégies du capital en Grèce, 1780–1930* (Paris: La Découverte, 1988).

19. Following the 1863 program, the School of Industrial Arts taught elementary mathematics, elements of static and mechanics (theoretical and applied), descriptive geometry, architecture and design, practical geometry, physics, and chemistry.

20. The construction of the canal of Corinthos, the development of railroads, the extension of roads, see: Vaxevanoglou, *I kinoniki ipodochi tis kenotomias*.

21. The Department of Architecture, the only 'Fine Arts' department at the Technical University of Athens educated architectural engineers who were familiar with science subjects like mechanics, higher mathematics, and physics. See, Proceedings of Administration Board Meeting, November 11, 1917 (*Praktika Sigglitou*), Archives of the National Technical University of Athens, Greece, Vol. 1917–1920. The Surveying Department offered a 3-year program which was extended to 4 years in 1930 and to 5 years in 1974.

22. The Chamber of Commerce and Industry (*Emporiko ke Viomichaniko Epimelitirio*) and The Chamber of Arts (*Eppaggelmatiko ke Viotechniko Epimelitirio*) were established in 1919 and 1925 respectively. Founded in 1898, the Greek Polytechnic Association (*Ellinikos Polytechnikos Sillogos*) organized university trained engineers, physics, and math graduates and non-formally trained craftsmen without an academic degree. In 1908, graduates from the Technical University of Athens separated and founded their own Association of Engineers of the Metsovio Polytechnio (*Sindesmos ton Michanikon tou Metsoviou Polytechniou*) in order to 'preserve the prestige of the Metsovio Polytechnio and to guarantee the professional dignity of its graduates.'

23. For Greek industrial development, see: Christos Hadziiossif, *I girea selini. I viomichania stin Ellada, 1830–1940* (Athens: Themelio, 1993); Dertilis, *Banquiers, usuriers et paysans*.

24. Aggelos Elefantis, *I epaggelia tis adinatis epanastasis. KKE ke astismos ston mesopolemo* (Athens: Themelio, 1979 [2nd ed.]).

25. Avdela, *Dimosii ipallili genous thilikou*, 33.

26. *Diavaso* 198 'I read,' (14 September 1988), Special Issue for a review dedicated to the women's press and feminist movement in Greece.

27. Elliniki Epitheorisi,' *Hellenic Review* 193 (November 1923): 13.

28. In 1923, 26 men graduated in civil engineering; 5 men (and 1 woman) in architecture, 12 in chemical engineering, 14 in mechanical and electrical engineering, 12 in surveying. For 1923–1950 period, see: Ethniko Metsovio Polytechnio, Registers of Graduate Students (*Diplomatouchi Anotaton Scholon*) (Athens: Ethniko Metsovio Polytechnio, 1950).

29. On the founder of the Society Alexander Zachariou (1869–1938), engineer of the Polytechnic University of Zurich, see, Helen Kalafati, 'Alexander Zachariou: a Main Engineer of the Greek Industry,' in Kalafati et al., *I palees silloges tis bibliothikis*, 119–44. For women engineers' careers until 1951, see the editions of the Greece's Engineering Association: Techniko Epimelitirio tis Ellados, *Technical Bulletin of Greece* (Techniki Epetiris tis Ellados), II (Athens: Ekdosis Technikou Epimelitiriou tis Ellados, 1934); *Name Index of Greek Engineers*, January 1939 (Onomastikos pinax ton melon tou Technikou Epimelitiriou, Ianouarios 1939; *Name Index of Greek Engineers*, January 1951 (Onomastikos pinax ton melon tou Technikou Epimelitiriou, Ianouarios 1951.

30. Among the seven graduates, only Chrisoula Troupkou (class of 1946) worked in a private firm in the city of Salonica. At Engineering Association, the names of the six others were either not registered (2) or did not include any professional address or affiliation (4). Two of them were daughters of engineers.

31. Ruth Oldenziel, 'Gender and the Meanings of Technology: Engineering in the United States, 1880–1945' (Yale University, diss, 1992); and her *Making Technology Masculine: Women, Men, and the Machine in America, 1880–1945* (Ann Arbor: University of Michigan Press, 1999), chapters 3, 5.

32. On the statute of the female employees in the public sector in the 1908–1955 period, see: Avdela, *Dimosii ipallili genous thilikou*.

33. Information comes from authors' interview (October 1998) with Eli Nicolaïdou-Vasilakioti (class of 1948), architect, classmate, and colleague of Roussou. Nicolaï dou, who was Director at the Ministry of Construction and Public Works, represented the Minister in International Organizations. She was awarded the Peace Medal by UNO.

34. John O. Iatrides, ed., *Greece in the 1940's: A Nation in Crisis* (Hanover, N.H.: University Press of New England, 1981).

35. On the path to dictatorship, see, Nicos Mouzelis, *Politics in the Semi-Periphery: Early Parliamentarism and Late Industrialization in the Balkans and Latin America* (London: MacMillan, 1986).

36. Sophie Antonopoulou, *O metapolemikos metaschimatismos tis Ellinikis ikonomias ke to ikistiko phenomeno, 1950–1980* (Athens: Papazisi, 1991); Nicholas G. Pirounakis, *The Greek Economy. Past, Present and Future* (London: MacMillan, 1997).

37. Stergios Babanasis, ed., *Ta characteristika apascholisis ton michanikon ke ton ikonomologon stin Ellada* (Athens: Techniko Epimelitirio tis Elladas, 1993), 47–48.

38. In the mid 1950s, children of liberal professionals, who often came from the most upper class, were 20 times more likely to enter the university than those of workers. In 1965, this factor became 5, but a decade later fell to 2,5. Georgia K. Polydorides, 'Equality of Opportunity in the Greek Higher Education System: The Impact of Reform Policies,' *Comparative Education Review* 22, 1 (February 1978): 80–93; Konstantinos Tsoukalas, *Kratos, kinonia, ergasia sti metapolemiki Ellada* (Athens: Themelio, 1986), 273–75.

39. According to some Greek scholars, this kind of upward social mobility through education had been taken place in Greece already during the 19th century Konstantinos Tsoukalas, *Exartisi ke anaparagogi. O Kinonikos rolos ton ekpedeftikon michanismon stin Ellada, 1830–1922* (Themelio: Athens, 1977) and his 'Formation de l'Etat moderne en Grèce,' *Peuples Méditarranéens* 27/28 (April-September 1984): 83–101.

40. Percentages of admissions to university was about 20 per cent for the 1973–1975 and 15 per cent for 1976–1981 period.

41. On the 1976 reform, see, *Comparative Education Review* 22, 1 (February 1978), Special Issue.

42. In 1972, Greece ranked second in the world as for the number of students abroad. From 1972 to 1982 this number had been multiplied by 2.5. In 1981, about 45,000 Greeks, or nearly half of the student population within Greek Universities registered for foreign universities. Babanasis, *Ta characteristika apascholisis ton michanikon ke ton ikonomologon stin Ellada*, 65, 153; Nikos Panayotopoulos, 'Les 'Grandes Ecoles' d'un petit pays. Les Études à l'Étranger: le cas de la Grèce,' *Actes de la Recherche en Sciences Sociales* 121/122 (March 1998): 77–91.

43. French women, who did not gain the right to vote until 1944, were the exception to this rule. Dimitra Samiou, 'Les femmes grecques à la conquête de l'Égalité politique: les luttes autour du droit du vote, 1864–1952' (University of Paris VII, Master Thesis, 1984).

44. Georgia Kontogiannopoulou-Polydorides, 'Greece,' in Maggie Wilson, ed., *Girls and Young Women in Education* (Oxford: Pergamon Press, 1991), 91–113; Marie Eliou, 'Equality of the Sexes in Education: and Now What,' *Comparative Education* 23, 1 (1987): 59–67; Michalis Kassotakis, 'The

Unequal Participation of the two Sexes in the Secondary Technical and Secondary Genaral Education,' in Papageorgiou, *Ekpedefsi ke Isotita Efkerion*, 36–47. In 1980–1981, female students consisted of 53.8 per cent of the total at upper general high schools and 19.9 per cent at secondary technical and vocational schools. The difference has been decreasing since then; in 1997–98, these figures were respectively 55 per cent and 38.9 per cent.

45. Georgia Polydorides, 'Women's Participation in the Greek Educational System,' *Comparative Education* 21, 3 (1985): 229–240; Kontogiannopoulou-Polydorides, 'Greece''; OECD, *Educational Policy and Planning. Educational Reform Policies in Greece* (Paris: OECD, 1980).

46. In 1981–1982, women students consisted of 38.1 per cent (8,366) of total at higher technical and vocational institutes and 42.5 per cent (37,187) in universities. In 1984–1985, these figures were 43.1 per cent (17,133) and 47.8 per cent (53,219); in 1989–1990, 46.7per cent (34,184) and 52.7 per cent (61,754).

47. Two reforms seemed to have had a positive impact on women education. In 1976, compulsory education increased from 6 years (primary school) to 9 years (primary school plus lower secondary-level school). Co-education was established for secondary schools by 1979 law. On the 1976 reform, see, *Comparative Education Review*, Special Issue.

48. In France, women represented 8 per cent of the engineering students in 1975, and 19.2 per cent in 1989–1990. These percentazges were 6.4 per cent and 11.9 per cent for Germany. See, Catherine Marry, 'Les ingénieurs: une profession encore plus masculine en Allemagne qu'en France?' *L'orientation scolaire et profesionelle* 21, 3 (September 1992): 245–267.

49. Jane Lambiri-Dimaki, *Social Stratification in Greece, 1962–1982* (Athens: Sakkoulas, 1983), 191.

50. The National Technical University of Athens was the only engineering school until the mid 1950s. The economic transformation of the country led to the creation of other engineering faculties within the various Greek universities. Babanasis, *Ta characteristika apascholisis ton michanikon ke ton ikonomologon stin Ellada*, 155–62.

51. Book of doctorate theses (*Biblion Didactorikon Diplomaton*), Archives of the National Technical University of Athens, Athens, Greece.

52. 9.2 per cent in 1960–1961; 12.4 per cent in 1980–1981. See, OCDE, Educational Policy and Planning, 96; Babanasis, *Ta characteristika apascholisis ton michanikon ke ton ikonomologon stin Ellada*, 66.

53. At the beginning of the 1970s, the Greek state employed more than half of all the University graduates and three quarters of the graduates receiving a salary. See, Tsoukalas, *Kratos, kinonia, ergasia sti metapolemiki Ellada*, 128. See, also, Stefanos Pesmazoglou, *Ekpedefsi ke anaptixi stin Ellada, 1948–1985* (Athens: Themelio, 1987).

54. From 1966 to 1972, over half of Greek engineers were self-employed (free-lance and heads of design offices) with a peak to 60 per cent in 1970. After the 1974 economic crisis, this percentage dropped down to 38 per cent in 1976 but increased again to 51 per cent in 1996. 'I epaggelmatiki apascholisi ton michanikon,' *Technika Chronika* (Jan.-March 1980), 56; Techniko Epimelitirio tis Elladas, *Erevna gia tin epaggelmatiki katastasi ke apascholisi ton diplomatouchon michanikon. Ekthesi apotelesmaton* (Athens: Techniko Epimelitirio tis Elladas, 1997), 24 and Appendix-Table G13.1 The study was based on a sample of 2,250 engineers registered with the Engineering Association since 1961.

55. This is an old phenomenon. From 1863 to 1890, when the Technical University of Athens was still a secondary-level institution, it awarded 380 diplomas in civil engineering, 140 in surveying and only 40 in mechanical engineering. Kalafati, *To Ethniko Metsovio Polytechnio*. From 1890 to 1910, the school granted 235 diplomas in civil engineering and 31 in mechanical engineering. *Ethniko Metsovio Polytechnio*, Registers of Graduate Students. In 1910, civil engineers represented only 37per cent of all engineers in Britain, 26.7per cent in the U.S.A. and 22.8per cent in Germany. Peter Lundgreen, Engineering Education in Europe and the USA, 1750–1930: The Rise to Dominance of School Culture and Engineering Profession *Annals of Science* 47 (180): 33–75, p. 70.

56. Antonopoulou, *Metaschimatismos tis Ellinikis ikonomias*.

57. The medium seize of such an area was 180 m2 for Athens and 146m2 for the port city of Piraeus. Antonopoulou, *Metaschimatismos tis Ellinikis ikonomias*, 194–196.

58. Mariella Doumanis, *Mothering in Greece* (London: Academic Press, 1983); Peter Loizos and Evthymios Papataxiarchis, eds., *Contested Identities. Gender and Kinship in Modern Greece* (Princeton: Princeton University Press, 1991). For the persisting set of cultural references within Greek society in the 1960–1980 years despite the major changes during those years, see, John K. Campbell, 'Traditional Values and Continuities in Greek Society,' in Richard Clogg, ed., *Greece in the 1980s* (London: MacMillan, 1983), 184–207.

59. During the sixties, young girls in small towns who sought to pursue a scientific apolitirion (high school diploma) in order to enter science and engineering, had to register for boys' schools because their schools only offered humanities programs.

60. For the 1970s and 1980s the percentage would have been higher if the large number of architects (especially men), who had failed the exams to enter university in Greece but studied abroad (mainly in Italy), had not been included in these figures.

61. *Erevna gia tin epaggelmatiki katastasi ke apascholisi ton diplomatouchon michanikon*, Appendix-Table G13.3. Another study on women engineers, who graduated between 1974–1979, showed that in Athens 37.2 per cent and Larissa 66.7 per cent were working in the public sector. Babanasis, *Ta characteristika apascholisis ton michanikon ke ton ikonomologon stin Ellada*, 199, 201.

62. Pirounakis, *The Greek Economy*; Konstantinos Drakatos, *O megalos kiklos tis Ellinikis ikonomias, 1945–1995* (Athens: Papazisi, 1997).

63. Until the 1980s, engineering students constituted a third of all Greek students studying abroad. Babanasis, *Ta characteristika apascholisis ton michanikon ke ton ikonomologon stin Ellada*, 150, 151.

64. *Erevna gia tin epaggelmatiki katastasi ke apascholisi ton diplomatouchon michanikon*, 15.

65. For a presentation of the present situation of the profession, see, Babanasis, *Ta characteristika apascholisis ton michanikon ke ton ikonomologon stin Ellada*; and *Erevna gia tin epaggelmatiki katastasi ke apascholisi ton diplomatouchon michanikon*.

66. *Erevna gia tin epaggelmatiki katastasi ke apascholisi ton diplomatouchon michanikon*, Appendix-Table G 13.3.

67. Notwithstanding the crisis and the devaluation of the engineer's status, the under-representation of the lower classes at the engineering faculties is noteworthy. At the same time, the professional crisis seems to accentuate the importance of family in engineer's career choice. During 1960–1990 period, 6 to 9 per cent of those registered with Greece's Engineering Association were children of engineers; this percentage rose to 13 per cent after 1990. *Erevna gia tin epaggelmatiki katastasi ke apascholisi ton diplomatouchon michanikon*, 14 and Appendix-Table A 11.5.

68. Ibid., Appendix-Tables B4.3 and A1.6; Aleka Vichert and Sophie Migadi, *Ifistameni epaggelmatiki katastasi ton ginekon* (Athens: Enosis Diplomatouchon Ellinidon Michanikon, 1997), 23.

69. A study funded by the Greek Women's Engineering Association, established in 1995, showed that very few young women engineers have children. On 73 women younger than 30.96 per cent do not have children: among those between 30 and 35, this rate is 78 per cent. Vichert and Migadi, *Ifistameni epaggelmatiki katastasi ton ginekon michanikon*, 41–42.

70. *Erevna gia tin epaggelmatiki katastasi ke apascholisi ton diplomatouchon michanikon*, Appendix-Table A8.3.

Moniko Greif

Epilogue
Women Engineers in Western Germany: Will We Ever Be Taken For Granted?

Sometimes one can get the impression that the situation of women engineers has hardly changed in Germany. There are still very few of us. Consider the topics raised at the first discussion forum for women engineers after the Second World War. Called Women in Engineering (*Ausschuss Frauen im Ingenieurberuf* or FIB), it was founded in 1965 within the German Society of Engineers (*Verein deutscher Ingenieure* or VDI). And at that first meeting, the agenda concerned the special problems of promotion and career advancement, of matching female role models and the inherently masculine image of the profession. Then there was the problem of balancing work, and what happens when having a family means interrupting a promising career – something with which men rarely have to deal.

All of these are still special problems for women. For more than twenty years, there have been government campaigns and projects to encourage young women to enter the technological professions. In civil engineering and architecture we now find female students representing twenty and fifty percent respectively of the first-year students. But in mechanical and electrical engineering the situation has been very slow to change. When I began to study mechanical engineering at the technical university in Darmstadt in 1971, female students were only 0.5 per cent of the field. Nearly thirty years later they are five percent of the students in that subject. Meanwhile, statistics show that female engineers are twice as likely to be unemployed as men, and much more so than women who work in professions that have been traditionally female dominated.

To be honest, some things have changed considerably. Very few counsellors at the national labor agency will still tell you that mechanical engineering is inappropriate for girls because foundries don't have ladies' toilets. Universities have become more aware of the needs and interests of female students, so easy to ignore when women are a minority population in the academic context. Because of this there are now many programs in place to help women in higher education. There are also associations of women engineers which offer all sorts of support, including special training courses for female managers and chances to network within the engineering community. Even if some managers and companies still dislike hiring women, they won't state their aversion openly, or say that women engineers are bad for production or for direct contact with the customers in service. And in fact, more and more employers look at women engineers favorably now.

Individually and collectively, women engineers and scientists have gained self-confidence, and brought new ideas to the discussion of women's participation in technology. This development is due not least to the contribution of feminist engineers. Many women engineers reject the label 'feminist,' and some are ambivalent about the idea of affirmative action. Only one of the associations of women engineers has 'feminist' in its name. Yet feminist ideas have definitely influenced the more traditional associations. When the FIB was reformed in the early eighties (after a hiatus in the seventies), feminist speakers were present at the first conference. In the sixties, the FIB sought to find appropriate fields of work for women, and tried to combat prejudice. In the seventies and eighties, its focus changed to claiming equal opportunities in all areas, and in the nineties debating the specific contribution that women engineers could make to product design and technology.

In so doing, they were picking up a discussion that had already begun at a yearly congress, Women in Science and Engineering, founded in Aachen in 1977. This discussion forum founded by feminist engineering students, continues to be a focal point for associations and activities (such as DIB, NUT, the journal Koryphäe, TUNIXEN, and FIT). Its initiators felt a double isolation: not only were they rare in the male engineering world of their workplace or university, but they were also seen as anomalous within the women's movement for doing a 'man's job.' This meant they were sometimes patronized or disdained for being 'male-identified.' Hence the need for a separate and supportive organization. Relations with other feminist groups and female social scientists were sometimes strained (something that has been true in other times and countries as well, as papers in this volume attest). As objects of sociological research, women engineers found themselves misunderstood, and sometimes confronted with what they felt were very traditional images of femininity. Most feminists, meanwhile, considered technology to be a masculine domain with women as its victims, a simplistic position that denied women's achievements as technical contributors or experts.

Nevertheless, at least the feminist critique of gender roles and the division of labor made women engineers generally less likely to accept discrimination as an individual problem. Participating in citizens' committees for the environment encouraged them to criticize the entrenched concepts of what technology can and should do. From the first meeting, masculinist technology (or that which was produced by or served to reinforce the patriarchy) was deplored; it was argued that the women's collaboration was crucial for creating a technology that took account of social and environmental concerns. In 1993, Women in Science and Technology (*Frauen in Naturwissenschaft und Technik* or

NUT) issued a memorandum explaining this position further. It stated that since men and women are socialized differently, and consequently have different needs, interests and experiences, this leads them to formulate methods and questions on different lines. The NUT made reference to the narrow perspective of mainstream engineering caused by the male monoculture, which in European countries is in addition mostly, white, and middle class. Without the input of women, it inferred, male engineers might be willing but unable to consider the needs and interests of other users of technology. They tend to neglect the social implications and consequences of design, when the problems to be solved always have a social as well as a technical dimension.

Many male engineers in Germany still consider it provocative to assert that gender has any impact whatsoever on science and technology. So, for a rather long time, it was impossible to have a reasonable discussion about this feminist perspective. But the way demand for technology and products has been changing, and recurring debates about what makes a good engineer, have brought the topic to the surface again and again. Employers now want their engineers to have new professional qualities to cope with project work; technical knowledge is no longer sufficient. Work is becoming more complex, it is now normal to co-operate across disciplines, and it is necessary to hold negotiations with, or presentations for, customers and users. Designers have to meet the requirements of customers very carefully to compete in a crowded market. So employers want ever-greater competence in communication and teamwork. German society is also no longer willing to accept new technology without question, as happened in the sixties. Instead, people want to be informed about risks and details. Engineers must be able to be 'technical translators,' communicating with those who do not have technical expertise.

How might these new conditions create opportunities for women? Some of the skills required are naturally learned by women in the process of socialization. Investigation has shown that female engineering students have a greater talent for communication and organization than their male colleagues. Female engineers find it easier to orient themselves to customers and understand what they want. There are hints that they are better at seeing the social implications of their work. A survey among IT workers showed that the women were much more willing than men to address the requirements of a project and conform to the wishes of the users. Male workers in IT tended to explore the technical limits of whatever was at hand, rather than focusing on what the customer wanted. No wonder that German products are often said to be 'over-engineered,' or that elderly people (very much a growing market) dislike products designed by 'techno-nerds'!

I would not want to claim that women are necessarily better engineers than men. But I think that increasing the participation of women is not only a question of equal opportunities within the profession, but looking too at the quality of results. Engineers are still a rather homogeneous group, but customers aren't. We need a wider spectrum of human talents and experiences to meet the challenges of the future. I am optimistic that more and more people will realize the role women can play in providing the new perspectives needed.

Index